AF566673

IET POWER AND ENERGY SERIES 82

Periodic Control of Power Electronic Converters

Other volumes in this series:

Volume 1 **Power Circuit Breaker Theory and Design** C. H. Flurscheim (Editor)
Volume 4 **Industrial Microwave Heating** A. C. Metaxas and R. J. Meredith
Volume 7 **Insulators for High Voltages** J. S. T. Looms
Volume 8 **Variable Frequency AC Motor Drive Systems** D. Finney
Volume 10 **SF_6 Switchgear** H. M. Ryan and G. R. Jones
Volume 11 **Conduction and Induction Heating** E. J. Davies
Volume 13 **Statistical Techniques for High Voltage Engineering** W. Hauschild and W. Mosch
Volume 14 **Uninterruptible Power Supplies** J. Platts and J. D. St Aubyn (Editors)
Volume 15 **Digital Protection for Power Systems** A. T. Johns and S. K. Salman
Volume 16 **Electricity Economics and Planning** T. W. Berrie
Volume 18 **Vacuum Switchgear** A. Greenwood
Volume 19 **Electrical Safety: A guide to causes and prevention of hazards** J. Maxwell Adams
Volume 21 **Electricity Distribution Network Design, 2nd Edition** E. Lakervi and E. J. Holmes
Volume 22 **Artificial Intelligence Techniques in Power Systems** K. Warwick, A. O. Ekwue and R. Aggarwal (Editors)
Volume 24 **Power System Commissioning and Maintenance Practice** K. Harker
Volume 25 **Engineers' Handbook of Industrial Microwave Heating** R. J. Meredith
Volume 26 **Small Electric Motors** H. Moczala *et al.*
Volume 27 **AC–DC Power System Analysis** J. Arrillaga and B. C. Smith
Volume 29 **High Voltage Direct Current Transmission, 2nd Edition** J. Arrillaga
Volume 30 **Flexible AC Transmission Systems (FACTS)** Y-H. Song (Editor)
Volume 31 **Embedded Generation** N. Jenkins *et al.*
Volume 32 **High Voltage Engineering and Testing, 2nd Edition** H. M. Ryan (Editor)
Volume 33 **Overvoltage Protection of Low-Voltage Systems, Revised Edition** P. Hasse
Volume 36 **Voltage Quality in Electrical Power Systems** J. Schlabbach *et al.*
Volume 37 **Electrical Steels for Rotating Machines** P. Beckley
Volume 38 **The Electric Car: Development and Future of Battery, Hybrid, and Fuel-Cell Cars** M. Westbrook
Volume 39 **Power Systems Electromagnetic Transients Simulation** J. Arrillaga and N. Watson
Volume 40 **Advances in High Voltage Engineering** M. Haddad and D. Warne
Volume 41 **Electrical Operation of Electrostatic Precipitators** K. Parker
Volume 43 **Thermal Power Plant Simulation and Control** D. Flynn
Volume 44 **Economic Evaluation of Projects in the Electricity Supply Industry** H. Khatib
Volume 45 **Propulsion Systems for Hybrid Vehicles** J. Miller
Volume 46 **Distribution Switchgear** S. Stewart
Volume 47 **Protection of Electricity Distribution Networks, 2nd Edition** J. Gers and E. Holmes
Volume 48 **Wood Pole Overhead Lines** B. Wareing
Volume 49 **Electric Fuses, 3rd Edition** A. Wright and G. Newbery
Volume 50 **Wind Power Integration: Connection and system operational aspects** B. Fox *et al.*
Volume 51 **Short Circuit Currents** J. Schlabbach
Volume 52 **Nuclear Power** J. Wood
Volume 53 **Condition Assessment of High Voltage Insulation in Power System Equipment** R. E. James and Q. Su
Volume 55 **Local Energy: Distributed generation of heat and power** J. Wood
Volume 56 **Condition Monitoring of Rotating Electrical Machines** P. Tavner, L. Ran, J. Penman, and H. Sedding
Volume 57 **The Control Techniques Drives and Controls Handbook, 2nd Edition** B. Drury
Volume 58 **Lightning Protection** V. Cooray (Editor)
Volume 59 **Ultracapacitor Applications** J. M. Miller
Volume 62 **Lightning Electromagnetics** V. Cooray
Volume 63 **Energy Storage for Power Systems, 2nd Edition** A. Ter-Gazarian
Volume 65 **Protection of Electricity Distribution Networks, 3rd Edition** J. Gers
Volume 66 **High Voltage Engineering Testing, 3rd Edition** H. Ryan (Editor)
Volume 67 **Multicore Simulation of Power System Transients** F. M. Uriate
Volume 68 **Distribution System Analysis and Automation** J. Gers
Volume 69 **The Lightning Flash, 2nd Edition** V. Cooray (Editor)
Volume 70 **Economic Evaluation of Projects in the Electricity Supply Industry, 3rd Edition** H. Khatib
Volume 72 **Control Circuits in Power Electronics: Practical issues in design and implementation** M. Castilla (Editor)
Volume 73 **Wide Area Monitoring, Protection and Control Systems: The enabler for Smarter Grids** A. Vaccaro and A. Zobaa (Editors)
Volume 74 **Power Electronic Converters and Systems: Frontiers and Applications** A. M. Trzynadlowski (Editor)
Volume 75 **Power Distribution Automation** B. Das (Editor)
Volume 76 **Power System Stability: Modelling, analysis and control** B. Om P. Malik
Volume 78 **Numerical Analysis of Power System Transients and Dynamics** A. Ametani (Editor)
Volume 79 **Vehicle-to-Grid: Linking electric vehicles to the smart grid** J. Lu and J. Hossain (Editors)
Volume 81 **Cyber-Physical-Social Systems and Constructs in Electric Power Engineering** Siddharth Suryanarayanan, Robin Roche, and Timothy M. Hansen (Editors)
Volume 86 **Advances in Power System Modelling, Control, and Stability Analysis** F. Milano (Editor)
Volume 88 **Smarter Energy: From Smart Metering to the Smart Grid** H. Sun, N. Hatziargyriou, H. V. Poor, L. Carpanini and M. A. Sánchez Fornié (Editors)
Volume 93 **Cogeneration and District Energy Systems: Modelling, Analysis, and Optimization** M. A. Rosen and S. Koohi-Fayegh
Volume 101 **Methane and Hydrogen for Energy Storage** R. Carriveau and David S-K. Ting
Volume 905 **Power System Protection, 4 volumes**

Periodic Control of Power Electronic Converters

Keliang Zhou, Danwei Wang,
Yongheng Yang and
Frede Blaabjerg

The Institution of Engineering and Technology

Published by The Institution of Engineering and Technology, London, United Kingdom

The Institution of Engineering and Technology is registered as a Charity in England & Wales (no. 211014) and Scotland (no. SC038698).

First published 2016

The Institution of Engineering and Technology
Michael Faraday House
Six Hills Way, Stevenage
Herts, SG1 2AY, United Kingdom

www.theiet.org

British Library Cataloguing in Publication Data
A catalogue record for this product is available from the British Library

ISBN 978-1-84919-932-2 (hardback)
ISBN 978-1-84919-933-9 (PDF)

Typeset in India by MPS Limited
Printed in the UK by CPI Group (UK) Ltd, Croydon.

Contents

Preface

Power electronics is entering into all kinds of energy processing systems – for many reasons – as it is a key technology to interface renewable energy sources into the grid and this field is booming; it is a technology, which, in many applications, ensures that an application is saving energy (e.g., in motor drives and lighting), the transportation is becoming electrified, and the industrial automation could not work without power electronics. It is the working horse in the modern society.

To master the power electronics technology we normally say that we have to understand the power, electronics, and control – it is a kind of multidisciplinary discipline – in the last years, more demands have also been on understanding the reliability of the system combined with an insight into the loading/environment of a power electronic system experience. The design of a power electronic system then becomes much more complicated and much information is needed in order to do a proper design. The control system is an important part of the power electronic system as mentioned above. As the "brain," the control technology enables the power electronic system to precisely and flexibly process power. Without a proper control design the system might be unstable and have poor reliability.

It is very common for power electronic systems to experience periodic signal processing, such as synchronous frame transformation, sinusoidal current regulation, sinusoidal voltage conditioning, and power harmonics mitigation. Classical controllers (e.g., the proportional integral controller) are not able to remove the dynamic periodic error completely. The residual periodic error will not only degrade the power quality but also even affect the stability and reliability of the electrical power systems. Thus it is a very interesting topic to investigate the internal model principle (IMP)-based periodic control solution for power electronic converters to tackle periodic signals. The name of "periodic control" clearly indicates that it is tailor-made for periodic signals. The foundation of the IMP was first championed by B. A. Francis and W. M. Wonham in 1976, which states that perfect asymptotic rejection/tracking of persistent inputs can be attained by replicating the signal generator in a stable feedback loop. The first IMP-based periodic control solution called *repetitive control*, which was presented in the 1980s, can achieve exact regulation of any periodic signal with a known period. Since then many scientists and engineers have made significant contributions to its both theoretical advancements and practical applications, especially its applications to power electronics. The IMP-based periodic control has yet been found to provide power electronic converters with a superior control solution to the compensation of periodic signals with high accuracy, fast dynamic response, good robustness, and cost-effective implementation. As an emerging topic, the periodic control has the great potential to be one of the best control solutions for power converters but not limited to, and to be a very popular standard industrial controller such as the proportional–integral–derivative (PID) control.

This book contributes to this discipline combined with power electronics applications/examples. It lays a foundation of the IMP-based periodic control theory with basic theory, derivation of applied equations, knowhow on the control synthesis, and some most recent progress. The book also gives several design and application examples, which can be fruitful in future controller designs, and the control methods are in some cases already applied in the industry. The book fills a gap in power converter control and it is the expectation that it will contribute both to student training and new research as well as being applied in many new products.

Have a nice reading in this exciting topic.

Chapter 1

Introduction

1.1 Background

Power electronics technology plays a pivotal role in improving the performance, efficiency, reliability, and security of the electrical power systems. It is a key enabling technology not only for the coming smart electrical power systems but also to create a more sustainable society. In this chapter, we will have a look at power, electronics, and control with an insight into the development of the power electronic systems and their control.

1.1.1 Trends in electrical power systems

Being evolved from 1896, the electrical power grid is the largest machine. Today, around 40% of all energy consumption is in electrical energy, but this will grow to 60% by 2040 [1], and maybe even more. A traditional power grid with unidirectional power flow and vertical operation and control is exemplified in Figure 1.1, which consists of central fossil-fueled (oil and coal), nuclear and/or large hydro power generation plants, a high voltage transmission system for carrying power closer to the consumers over long-distance transmission lines, and low voltage distribution networks for transferring power to the consumers over distribution lines. Less energy efficiency and more expensive cost due to long-distance power transmission, poor power security due to the centralized generation structure, poor reliability due to slow response, and poor visibility-caused blackouts due to limited real-time grid information, hinder the traditional power grid from offering a clean, sustainable, affordable, and reliable power supply for our lives.

Growing electricity demand, rising cost of limited fossil fuel, and increasing pressure to reduce greenhouse gas emission drive the electricity network toward a more distributed, greener and smarter power grid. The electricity network is undergoing the natural evolution from a conventional network with centralized large-scale power generation plants, long transmission lines, and passive distributed systems to a smart grid structure with large-scale deployment of small-size renewable distributed generators (DGs) within distribution systems in many countries [1–3]. For example, there are about 12 million DG units installed across the United States, with a total capacity of about 200 gigawatts [3]. The advent of large amounts of clean renewable energy (such as wind and solar) DGs gives rise to a series of new challenging problems, such as the mitigation of intermittency due to renewables, bidirectional power flow regulation within the distribution networks

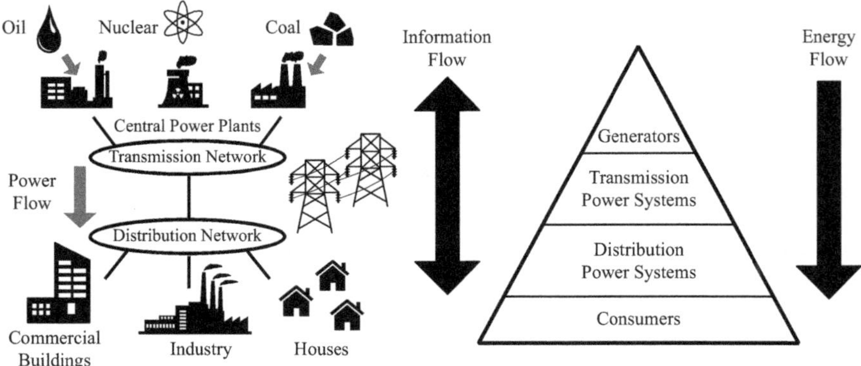

Figure 1.1 Architecture of a traditional electrical power grid based on oil, nuclear, and coal

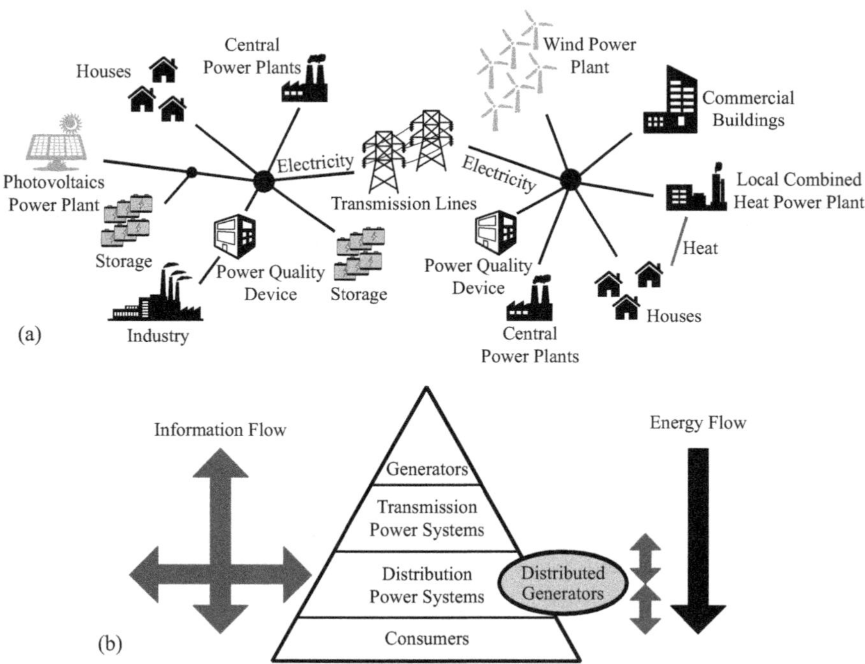

Figure 1.2 A possible more distributed, greener, and smarter power grid: (a) system architecture with renewables and (b) energy and information flows within the grid

and active consumer participation in energy management, and so on. Information and communication technology (ICT) for sensing, communication, and control, and power electronics for active power processing are the two key enabling technologies for smart grids. Figure 1.2 shows a possible case of a more distributed, greener and smarter power grid.

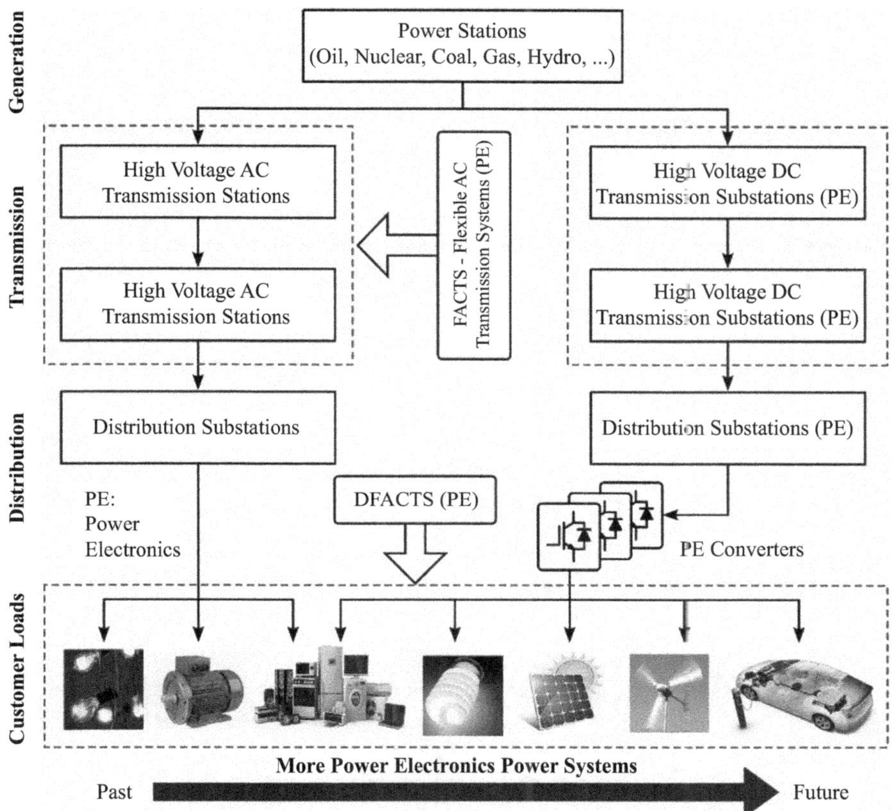

Figure 1.3 Power electronics in electrical power systems

1.1.2 Role of power electronics

Generally, power electronics is the process of using semiconductor switching devices to control and convert electrical power flow from one form to another to meet a specific need. As shown in Figure 1.3, it enables efficient, flexible conversion and conditioning of electrical power for the needs of different loads and for the purpose of energy savings and efficiency improvements in the electrical power systems. Modern power electronics started with the development of the first half-controllable semiconductor switch – the thyristor or silicon controlled rectifier (SCR) in Bell Laboratory in 1957 [3]. As shown in Figure 1.4, toward low power losses, fast switching, easy control, and high operation temperature, the evolution of switching power devices, which is the muscle of power electronic systems, benchmarks the history of power electronics. At the same time, significant achievements in the fields of micro-electronics, control theory, and informatics equip modern power electronic systems with "brain and nerves" as shown in Figure 1.5 [3], and significantly stimulate the development of power electronics. Nowadays it is widely used to process power from milliwatts customer electronics to gigawatts high voltage DC (HVDC) transmission, with conversion efficiencies typically in the excess

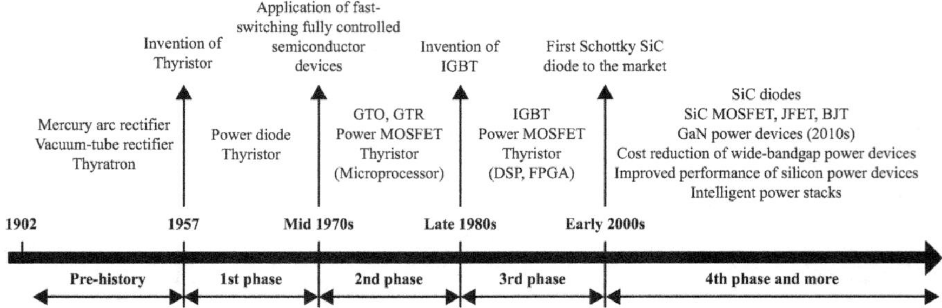

Figure 1.4 Evolution of the power electronics semiconductor technology

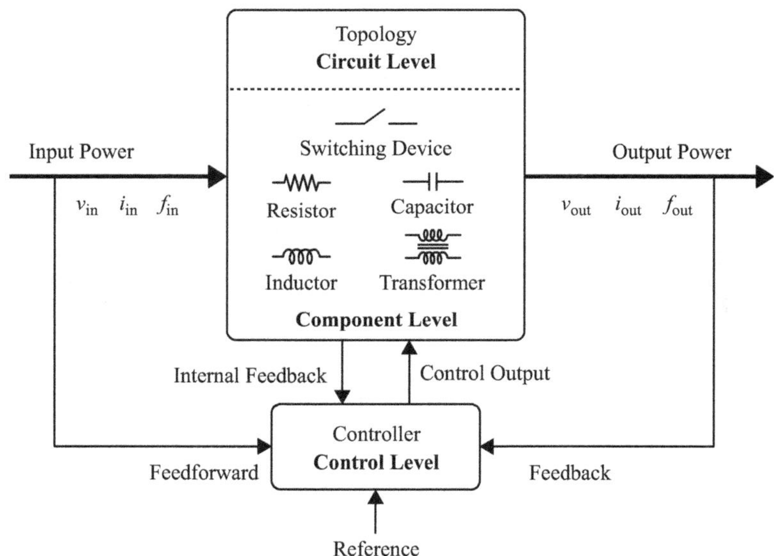

Figure 1.5 Power electronic converter system, where v_{in}, i_{in}, f_{in} and v_{out}, i_{out}, f_{out} are the input and output variables (voltage, current, and frequency)

of 90% and up to 98% for large systems. Today, wide-bandgap devices (e.g., silicon-carbide and gallium-nitride devices) seem to be able to achieve efficiencies higher than 99% [2]. More than 40% of the electrical energy is processed by power electronic systems in 2008. The number of electrical devices by controlling power electronics keeps increasing in electrical power systems, especially a large amount of power converter-interfaced DGs, energy storage systems (ESS, e.g., electric vehicles), loads, and distributed flexible AC transmission systems (DFACTS) being integrated into the distribution network [4–7].

Many issues come into the spotlight with the advent of DGs into distribution networks, such as power quality, voltage profile, reactive power and voltage

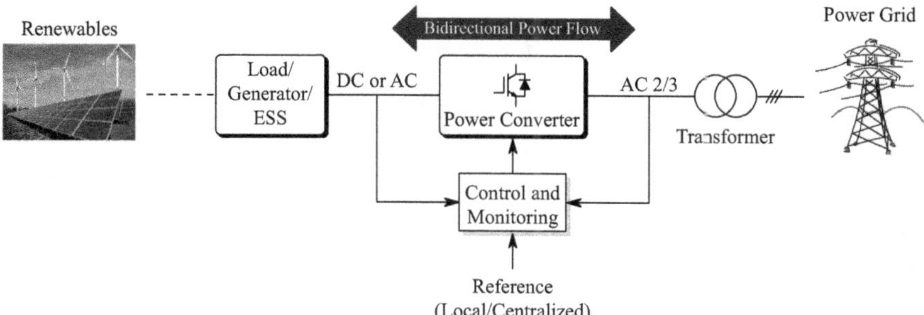

Figure 1.6 A general system structure for power converter-interfaced DGs

control, contribution to ancillary grid services, ride-through capability of DGs to withstand faults or disturbances, protection aspects, islanding and islanded operation, and so on [8,9]. These power converter interfaces electrically decouple DGs from the grid, and render versatile power control functions to improve the overall system efficiency, quality of the electricity supply, and operating conditions. It is impossible to achieve a massive integration of DGs, ESS, and loads without power electronic interfaces (i.e., power converters) into the grids.

Figure 1.6 gives a general system structure for DGs (including ESS/loads/DFACTS) with a power converter interface, which would provide power conditioning functionalities for grid forming, grid feed-in, or grid supporting upon grid-side demands. For examples, as shown in Figure 1.7, grid-connected wind power generators or PV power plants mainly feed active power into the grid, and would inject reactive power to shape voltage profile at the point of common coupling (PCC) or help to sustain riding through grid faults. In the case of an ESS and vehicle to grid (V2G) plug-in electric vehicles, peak load leveling by "valley filling" (charging when the demand is low), and "peak shaving" (discharging power back to the grid when the demand is high) can be achieved [2,3]. At the PCC, nonlinear loads, e.g., diode rectifier interfaced compact fluorescent lamps (CFL) and/or adjustable speed drive (ASD) systems, produce high current harmonics distortions, which may cause equipment malfunction, overheating transformers, efficiency degradation, and resonance. Consequently, DFACTS devices, e.g., active power filters, are exclusively employed to mitigate harmonics distortions and compensate reactive power. With a wide deployment of DGs, power converter-interfaced DGs will become major interactive players in the distribution network, and in many cases, it is already the case. The control of these grid interactive power converters will significantly affect power quality and even determine the stability of the existing and future grids.

1.1.3 Control of power electronic systems

Being the "brain," control technology enables power converters to precisely and flexibly convert and regulate the electrical power for the needs of different DGs, loads, and transmission devices, underpins power conditioning functionalities of

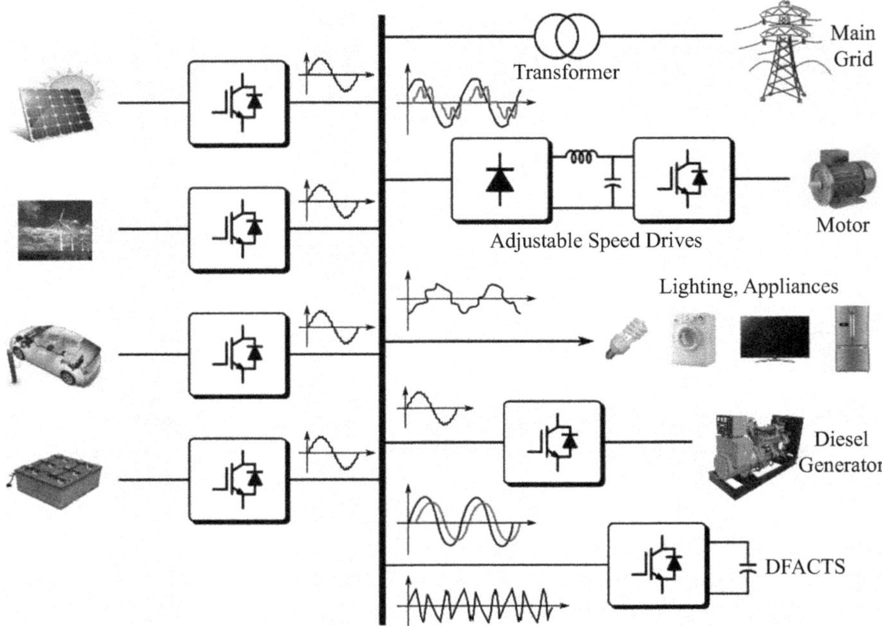

Figure 1.7 A power electronics-based micro-grid system with a connection to the main grid

grid interactive converters, and consequently influences the power quality and the stable operation of the electrical networks. For example, under unbalanced grid voltages caused by grid faults or severe load imbalances, for a grid-connected wind turbine system with power converter interfaces, one synchronous reference frame (SRF) current controller will lead to unexpected power flow oscillations due to the interaction between positive-sequence and negative-sequence voltages. Therefore, dual SRF current controllers [10], one for the positive-sequence and the other for the negative-sequence reference frames, are needed to independently regulate the positive- and negative-sequence currents to avoid unexpected power flow oscillations and ride through grid faults/disturbances. In addition, for three-phase DC–AC power converters [11–13], the interaction between fundamental frequency feed-in current and inherent sixth-order harmonic ripples in the DC-link voltage may degrade the power quality by injecting $6k \pm 1$ order harmonic pollutions into the feed-in currents. Similarly, for single-phase DC–AC power converters, the voltage ripples with twice the grid fundamental frequency in the DC-link may deteriorate the quality of the feed-in current.

To further illustrate how the control method for power converters poses the issue of power quality, an example of a single-phase grid-connected DG converter is examined in detail. One of the most important requirements for the inverter is to feed good quality AC power with less than 5% total harmonic distortion (THD) of its current at the PCC in accordance with the IEEE 519 standard [14,15].

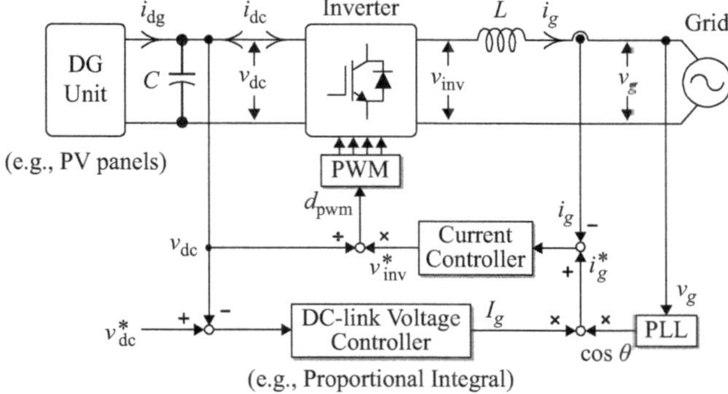

Figure 1.8 Overall structure and control of a grid-connected single-phase PWM inverter system

For the single-phase grid-connected DG system shown in Figure 1.8, it can be described as:

$$\begin{cases} \dfrac{di_g}{dt} = \dfrac{1}{L}\left(v_{\text{inv}} - v_g\right) \\ \dfrac{dv_{\text{dc}}}{dt} = \dfrac{1}{C}\left(i_{\text{dg}} - i_{\text{dc}}\right) \end{cases} \tag{1.1}$$

where i_g is the feed-in current into the grid, v_{inv} is the PV inverter output voltage, v_g is the grid voltage at the PCC, i_{dg} is the output current from the DG unit, and i_{dc} is the DC-link current. If the inverter output voltage v_{inv} or the grid voltage v_g contains any low-order harmonics, the injected feed-in current in the AC side current i_g can be written as:

$$\begin{aligned} i_g &= \frac{1}{L}\int\left[\left(v^1_{\text{inv}} + \sum_{h=2}^{n} v^h_{\text{inv}}\right) - \left(v^1_g + \sum_{h=2}^{n} v^h_g\right)\right]dt \\ &= \underbrace{\frac{1}{L}\int\left(v^1_{\text{inv}} - v^1_g\right)dt}_{i^1_g} + \sum_{h=2}^{n} i^h_{\text{g,C}} - \sum_{h=2}^{n} i^h_{\text{g,G}} \end{aligned} \tag{1.2}$$

with

$$i^h_{\text{g,C}} = \frac{1}{L}\int v^h_{\text{inv}}dt \quad \text{and} \quad i^h_{\text{g,G}} = \frac{1}{L}\int v^h_g dt \tag{1.3}$$

being the feed-in current harmonics induced by the inverter output voltage harmonics v^h_{inv} and the grid voltage distortions v^h_g respectively, i^1_g, v^1_{inv}, and v^1_g, representing the fundamental components of the feed-in current i_g, the inverter output voltage v_{inv}, the grid voltage v_g, and h being the harmonic order. It should be pointed out that the grid voltage harmonics are normally uncontrolled background distortions, which are induced by the nonlinear loads and nonlinear behaviors of the components. Fortunately, (1.2) implies that the current harmonics $i^h_{\text{g,C}}$ induced by the grid voltage distortions can be eliminated by the controllable inverter output voltage v_{inv} if a proper control method and/or modulation technique is applied to the grid-connected inverter. The following will illustrate how the inverter output voltage harmonics are generated, which in turn affects the feed-in grid current quality seen from (1.2) and (1.3).

Without loss of generality, neglecting high-switching frequency harmonics, the average pulse-width modulation (PWM) output voltage v_{inv} of the inverter can be written as:

$$\begin{aligned} v_{\text{inv}} = d_{\text{pwm}} v_{\text{dc}} &= \left(d^1_{\text{pwm}} + \sum_{h=2}^{n} d^h_{\text{pwm}} \right) (V_{\text{dc}} + \tilde{v}_{\text{dc}}) \\ &= d^1_{\text{pwm}} V_{\text{dc}} + d^1_{\text{pwm}} \tilde{v}_{\text{dc}} + V_{\text{dc}} \sum_{h=2}^{n} d^h_{\text{pwm}} + \tilde{v}_{\text{dc}} d^h_{\text{pwm}} \\ &= v^1_{\text{inv}} + \sum_{h=2}^{n} v^h_{\text{inv}} \end{aligned} \tag{1.4}$$

where $-1 \leq d_{\text{pwm}} \leq 1$ is the normalized output of current controller in the linear modulation region and the input of the PWM modulator shown in Figure 1.8, d^1_{pwm} and d^h_{pwm} denote the fundamental component and harmonics of d_{pwm}, respectively, V_{dc} and $\tilde{v}_{\text{dc}}$ are the DC and AC components of the DC-link voltage v_{dc} (i.e., DC-link capacitor voltage). Note that the high-order term $\tilde{v}_{\text{dc}} d^h_{\text{pwm}}$ can be ignored for simplicity, since it also can be filtered out by the low-pass inductor filter.

It is demonstrated by (1.4) that the harmonic components v^h_{inv} of the inverter output voltage v_{inv} are induced by the control output harmonics d^h_{pwm} and the DC-link voltage variation $\tilde{v}_{\text{dc}}$. As a consequence, according to (1.3), the injected current will inevitably inherit the harmonics, and thus the power quality will be affected. Furthermore, it can also be observed in (1.4) that it is possible to apply appropriate $V_{\text{dc}} d^h_{\text{pwm}}$ to eliminate the corresponding harmonics distortions in $d^1_{\text{pwm}} \tilde{v}_{\text{dc}}$ if the inverter controller is properly designed; otherwise $V_{\text{dc}} d^h_{\text{pwm}}$ would even deteriorate the feed-in current quality. The above analysis reveals how the inverter output voltage will affect the grid current quality.

Regarding the DC-link voltage variation $\tilde{v}_{\text{dc}}$, prior-art study has shown that $\tilde{v}_{\text{dc}}$ mainly contains a double-grid frequency component [14,15]. It should be highlighted that the DC-link voltage variation $\tilde{v}_{\text{dc}}$ is inherently induced by the single-phase DC–AC conversion; in other words, it comes from the energy conservation principle – the instantaneous power balance between the DC side and AC side of the single-phase inverter. For a better understanding of the origin mechanism of the DC-link voltage variation, let us assume the fundamental grid voltage v_g^1, the fundamental grid current i_g^1, and the inverter output voltage v_{inv}^1 as:

$$\begin{cases} v_g^1 = \sqrt{2}V_g^1 \cos(\omega_0 t + \varphi) \\ i_g^1 = \sqrt{2}I_g^1 \cos(\omega_0 t) \\ v_{\text{inv}}^1 = \sqrt{2}V_g^1 \cos(\omega_0 t + \varphi - \varphi_1) \end{cases} \tag{1.5}$$

with ω_0 and φ being the grid fundamental angular frequency and the power angle, respectively, φ_1 being the phase angle between the input voltage and the inverter output voltage, and $\theta = \omega_0 t$ being the phase of the input voltage.

From Figure 1.8, the inverter input instantaneous power p_{dc} at the DC side and the inverter output instantaneous power p_{inv} can be respectively obtained as:

$$p_{\text{dc}} = v_{\text{dc}} i_{\text{dc}} = p_{\text{dg}} - v_{\text{dc}} C \frac{dv_{\text{dc}}}{dt} = \underbrace{v_{\text{dc}} i_{\text{dg}}}_{\text{active power}} - \left(C V_{\text{dc}} \frac{d\tilde{v}_{\text{dc}}}{dt} + C\tilde{v}_{\text{dc}} \frac{d\tilde{v}_{\text{dc}}}{dt} \right) \tag{1.6}$$

$$\begin{aligned} p_{\text{inv}} = v_{\text{inv}} i_g &= \underbrace{V_{\text{inv}}^1 I_g^1 \cos(\varphi - \varphi_1)}_{\text{active power}} + V_{\text{inv}}^1 I_g^1 \cos(2\omega_0 t + \varphi_1 - \varphi) \\ &+ \left(i_g^1 \sum_{h=2}^{n} v_{\text{inv}}^h + v_{\text{inv}}^1 \sum_{h=2}^{n} i_g^h \right) + O(h) \end{aligned} \tag{1.7}$$

Being used for power decoupling between the DC side and the AC side, the DC-link capacitor C is to handle the reactive power flow with DC-link voltage variations. Neglecting the inverter losses gives $p_{\text{inv}} = p_{\text{dc}}$ and $p_{\text{dg}} = v_{\text{dc}} i_{\text{dg}} \approx V_{\text{inv}}^1 I_g^1 \cos(\varphi - \varphi_1)$. In most cases, due to $|V_{\text{dc}}| \gg |\tilde{v}_{\text{dc}}|$, $|i_g^1| \gg |i_g^h|$, and $|v_{\text{inv}}^1| \gg |v_{\text{inv}}^h|$, the high-order terms of $C\tilde{v}_{\text{dc}} d\tilde{v}_{\text{dc}}/dt$ in (1.6) and $v_{\text{inv}}^j i_g^k$ $(j, k = 2, 3, \ldots, n)$ in (1.7) can be ignored, and then $p_{\text{inv}} = p_{\text{dc}}$ will lead to:

$$\tilde{v}_{\text{dc}} \approx -\int \left(\frac{V_{\text{inv}}^1 I_g^1}{C V_{\text{dc}}} \cos(2\omega_0 t + \varphi - \varphi_1) \right) dt \tag{1.8}$$

which confirms that a major part of the DC-link voltage variation for single-phase inverters is a double-grid frequency component.

Equation (1.8) implies that, for a given inverter, the amplitude of $\tilde{v}_{\text{dc}}$ is proportional to the grid voltage level V_{inv}^1 and the feed-in current level I_g^1, and varies with the power angle φ. Moreover, a larger DC-link capacitor value C and the DC-link voltage V_{dc} will lead to a smaller DC-link voltage variation $\tilde{v}_{\text{dc}}$ [14,15]. The phasor diagram in Figure 1.9 illustrates the relationship among the fundamental frequency grid voltage v_g^1, feed-in current i_g^1, and inverter output voltage v_{inv}^1. It can be seen in Figure 1.9 that the amplitude of $\tilde{v}_{dc}$ will reach its maximum value at $\varphi = 90°$.

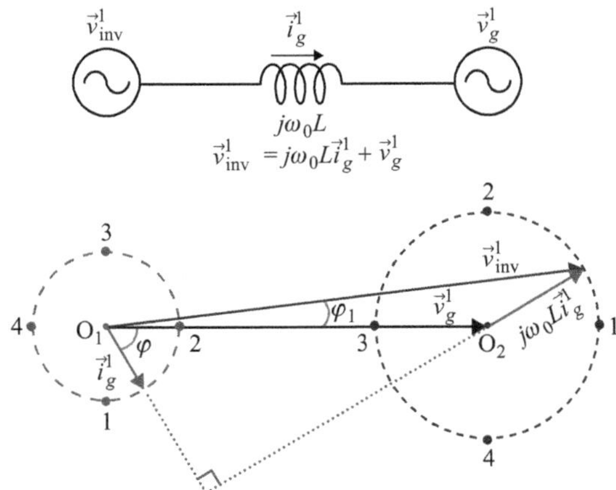

Figure 1.9 Fundamental frequency phasor diagrams for a grid-connected single-phase PWM inverter system

Furthermore, according to (1.2)–(1.4), it can be concluded that the product of $d^1_{\text{pwm}}\tilde{v}_{\text{dc}}$ will introduce harmonics into the inverter output voltage v_{inv}, and further bring harmonics into the feed-in current i_g. Equation (1.8) shows that $\tilde{v}_{\text{dc}}$ mainly contains harmonics with a double-grid frequency of $2\omega_0$. Therefore, $d^1_{\text{pwm}}\tilde{v}_{\text{dc}}$ will produce third-order harmonics with a frequency of $3\omega_0$ in v_{inv}, and consequently results in third-order harmonics appearing in the feed-in current. That explains why the third-order harmonics distortion dominates in the feed-in current from single-phase grid-connected inverters.

The previous example illustrates the mechanism of harmonic current injection from grid-connected single-phase converter. It is also clearly demonstrated that, for a given power converter, to a great extent, the associated power quality is determined by the control strategy. In the following section, we will discuss how to develop desired control strategies for power converters.

1.2 Periodic control of power converters

As shown in Figure 1.7, both sinusoidal grid voltages from the main grid and synchronized feed-in currents from all grid-connected power converter-interfaced DGs are periodic signals. Moreover, the majority of the current/voltage disturbances from grid-connected loads and the current/voltage compensations from DFACTS are also periodic signals. The regulation of periodic voltages/currents for grid-interactive power converters will significantly affect power quality and even determine the stability of the electrical power systems. It is a challenging but very interesting job to investigate high-performance control solutions for power converters to tackle periodic signals with high accuracy, fast dynamic response, good robustness, and cost-effective implementation.

In this section, we will examine a class of disturbance rejection and reference tracking methods in which the disturbances or reference signals are periodic signals. In this scenario, the controller can reject these disturbances or track the references by incorporating the model of the disturbance or the reference signal within itself. This approach is known as internal model principle (IMP), first championed by B. A. Francis and W. M. Wonham (1976) [16,17]. Hereafter, all IMP-based control methods for rejection and tracking of periodic signals are named *Periodic Control* in this book. Periodic control of power converters will be discussed.

1.2.1 Basic control problem for power converters

The very basic control problem for a control system is how to force its output signal to accurately, fast, and robustly track the reference input signal. Figure 1.10 shows the diagram of a typical feedback control system, where $G_c(s)$ is the feedback controller, $G_p(s)$ is the control plant, $d(s)$ is the disturbance, $r(s)$ is the reference voltage/current, and $y(s)$ is the output voltage/current. The transfer functions for such a power control system can be written as follows:

$$y(s) = \frac{G_c(s)G_p(s)}{1 + G_c(s)G_p(s)} r(s) + \frac{1}{1 + G_c(s)G_p(s)} d(s) \tag{1.9}$$

and

$$e(s) = r(s) - y(s) = \frac{1}{1 + G_c(s)G_p(s)} r(s) - \frac{1}{1 + G_c(s)G_p(s)} d(s) \tag{1.10}$$

Theoretically, according to (1.10), for a given control plant $G_p(s)$, if the feedback controller $G_c(s) \to \infty$, we will have $y(s) \to r(s)$ even in the presence of disturbance $d(s)$. It means that $G_c(s) \to \infty$ will enable the closed-loop control system to exactly track a signal of any frequency. However, it is an impossible mission to stabilize such a closed-loop control system with such a controller $G_c(s) \to \infty$ in practical applications.

Fortunately, for many power converter systems, their reference input signals $r(s)$ and disturbances $d(s)$ are periodic signals, e.g., sinusoidal feed-in current reference signal for grid-connected converter-interfaced DGs, sinusoidal output voltage reference signal for stand-alone converter-interfaced uninterruptible power supply (UPS), and targeted current harmonics distortions for active power filters. According to the Fourier series analysis, any periodic signal can be decomposed into the sum of a (possibly infinite) set of simple sine and cosine harmonics

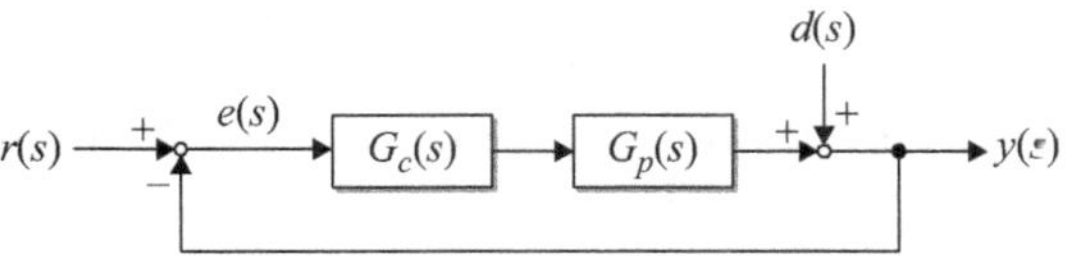

Figure 1.10 A typical feedback control system with a disturbance

(including a DC signal). This implies that, for power converter control systems, the simplified control problem is how to produce $G_c(s) \to \infty$ only at the interested harmonic frequencies for forcing $y(s) \to r(s)$ while still maintaining a fast transient response, guaranteeing robustness, and being feasible to implement. Such a problem is called *periodic control problem of power converters.*

1.2.2 Internal model principle (IMP)

If the reference signal, or deterministic disturbance $x(t)$, is generated by an external system or satisfies a known linear differential or difference equation [18], e.g.:

$$\frac{d^n}{dt^n}x(t) + \alpha_{n-1}\frac{d^{n-1}}{dt^{n-1}}x(t) + \ldots + \alpha_1\frac{d}{dt}x(t) + \alpha_0 x(t) = 0 \tag{1.11}$$

with α_i ($i = 0, 1, 2, \ldots, n-1$) being the differential equation coefficients and n being the equation order, the *Laplace* transform can be obtained [18] as:

$$X(s) = \mathcal{L}\{x(t)\} = \frac{f_x(0,s)}{s^n + \alpha_{n-1}s^{n-1} + \ldots + \alpha_0} = \frac{f_x(0,s)}{\Gamma_x(s)} \tag{1.12a}$$

where $f_x(0,s)$ is a polynomial function of s because of initial conditions (i.e., $x(0)$, $\dot{x}(0)$, $\ddot{x}(0)$, etc.), and $\Gamma_x(s)$ is the denominator of the function $X(s)$.

The denominator $\Gamma_x(s)$ that contains all complex poles p_i of the function $X(s)$, can be rewritten as:

$$\Gamma_x(s) = \prod\nolimits_i^n (s - p_i)$$

It is well known that the inverse *Laplace* transform of $X(s)$ in (1.12a) can be represented as a linear combination of $e^{p_i t}$:

$$x(t) = \mathcal{L}^{-1}\{X(s)\} = \sum_{i=1}^{n} C_i e^{p_i t} \tag{1.12b}$$

in which the complex coefficient C_i is determined by the initial condition function $f_x(0,s)$. Apparently, (1.12b) indicates that $e^{p_i t}$ can be treated as the general basis function for the signal $x(t)$ in the time domain.

Correspondingly, in the s-domain, the signal $X(s)$ can also be represented as a linear combination of its general basis functions in the following:

$$X(s) = \mathcal{L}\{x(t)\} = \mathcal{L}\left\{\sum_{i=1}^{n} C_i e^{p_i t}\right\} = \sum_{i=1}^{n} \mathcal{L}\{C_i e^{p_i t}\} = \sum_{i=1}^{n}\left(C_i \frac{1}{s - p_i}\right) \tag{1.12c}$$

in which $(s - p_i)^{-1}$ ($i = 1, 2, \ldots, n$) constitutes the basis function for the signal $X(s)$.

As shown in Figure 1.11, the type of the signal $x(t)$ is defined by the location of the poles p_i on the s-plane. Specifically, all left half-plane (LHP) poles indicate that $x(t)$ is an asymptotically decayed function; any right half-plane (RHP) pole will lead $x(t)$ to be divergent; the imaginary part of a complex pole p_i denotes its oscillation frequency; the real part of a complex pole p_i denotes its decaying rate.

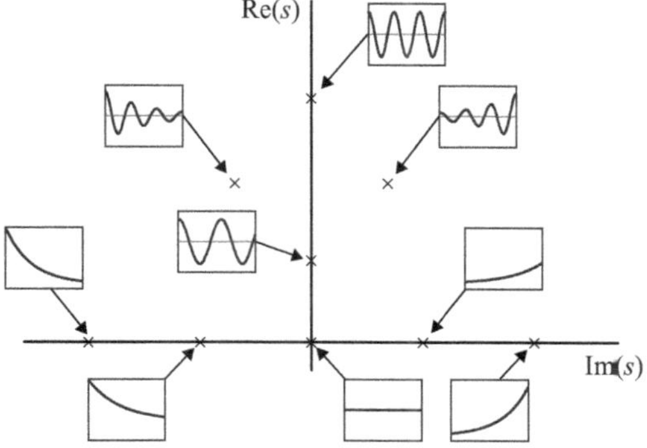

Figure 1.11 Relationship between the pole location on the s-plane and the type of the signal in the time domain

Since $\Gamma_x(s)$ contains all the poles p_i, it can determine the type of the signal $x(t)$, and thus it is defined as the *signal generating polynomial*. More generally, any function that takes the same set of basis functions of $(s - p_i)^{-1}$ (the subscript $i = 1, 2, \ldots, n$), gives a reduplicated model of the dynamic structure of the disturbance and reference signal $x(t)$ [16], and it is called an *internal model* or a signal generator of the signal $x(t)$. The function $X(s)$ in (1.12c) with various coefficients C_i gives a general internal model of the signal $x(t)$.

For example:

1. For $x(t) = \sin(\omega t)$, we have

$$X(s) = \frac{\omega}{(s - j\omega)(s + j\omega)} = \frac{\omega}{s^2 + \omega^2} = \frac{\omega}{\Gamma_x(s)} \text{ with } \Gamma_x(s) = s^2 + \omega^2$$

2. For $x(t) = x_0$, we have

$$X(s) = \frac{x_0}{s - 0} = \frac{x_0}{s} = \frac{x_0}{\Gamma_x(s)} \text{ with } \Gamma_x(s) = s$$

3. For $x(t) = x_0 + \sin(\omega t)$, we have

$$X(s) = \frac{x_0(s^2 + \omega^2) + \omega s}{s(s^2 + \omega^2)} = \frac{1}{\Gamma_x(s)} \text{ with } \Gamma_x(s) = s(s^2 + \omega^2)$$

If the generating polynomial $\Gamma_d(s)$ of the disturbance $d(s)$ or the generating polynomial $\Gamma_r(s)$ of the reference $r(s)$ is included as a part of the controller, asymptotic disturbance rejection or reference tracking can be attained. In other words, asymptotic rejection/tracking of a persistent input signal $x(t)$ can only be attained by suitably replicating its internal model $X(s)$ in a stable feedback loop. The necessity of this control structure constitutes the IMP [16]. Moreover, it should be pointed out that the coefficient C_i for the internal model in (1.12c) should be properly designed to stabilize the closed-loop IMP-based control system.

For a typical control system as shown in Figure 1.10, let the plant model be:

$$G_p(s) = \frac{B_p(s)}{A_p(s)} \tag{1.13}$$

and assuming that $\Gamma_d(s)$ and $\Gamma_r(s)$ are the signal generating polynomials for the disturbance $d(s)$ and the reference $r(s)$, respectively, and either $\Gamma_d(s)$ or $\Gamma_r(s)$ is not a factor of $A_p(s)$. The disturbance $d(s)$ and the reference $r(s)$ can be written as:

$$d(s) = \frac{f_d(0,s)}{\Gamma_d(s)} \quad \text{and} \quad r(s) = \frac{f_r(0,s)}{\Gamma_r(s)}$$

If the feedback controller takes the following form:

$$G_c(s) = \frac{B_c(s)}{A_c(s)\Gamma_d(s)\Gamma_r(s)} \tag{1.14}$$

Equation (1.10) can be rewritten as:

$$e(s) = r(s) - y(s) = G(s)r(s) - G_d(s)d(s) \tag{1.15}$$

with

$$G(s) = \frac{A_c(s)A_p(s)\Gamma_d(s)\Gamma_r(s)}{A_c(s)A_p(s)\Gamma_d(s)\Gamma_r(s) + B_c(s)B_p(s)}$$

$$G_d(s) = \frac{B_p(s)}{A_p(s)}\frac{A_c(s)A_p(s)\Gamma_d(s)\Gamma_r(s)}{A_c(s)A_p(s)\Gamma_d(s)\Gamma_r(s) + B_c(s)B_p(s)}$$

Supposing that $B_c(s)$ and $A_c(s)$ have been selected such that the closed-loop characteristic equation is:

$$A_{CE}(s) = A_c(s)A_p(s)\Gamma_d(s)\Gamma_r(s) + B_c(s)B_p(s) \tag{1.16}$$

whose roots locate on the left half s-plane, i.e., the closed-loop system is asymptotically stable.

Therefore, (1.15) can be rewritten as:

$$e(s) = r(s) - y(s) = \frac{A_c(s)A_p(s)\Gamma_d(s)}{A_{CE}(s)}f_r(0,s) - \frac{B_p(s)}{A_p(s)}\frac{A_c(s)A_p(s)\Gamma_r(s)}{A_{CE}(s)}f_d(0,s) \tag{1.17}$$

Since $A_{CE}(s)$ is assumed to have stable roots, the inverse *Laplace* transform of $E(s)$ converges to 0, i.e., $e(t \to \infty) = 0$ and $y(t) \to r(t)$ in the presence of the disturbance $d(t)$. It is shown that, in the frequency domain, the purpose of the internal model is to provide closed-loop zeros that cancel the poles of the disturbances and reference signals [16].

1.2.3 Internal model principle-based periodic control

According to the Fourier series analysis, a periodic reference or disturbance signal $x(t)$ with a fundamental frequency $\omega_0 = 2\pi/T_0$ can be decomposed into a linear combination of sine and cosine functions, or exponential functions as:

$$x(t) = \frac{a_0}{2} + \sum_{h=1}^{\infty} [a_h \cos(h\omega_0 t) + b_h \sin(h\omega_0 t)] = \sum_{h=0}^{\infty} c_h e^{jh\omega_0 t} \tag{1.18}$$

where

$$a_0 = \frac{1}{T}\int_{t_0}^{t_0+T_0} x(t)dt$$

$$a_h = \frac{1}{T}\int_{t_0}^{t_0+T_0} x(t)\cos(h\omega_0 t)dt$$

$$b_h = \frac{1}{T}\int_{t_0}^{t_0+T_0} x(t)\sin(h\omega_0 t)dt$$

$$c_h = \frac{1}{T}\int_{t_0}^{t_0+T_0} x(t)e^{-jh\omega_0 t}dt$$

Thus a general internal model for the periodic signal $x(t)$ can be written as:

$$X(s) = \frac{a_0}{2}\frac{1}{s} + \sum_{h=1}^{\infty}\left(\frac{a_h s}{s^2 + (h\omega_0)^2} + \frac{b_h(h\omega_0)}{s^2 + (h\omega_0)^2}\right) = \sum_{h=-\infty}^{\infty}\left(\frac{c_h}{s - j(h\omega_0)}\right) \tag{1.19}$$

The internal model $X(s)$ is actually a combination of a series of resonant controllers (including an integral controller) at harmonic frequencies of $h\omega_0$ with $h = 0, 1, 2, \ldots$.

For the typical control system shown in Figure 1.10, the internal model $X(s)$ is embedded into the feedback controller as:

$$G_c(s) = G_f(s)X(s) \tag{1.20}$$

where $G_f(z)$ is a compensator to stabilize the feedback system.

It can be found that if $s = \pm jh\omega_0$ with $h = 0, 1, 2, \ldots$, the controller $G_c(s) \to \infty$ at harmonic frequencies of $h\omega_0$, and then (1.9) will lead to:

$$y(jh\omega_0) = \frac{\infty \cdot G_p(jh\omega_0)}{1 + \infty \cdot G_p(jh\omega_0)} r(jh\omega_0) + \frac{1}{1 + \infty \cdot G_p(jh\omega_0)} d(jh\omega_0)$$

Thus $y(s) \to r(s)$ at harmonic frequencies of $h\omega_0$, i.e., zero-error tracking of periodic signal $y(t)$ with known frequency of $h\omega_0$ is achieved in the presence of disturbance $d(t)$. Hereafter, IMP-based controller $G_c(s)$ in (1.20) is called *periodic controller* in this book.

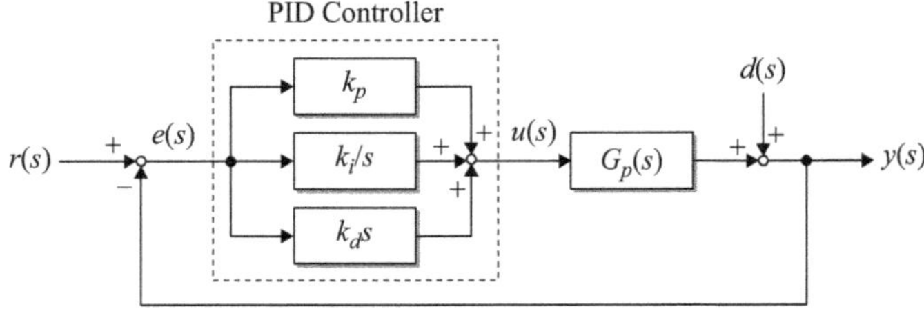

Figure 1.12 A typical proportional–integral–derivative (PID) control system

The proportional–integral–derivative (PID) controller shown in Figure 1.12 is the most commonly used IMP-based periodic controller in industrial applications. The PID controller in the *Laplace* domain can be written as:

$$G_{\mathrm{PID}}(s) = k_p + k_i \frac{1}{s} + k_d s \tag{1.21}$$

where k_p is the proportional gain, k_i is the integral gain, and k_d is the derivative gain. However, the derivative term is rarely used due to the increased noise sensitivity and stability problems associated with high-frequency resonance modes. Obviously, the integral control term $1/s$ in the PID controller (1.21) is the internal model for DC signals. It (i.e., the integral control term $1/s$) enables zero-error tracking of DC signals. Meanwhile P (i.e., 1) and D (i.e., s) terms are used to improve the system stability and transient responses (such as overshoot, settling time, rise rime, and so on). Thus the three-term-type structure PID can achieve accurate, fast, and robust control of DC signals.

Likewise, considering a sinusoidal signal, $x(t) = \sin(h\omega_0 t + \varphi)$ with ω_0 being its fundamental frequency, φ being its phase and h being the harmonic order, the internal model for $x(t)$ can be written as:

$$\hat{G}_h(s) = \frac{s \sin \varphi + (h\omega_0)\cos \varphi}{s^2 + (h\omega_0)^2} \tag{1.22}$$

which also provides an IMP-based periodic control scheme for any sine/cosine signal with a specific frequency $h\omega_0$, called *resonant controller* (RSC) [19–29].

Moreover, a delay-based repetitive controller (RC) [30–50] can be written as:

$$\hat{G}_{\mathrm{rc}}(s) = \frac{e^{-sT_0}}{1 - e^{-sT_0}} = -\frac{1}{2} + \frac{1}{T_0 s} + \frac{1}{T_0} \sum_{h=1}^{\infty} \frac{2s}{s^2 + (h\omega_0)^2} \tag{1.23}$$

which also contains the internal models for all harmonics. Thus, the RC offers an effective IMP-based periodic controller for any periodic signal with a known fundamental frequency ω_0.

From the above analysis, it can be seen that that the IMP-based periodic control provides a simple but accurate control solution to power converters.

1.2.4 Periodic control of power converters

So far, vast progresses have been achieved in periodic control of power converters. Over the past decade, the periodic control technology has been significantly improved to become an attractive high-performance control solution for the power converters to regulate voltages/currents with higher accuracy, faster dynamic response, better robustness, and more feasibility. The RSC and RC are two most popular periodic controllers, which have been widely applied to power converters.

For any sine/cosine signal, IMP-based RSC can achieve zero steady-state error tracking/rejection by including an internal model of the cosine signal at its resonant frequency [21–23]. Owing to its simplicity and effectiveness, RSC becomes a popular current controller for grid converters. A stationary-frame-based RSC is actually equivalent to a synchronous-frame-based proportional–integral (PI) controller [21,24], but it provides several advantages over the latter, such as much less computational burden and complexity due to its lack of the Park transformations, less sensitiveness to noise and error in synchronization. RSC is proposed to control grid-tied PWM rectifiers to produce high-quality sinusoidal input currents by Sato *et al.* in 1998 [22]. Various improvements have been made in RSC:

- Proportional plus RSC and PI plus RSC schemes [24,25] have been applied to grid converters. These controllers behave like a general PI control – the RSC acts as a general integrator for the sinusoidal signal and the proportional/PI acts as a general proportional control.
- Several discretization methods for RSC have been investigated to improve the control system stability and immunity against discretization errors [26].
- Phase-lead compensation RSC has been proposed to improve the transient response and stability of the systems [27–29].
- Parallel combination of multiple RSC (MRSC) controllers has been used to eliminate more harmonics for the low total harmonic distortions. A parallel structure enables the MRSC to independently choose the gain of each RSC component to achieve quite fast transient response. However, if the number of RSC components is large, MRSC [25,27] may cause heavy parallel computational burden and parameter tuning difficulties.

For any arbitrary periodic signal with a known period, the IMP-based RC could achieve zero steady-state errors [23] by including a parallel combination of the internal models for its all harmonics components (including the DC component) [55,56]. These internal models – integral controller for the DC component and RSC components for harmonics, enable the RC to reject all harmonics and DC bias. A delay-based compact infinite impulse response (IIR) form enables the RC to consume much less computation than what the MRSC does. However, since the gains for all RSC components are identical, it is impossible for RC systems to optimize their dynamic response. That is to say, RC usually yields much slower transient response than what the MRSC does. It is initially developed to accurately control a periodic magnet power supply curve for proton synchrotron by Inoue *et al.* in 1981 [30]. Middleton, Goodwin, and Longman applied the RC to Robotics in 1985 [31]. Tomizuka *et al.* [32] proposed a complete synthesis and design method for the digital RC schemes, such as the modified RC, zero-phase compensation for the RC, and so on. RC was first applied to single-phase PWM inverters by Haneyoshi, Kawamura, and Hoft in 1988 [33]. The first patent application of the RC in PWM inverters was filed by Fuji Electric in 1990. Tzou applied the RC to single-phase PWM inverters in 1997 [34]. Over the past decade, a series of RC schemes have been successfully developed and employed in power converter systems:

- PID-like plug-in RC has been applied to PWM inverters [35–49] and rectifiers [50,51] for accurate, fast and robust control.
- Phase-lead compensation method [52,53] has been developed for the RC to improve the stability and transient response.
- "Tailor-made" selective harmonic control (SHC) [54–63] (including the odd harmonic RC, dual module, parallel structure RC, etc.) has been introduced to deal with unevenly distributed power harmonics. It makes a good trade-off between the slow but accurate RC and the fast but less feasible MRSC.
- Frequency adaptive RC [64–74] has been proposed to maintain high control accuracy in the presence of grid frequency variations.

Being the "bread and butter" of control engineering, the IMP-based PID controller is the most common and successful industrial controller due to its simplicity, effectiveness, and maturity. In practical applications, the IMP-based periodic controllers usually take PID-like hybrid structure – the internal model control term (e.g., the RSC, RC, and/or integral controller) is combined with complementary feedback control term (e.g., proportional, proportional–derivative). Taking the periodic control (i.e., internal model) term as a *general integrator* [21], such a hybrid structure periodic controller can be treated as *general proportional–integral–derivative (GPID) control* for periodic signals, where the *general integrator* $G_I(s)$ is used to ensure accurate tracking/elimination of periodic signals and a complementary feedback control term $G_c(s)$ is used to improve the system stability and dynamic response, as shown in Figure 1.13. Like the PID, the hybrid structure enables simple but effective GPID control that can achieve accurate, fast, and robust control of periodic signals. Plug-in RC and PI plus RSC are two examples of typical general PI control schemes for power converters. GPID provides a universal framework for housing various IMP-based periodic controllers. In summary, taking the internal model term as *general integrator*, GPID provides a universal framework for housing various IMP-based periodic controllers to solve different problems in practical applications.

Perfect tracking or rejection of periodic signals plays a critical role not only in electrical power processing for better power quality, but also in many other engineering practices where periodic disturbances are dominant or tight performances are demanded [74–82], such as the track-following servo system of disk drives, steel casting, power electronics-enabled active power filters, active air-bearing systems, active noise control, satellite attitude stabilization, peristaltic pumps used in medical devices and robotized laparoscopic surgery, and robots performing

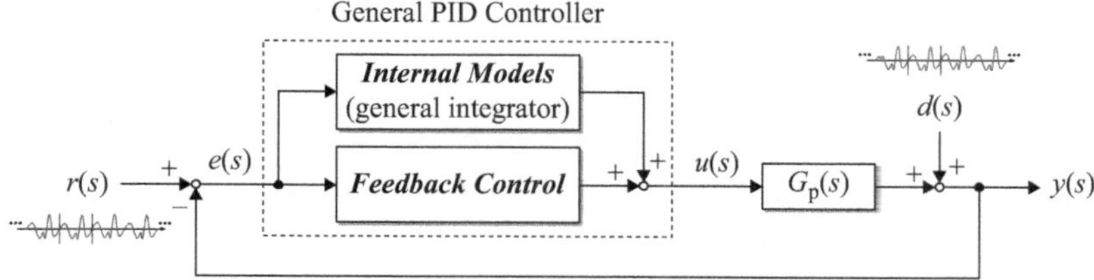

Figure 1.13 General PID control system for periodic signals

repetitive tasks. The IMP-based periodic control has the potential to be the simplest, the most efficient and easily tunable high-performance control solution for these real-world problems and to be a very popular standard industrial controller such as the PID control.

1.3 What is in this book

Advanced power electronic converters, which can precisely and efficiently convert, control, and condition electricity, play a key role in the successful grid integration of different distributed generators, loads, and transmission devices. The global electricity processed by power electronic converters would be up to 80% in the very near future. The power quality and even the stability of electrical power systems would be affected and even determined by massive interfacing power converters. As a consequence, power converters highly demand optimal control strategies for periodic voltages/currents compensation to assure good power quality and stable power system operation. Simple but very effective IMP-based periodic controllers offer attractive control solutions to power converters.

Addressing the emerging issues discussed in this chapter, this book will systematically and comprehensively present the most recent IMP-based periodic control technology, which is dedicated to offering optimal and even perfect periodic control solution for power electronic conversion systems – high control accuracy, fast transient response, good robustness, and easy implementation. It will mainly include the following:

- IMP-based periodic control in Chapter 1, which investigates the demand of periodic control, IMP, internal model and periodic control, and proposes a general PID framework.
- Fundamental periodic control technology in Chapter 2, which explores the RC and RSC schemes in more detail.
- Advanced periodic control technology in Chapter 3, which involves the parallel structure, selective harmonic elimination, and optimal periodic control.
- Application examples of the fundamental and advanced periodic control of various power converters in Chapter 4, which includes the periodic control of voltages for constant-voltage constant-frequency (CVCF) converters and the periodic control of currents for grid-connected converters.

- Frequency-adaptive periodic control in Chapter 5, which presents two fractional order approaches to achieve high-performance periodic control in response to frequency variations.
- Application examples of the frequency-adaptive periodic control (FAPC) of various power converters in Chapter 6, which includes the FAPC of voltages for CVCF converters and the FAPC of currents for grid-connected converters.
- Continuing development of the periodic control technology in Chapter 7, which discusses the multi-period periodic control and periodic signal processing for the power converter control.

More important, besides complete analysis and synthesis methods for periodic systems, plenty of practical application examples will be provided to demonstrate the validity of the proposed periodic control (PC) technology for power converters e.g., voltage control of power inverters for UPS, current control for grid-connected converters for PV generators or active power filters. Periodic control provides an excellent control solution to periodic signal compensation in extensive engineering applications, such as ultrahigh accuracy nano-positioning, grid integration of renewable generation via power converters, power quality systems, and so on.

References

[1] Prem, E. "Electronics enabling efficient energy usage – Results from the E4U project, Vienna, Austria," 2009. http://www.e4efficiency.eu/fileadmin/downloads/Final_E4U_publication_V13.pdf

[2] Tolbert, L. M., King, T. J., Ozpineci, B. *et al.*, "Power electronics for distributed energy systems and transmission and distribution applications: Assessing the technical needs for utility applications," ORNL/TM-2005/230, Oak Ridge National Laboratory, Dec. 2005.

[3] Strzelecki, R. and Benysek, G., Editors, *Power Electronics in Smart Electrical Energy Networks*, Springer-Verlag, 2008.

[4] Blaabjerg, F., Chen, Z., and Kjaer, S. B., "Power electronics as efficient interface in dispersed power generation systems," *IEEE Trans. Power Electron.*, vol. 19, no. 5, pp. 1184–1194, 2004.

[5] Carrasco, J.M., Franquelo, L.G., and Bialasiewicz, J.T. *et al.*, "Power-electronic systems for the grid integration of renewable energy sources: A survey," *IEEE Trans. Ind. Electron.*, vol. 53, no. 4, pp. 1002–1016, 2006.

[6] Divan, D., Moghe, R., and Prasai, A., "Power electronics at the grid edge – The key to unlocking value from the smart grid," *IEEE Power Electron. Mag.*, pp. 16–22, Dec. 2014.

[7] Teodorescu, R., Liserre, M., and Rodríguez, P., *Grid Converters for Photovoltaic and Wind Power Systems*, Wiley-IEEE, 2011.

[8] Vasquez, J. C., Guerrero, J. M., Luna, A., Rodriguez, P., and Teodorescu, R., "Adaptive droop control applied to voltage-source inverters operating in grid-connected and islanded modes IEEE 4088–4096," *IEEE Trans. Ind. Electron.*, vol. 56, no. 10, pp. 4088–4096, 2009.

[9] Vasquez, J.C., Mastromauro, R.A., Guerrero, J.M., and Liserre, M., "Voltage support provided by a droop-controlled multifunctional inverter," *IEEE Trans. Ind. Electron.*, vol. 56, no. 11, pp. 4510–4519, 2009.

[10] Song, H.-S. and Nam, K., "Dual current control scheme for PWM converter under unbalanced input voltage conditions," *IEEE Trans. Ind. Electron.*, vol. 46, no. 5, pp. 953–959, 1999.

[11] Zhou, K., Yang, Y., Blaabjerg, F., and Wang, D., "Optimal selective harmonic control for power harmonics mitigation," *IEEE Trans. Ind. Electron.*, vol. 62, no. 2, pp. 1220–1230, 2015.

[12] Lu, W., Zhou, K., Wang, D., and Cheng, M., "A generic digital nk +/-m order harmonic repetitive control scheme for PWM converters," *IEEE Trans. Ind. Electron.*, vol. 61, no. 3, pp. 1516–1527, 2013.

[13] Zhou, K., Qiu, Z. and Yang, Y., Current Harmonics Suppression of Single-Phase PWM Rectifiers. *IEEE PEDG*, Jun. 25–28, 2012.

[14] Zhou, K., Qiu, Z., Watson, N. R., and Liu, Y., "Mechanism and elimination of harmonic current injection from single-phase grid-connected PWM converters," *IET Power Electron.*, vol. 6, no. 1, pp. 88–95, 2013.

[15] Yang, Y., Zhou, K., and Blaabjerg, F., "Current harmonics from single-phase grid-connected inverters – examination and suppression," *IEEE J. Emerging and Selected Topics in Power Electron.*, vol. 49, no. 1, pp. 221–233, 2016.

[16] Francis, B. A. and Wonham, W. M., "The internal model principle of control theory," *Automatica*, vol. 12, no. 5, pp. 457–465, 1976.

[17] Davison, E. J., "The robust control of a servomechanism problem for time-invariant multivariable systems," *IEEE Trans. Automatic Control*, vol. 21, no. 2, pp. 25–34, 1976.

[18] Goodwin, G.C., Graebe, S.F., and Salgado, M.E., *Control System Design*, Prentice Hall, 2000.

[19] Rowan, T.M. and Kerkman, R.J., "A new synchronous current regulator and an analysis of current regulated PWM inverters," *IEEE Trans. Ind. Appl.*, vol. 22, no. 3, pp. 678–690, 1986.

[20] Draou A., Sato Y., and Kataoka T., "A new state feedback-based transient control of PWM AC to DC voltage type converters," *IEEE Trans. Power Electron.*, vol. 10, no. 6, pp. 716–724, 1995.

[21] Yuan, X., Merk, W., Stemmler, H., and Allmeling, J., "Stationary frame-generalized integrators for current control of active power filters with zero steady-state error for current harmonics of concern under unbalanced and distorted operating conditions," *IEEE Trans. Ind. Appl.*, vol. 38, no. 2, pp. 523–532, 2002.

[22] Sato, Y., Ishizuka, T., Nezu, K., and Kataoka, T., "A new control strategy for voltage-type PWM rectifiers to realize zero steady-state control error in input current," *IEEE Trans. Ind. Appl.*, vol. 34, no. 3, pp. 480–486, 1998.

[23] Fukuda, S. and Yoda, T., "A novel current-tracking method for active filters based on a sinusoidal internal model," *IEEE Trans. Ind. Appl.*, vol. 37, no. 3, pp. 888–895, 2001.

[24] Liserre, M., Teodorescu, R., and Blaabjerg, F., "Multiple harmonics control for three-phase grid converter systems with the use of PI-RES current controller in a rotating frame," *IEEE Trans. Power Electron.*, vol. 21, no. 3, pp. 836–841, 2006.

[25] Teodorescu, R., Blaabjerg, F., Liserre, M., and Loh, P. C., "Proportional-resonant controllers and filters for grid-connected voltage-source converters," *IEE Proc. – Electric Power Appl.*, vol. 153, no. 5, pp. 750–762, 2006.

[26] Yepes, G., Freijedo, F. D., Lopez, O., and Doval-Gandoy, J., "High performance digital resonant controllers implemented with two integrators," *IEEE Trans. Power Electron.*, vol. 6, no. 2, pp. 563–576, 2011.

[27] Yang, Y., Zhou, K., and Cheng, M., "Phase compensation multi-resonant control of CVCF PWM converters," *IEEE Trans. Power Electron.*, vol. 28, no. 8, pp. 3923–3930, 2013.

[28] Yang, Y., Zhou, K., and Cheng, M., "Phase compensation resonant controller for PWM converters," *IEEE Trans. Ind. Informatics*, vol. 9, no. 2, pp. 957–964, 2013.

[29] Cárdenas, R., Juri, C., Peña, R., Wheeler, P., and Clare, J., "The application of resonant controllers to four-leg matrix converters feed-in unbalanced or nonlinear loads," *IEEE Trans. Power Electron.*, vol. 27, no. 3, pp. 1120–1129, 2012.

[30] Inoue, T., Nakano, M., Kubo, T., Matsomoto, S., and Baba, H., "High accuracy control of a proton synchrotron magnet power supply," in Proceedings of the 8th IFAC World Congress, pp. 3137–3142, 1981.

[31] Middleton, R. H., Goodwin, G. C., and Longman, R. W., "A method for improving the dynamic accuracy of a robot performing a repetitive task," *Int. J. Robotics Research*, vol. 8, no. 5, pp. 67–74, 1989.

[32] Tomizuka, M., Tsao, T.-C., and Chew, K.-K., "Analysis and synthesis of discrete-time repetitive controllers," *ASME: J. Dyn. Syst. Meas. Control*, vol. 111, pp. 353–358, 1989.

[33] Haneyoshi, T., Kawamura, A., and Hoft, R. G., "Waveform compensation of PWM inverter with cyclic fluctuating loads," *IEEE Trans. Ind. Appl.*, vol. 24, no. 4, pp. 582–589, 1988.

[34] Tzou, Y.-Y., Ou, R.-S., Jung, S.-L., and Chang, M.-Y., "High-performance programmable AC power source with low harmonic distortion using DSP-based repetitive control technique," *IEEE Trans. Power Electron.*, vol. 12, no. 4, pp. 715–725, 1997.

[35] Zhou, K. and Wang, D., "Zero tracking error controller for three-phase CVCF PWM inverter," *Electronics Letters*, vol. 36, no. 10, pp. 864–865, 2000.

[36] Zhou, K., Wang, D., Low, K. S., "Periodic errors elimination in CVCF PWM DC/AC converter systems: A repetitive control approach," *IEE Proc. – Control Theory and Appl.*, vol. 147, no. 6, pp. 694–700, 2000.

[37] Zhou, K. and Wang, D., "Digital repetitive learning controller for three-phase CVCF PWM inverter," *IEEE Trans. Ind. Electron.*, vol. 48, no. 4, pp. 820–830, 2001.

[38] Zhou, K. and Wang, D., "Unified robust zero error tracking control of CVCF PWM converters," *IEEE Trans. Circuits and Systems (I)*, vol. 49, no. 4, pp. 492–501, 2002.
[39] Zhang, K., Kang, Y., Xiong, J., and Chen, J., "Direct repetitive control of SPWM inverter for UPS purpose," *IEEE Trans. Power Electron.*, vol. 18, no. 3, pp. 784–792, 2003.
[40] Zhao, W., Lu, D. D.-C, and Agelidis, V. G., "Current control of grid-connected boost inverter with zero steady-state error," *IEEE Trans. Power Electron.*, vol. 26, no. 10, pp. 2825–2834, 2011.
[41] Jiang, S., Cao, D., Li, Y., Liu, J., and Peng, F. Z., "Low THD, fast transient, and cost-effective synchronous-frame repetitive controller for three-phase UPS inverters," *IEEE Trans. Power Electron.*, vol. 27, no. 6, pp. 2994–3005, 2012.
[42] Cho, Y. and Lai, J., "Digital plug-in repetitive controller for single-phase bridgeless PFC converters," *IEEE Trans. Power Electron.*, vol. 28, no. 1, pp. 165–175, 2013.
[43] Ouyang, H., Zhang, K., Zhang, P., Kang, Y., and Xiong, J., "Repetitive compensation of fluctuating DC link voltage for railway traction drives," *IEEE Trans. Power Electron.*, vol. 26, no. 8, pp. 2160–2171, 2011.
[44] Ye, Y., Zhang, B., Zhou, K., Wang, D., and Wang, Y., "High-performance cascade type repetitive controller for CVCF PWM inverter: Analysis and design," *IET Electrical Power Appl.*, vol. 1, no. 1, 112–118, 2007.
[45] Ye, Y., Zhou, K., Zhang, B., Wang, D., and Wang, J., "High performance repetitive control of PWM DC-AC converters with real-time phase-lead FIR filter," *IEEE Trans. Circuits and Systems (II)*, vol. 53, no. 8, pp. 768–772, 2006.
[46] Yang, Y., Zhou, K., and Lu, W., "Robust repetitive control scheme for three-phase CVCF PWM inverters," *IET Power Electron.*, vol. 5, no. 6, pp. 669–677, 2012.
[47] Hornik, T. and Zhong, Q.-C., "A current-control strategy for voltage-source inverters in microgrids based on H∞ and repetitive control," *IEEE Trans. Power Electron.*, vol. 26, no. 3, pp. 943–952, 2011.
[48] Zhang, B., Zhou, K., and Wang, D., "Multirate repetitive control for PWM DC/AC converters," *IEEE Trans. Ind. Electron.*, vol. 61, no. 6, pp. 2883–2890, 2014.
[49] Zhu, W., Zhou, K., and Cheng, M., "A bidirectional high-frequency-link single-phase inverter: Modulation, modeling, and control," *IEEE Trans. Power Electron.*, vol. 29, no. 8, pp. 4049–4057, 2014.
[50] Zhou, K. and Wang, D., "Digital repetitive controlled three-phase PWM rectifier," *IEEE Trans. Power Electron.*, vol. 18, no. 1, pp. 309–316, 2003.
[51] Zhu, W., Zhou, K., Cheng, M., and Peng, F., "A high-frequency-link single-phase PWM rectifier," *IEEE Trans. Ind. Electron.*, vol. 62, no. 1, pp. 289–298, 2015.
[52] Zhang, B., Wang, D., Zhou, K., Wang, Y., "Linear phase-lead compensated repetitive control of PWM inverter," *IEEE Trans. Ind. Electron.*, vol. 55, no. 4, pp. 1595–1602, 2008.

[53] Zhang, B., Zhou, K., Wang, Y., and Wang, D., "Performance improvement of repetitive controlled PWM inverters: A phase-lead compensation solution," *Int. J. Circuit Theory and Appl.*, vol. 38, no. 5, pp. 453–469, 2010.

[54] Costa-Castelló, R., Grinó, R., and Fossas, E., "Odd-harmonic digital repetitive control of a single-phase current active filter," *IEEE Trans. Power Electron.*, vol. 19, no. 4, pp. 1060–1068, 2004.

[55] Griñó, R. and Costa-Castelló, R., "Digital repetitive plug-in controller for odd-harmonic periodic references and disturbances," *Automatica*, vol. 41, no. 1, pp. 153–157, 2005.

[56] Escobar, G., Hernandez-Briones, P. G., Martinez, P. R., Hernandez-Gomez, M., and Torres-Olguin, R. E., "A repetitive-based controller for the compensation of $6l\pm1$ harmonic components," *IEEE Trans. Ind. Electron.*, vol. 55, no. 8, pp. 3150–3158, 2008.

[57] Zhou, K., Low, K. S., Wang, D., Luo, F.-L., Zhang, B., and Wang, Y., "Zero-phase odd-harmonic repetitive controller for a single-phase PWM inverter," *IEEE Trans. Power Electron.*, vol. 21, no. 1, pp. 193–201, 2006.

[58] Zhou, K., Wang, D., Zhang, B., Wang, Y., Ferreira, J. A., and de Haan, S. W. H., "Dual-mode structure digital repetitive control," *Automatica*, vol. 43, no. 3, pp. 546–554, 2007.

[59] Zhou, K., Wang, D., Zhang, B., Wang, Y., "Plug-in dual-mode structure digital repetitive controller for CVCF PWM inverters," *IEEE Trans. Ind. Electron.*, vol. 56, no. 3, pp. 784–791, 2009.

[60] Lu, W., Zhou, K., and Wang, D., "General parallel structure digital repetitive control," *Int. J. Control.* vol. 86, no. 1, pp. 70–83, 2013.

[61] Zhou K., Yang Y., Blaabjerg F., and Wang D., "Optimal selective harmonic control for power harmonics mitigation," *IEEE Trans. Ind. Electron.*, vol. 62, no. 2, pp. 1220–1230, 2015.

[62] Lu, W., Zhou, K., Wang, D., and Cheng, M., "A general parallel structure repetitive control scheme for multiphase DC-AC PWM converters," *IEEE Trans. Power Electron.*, vol. 28, no. 8, pp. 3980–3987, 2013.

[63] Mattavelli, P., and Marafão, F. P., "Repetitive-based control for selective harmonic compensation in active power filters," *IEEE Trans. Ind. Electron.*, vol. 51, no. 5, pp. 1018–1024, 2004.

[64] Wang, Y., Wang, D., Zhang, B., and Zhou, K., "Fractional delay-based repetitive control with application to PWM DC/AC converters," *IEEE Multi-Conference on Systems and Control*, Singapore, Oct. 1–3, 2007.

[65] Longman, R., "On the theory and design of linear repetitive control systems," *Eur. J. Control*, vol. 5, pp. 447–496, 2010.

[66] Rashed, M., Klumpner, C., and Asher, G., "Repetitive and resonant control for a single-phase grid-connected hybrid cascaded multilevel converter," *IEEE Trans. Power Electron.*, vol. 28, no. 5, pp. 2224–2234, 2013.

[67] Chen, D., Zhang, J., and Qian, Z., "An improved repetitive control scheme for grid-connected inverter with frequency-adaptive capability," *IEEE Trans. Ind. Electron.*, vol. 60, no. 2, pp. 814–823, 2013.

[68] Zou, Z., Zhou, K., Wang, Z., and Cheng, M., "Fractional-order repetitive control of programmable AC power sources," *IET Power Electron.*, vol. 7, no. 2, pp. 431–438, 2014.

[69] Liu, T., Wang, D., and Zhou, K., "High-performance grid simulator using parallel structure fractional repetitive control," *IEEE Trans. Power Electron.*, vol. 31, no. 3, pp. 2669–2679, 2016.

[70] Yang, Y., Zhou, K., and Blaabjerg, F., "Enhancing the frequency adaptability of periodic current controllers with a fixed sampling rate for grid-connected power converters," *IEEE Trans. Power Electron.*, vol. 31, no. 10, pp. 7273–7285, 2016.

[71] Yang, Y., Zhou, K., Wang, H., Blaabjerg, F., Wang, D., and Zhang, B., "Frequency-adaptive selective harmonic control for grid-connected inverters," *IEEE Trans. Power Electron.*, vol. 30, no. 7, pp. 3912–3924, 2015.

[72] Yang, Y., Zhou, K., and Blaabjerg, F., "Frequency adaptability of harmonics controllers for grid-interfaced converters," *Int. J. Control*, pp. 1–12, DOI: 10.1080/00207179.2015.1022957, 2015 (Early Online Publication).

[73] Zou, Z.-X., Zhou, K., Wang, Z., and Cheng, M., "Frequency adaptive fractional order repetitive control of shunt active power filters," *IEEE Trans. Ind. Electron.*, vol. 62, no. 3, pp. 1659–1668, 2015.

[74] Ramos, G. A., Costa-Castelló, R., Olm, J. M., *Digital Repetitive Control under Varying Frequency Conditions*, Springer-Verlag, 2013.

[75] Bhikkaji B. and Moheimani S. O., "Integral resonant control of a piezoelectric tube actuator for fast nanoscale positioning," *IEEE/ASME Trans. Mechatronics*, vol. 13, no. 5, pp. 530–537, 2008.

[76] Lee, H. S., "Controller optimization for minimum position error signals of hard disk drives," *IEEE Trans. Ind. Electron.*, vol. 48, no. 5, pp. 945–950, 2001.

[77] Ramesh, R., Mannan, M. A., and Poo, A. N., "Tracking and contour error control in CNC servo systems," *Int. J. Machine Tools and Manufacture*, vol. 45, no. 3, pp. 301–326, 2005.

[78] Bodson, M., Jensen, J. S., and Douglas, S. C., "Active noise control for periodic disturbances," *IEEE Trans. Control Syst. Technol.*, vol. 9, no. 1, pp. 200–205, 2001.

[79] Manayathara, T. J., Tsao, T.-C., and Bentsman, J., "Rejection of unknown periodic load disturbances in continuous steel casting process using learning repetitive control approach," *IEEE Trans. Control Syst. Technol.*, vol. 4, no. 3, pp. 259–265, 1996.

[80] Hillerstrom, G., "Adaptive suppression of vibrations – a repetitive control approach," *IEEE Trans. Control Syst. Technol.*, vol. 4, no. 1, pp. 72–78, 1996.

[81] Pipeleers, G., Demeulenaere, B., Swevers, J., *Optimal Linear Controller Design for Periodic Inputs*, Springer-Verlag, 2009.

[82] Rocabert, J., Luna, A., Blaabjerg, F., and Rodríguez, P., "Control of power converters in AC microgrids," *IEEE Trans. Power Electron.*, vol. 27, no. 11, pp. 4734–4748, Nov. 2012.

Chapter 2
Fundamental periodic control

Abstract

The internal model principle (IMP) states that perfect asymptotic rejection/tracking of persistent inputs can only be attained by replicating the signal generator in a stable feedback loop [1]. The signal generator is also called "internal model" of the inputs. W. M. Wonham summarized IMP as "Every good regulator must incorporate a model of the outside world." Based on IMP [1,2], this chapter presents the fundamental periodic controllers for providing zero steady-state error compensation for periodic signals and elaborates their general design methodology. These IMP-based periodic controllers include repetitive control (RC) [3–21], multi-resonant control (MRSC) [22–34], and discrete Fourier transformation (DFT)-based RC [26,35–36]. The general design methodology comprises a standard internal model for periodic signals and the synthesis methods for universal plug-in structure periodic control (PC) systems. The relationship among these three fundamental periodic controllers will also be demonstrated.

2.1 Repetitive control (RC)

2.1.1 Internal model of any periodic signal

As shown in Figure 2.1, a periodic signal with a period of T_0 can be generated by a time-delay-based positive feedback loop in the continuous-time domain. According to Figure 2.1(a), the transfer function of the periodic signal generator can be written as [3]:

$$\hat{G}_{\rm rc}(s) = \frac{u_{\rm rc}(s)}{e(s)} = \frac{e^{-sT_0}}{1 - e^{-sT_0}} \tag{2.1}$$

where $T_0 = 2\pi/\omega_0 = 1/f_0$ is the signal period with f_0 being the signal frequency and ω_0 being the angular signal frequency. The periodic signal generator $\hat{G}_{\rm rc}(s)$ can be expanded as [34]:

$$\hat{G}_{\rm rc}(s) = \frac{e^{-sT_0}}{1 - e^{-sT_0}} = \overbrace{-\frac{1}{2}}^{\text{Impulse}} + \overbrace{\frac{1}{sT_0}}^{\text{Step}} + \overbrace{\frac{1}{T_0}\sum_{n=1}^{\infty}\frac{2s}{s^2 + (n\omega_0)^2}}^{\text{Harmonics}} \tag{2.2}$$

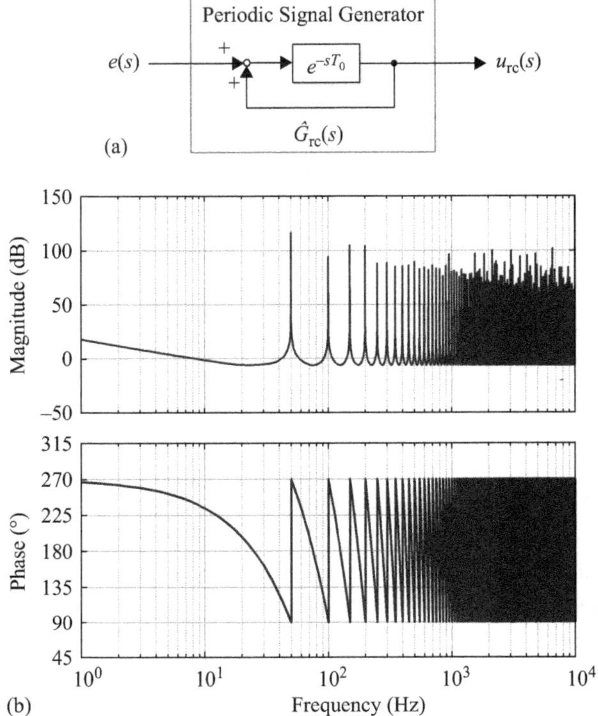

Figure 2.1 Periodic signal generator: (a) basic structure and (b) Bode plot characteristics, where $T_0 = 0.02$ s

which indicates that the periodic signal generator $\hat{G}_{rc}(s)$ is equivalent to the parallel combination of the signal generators for an impulse signal, a step signal, and all harmonics. Notice that the signal generator $\hat{G}_{rc}(s)$ has poles at $s = \pm jn\omega_0$, $n \in \mathbb{N}$. Hence, $\hat{G}_{rc}(s)$ exhibits infinite gains at the harmonic frequencies of $n\omega_0$, i.e., $\hat{G}_{rc}(s)|_{s=\pm jn\omega_0} \to \infty$. The infinite gains lead to zero-tracking errors at these frequencies if $\hat{G}_{rc}(s)$ is included into the closed-loop system. It should be pointed out that the periodic signal generator $\hat{G}_{rc}(s)$ is actually an *internal model* of the periodic signal of the fundamental frequency f_0.

2.1.2 Classic RC scheme

Based on the internal model $\hat{G}_{rc}(s)$ in (2.1), a general structure of the IMP-based RC is shown in Figure 2.2(a).

The corresponding transfer function of the classic RC in the continuous-time domain is [34]:

$$G_{rc}(s) = \frac{u_{rc}(s)}{e(s)}$$

$$= k_{rc}\frac{Q(s)e^{-sT_0}}{1-Q(s)e^{-sT_0}}e^{sT_c} \overset{Q(s)=1}{\longrightarrow} k_{rc}e^{sT_c}\left[-\frac{1}{2}+\frac{1}{sT_0}+\frac{1}{T_0}\sum_{n=1}^{\infty}\frac{2s}{s^2+(n\omega_0)^2}\right] \tag{2.3}$$

where T_0 is the period of the periodic signal as defined previously, k_{rc} is the control gain for tuning the error convergence rate, $Q(s)$ with $|Q(j\omega)| \leq 1$ is usually a low-pass filter (LPF) for compensating the harmonics of interest, and e^{sT_c} with T_c being the compensation time is a phase-lead compensator for the entire system delay compensation. In practical applications, like the coefficients C_i in (1.12c), k_{rc}, $Q(s)$, and e^{sT_c} are employed to enhance the RC system performance, such as system robustness and dynamic response.

Note that, from (2.3), it is known that:

- When $Q(s) \rightarrow 1$, the classic RC $G_{rc}(s) \rightarrow \infty$ at poles $s = \pm jn\omega_0$, and thus it offers exact tracking control of the periodic signals with an interested frequency f_0. That is to say, the RC offers an accurate control solution to the compensation of periodic signals.
- Since both the coefficients and phase-lead compensation for all harmonic generator components in (2.3) are identical, the RC $G_{rc}(s)$ does not have the freedom of coefficient-tuning and/or phase-lead compensation independently at each harmonic frequency in order to optimize its dynamic response.

Most modern controllers are implemented in their digital forms in practical applications. In the discrete-time domain, the internal model $\hat{G}_{rc}(s)$ in (2.1) can be discretized as:

$$\hat{G}_{rc}(z) = \frac{z^{-N}}{1-z^{-N}} \tag{2.4}$$

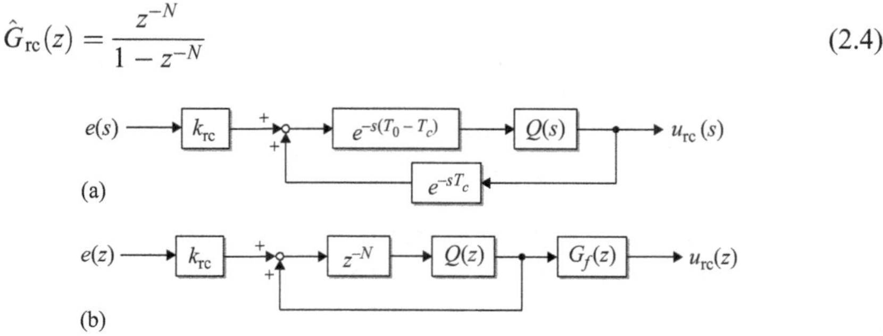

Figure 2.2 Classic repetitive controller: (a) in the continuous-time domain and (b) in the discrete-time domain

where $N = T_0/T_s \in \mathbb{N}$ is the fundamental period with T_s being the sampling period. Notice that the finite-impulse-response (FIR) filter z^{-N} is an all-pass filter for all harmonics up to the Nyquist frequency. The signal generator $\hat{G}_{rc}(z)$ is constructed by the FIR filter z^{-N} with a positive feedback loop, and it contains the internal models of all harmonics up to the Nyquist frequency.

As shown in Figure 2.2(b), the general form of the digital classic RC can be written as [4]:

$$G_{rc}(z) = k_{rc}\frac{Q(z)z^{-N}}{1 - Q(z)z^{-N}}G_f(z) \tag{2.5}$$

in which $G_f(z)$ is the phase-lead compensator, and the LPF $Q(z)$ usually employs an FIR LPF as follows:

$$Q(z) = \left(\sum_{i=1}^{m} a_i z^i + a_0 + \sum_{i=1}^{m} a_i z^{-i}\right) \Big/ \left(2\sum_{i=1}^{m} a_i + a_0\right) \tag{2.6}$$

with $a_i > 0$ and $m \in \mathbb{N}$ with $m \leq (N-1)/2$. It is clear that $|Q(e^{j\omega})| \leq 1$ and $\angle Q(e^{j\omega}) = 0$.

From (2.5), it is known that:

- $1 - Q(z)z^{-N} \to 0$ when $Q(z) \to 1$, and thus $G_{rc}(z) \to \infty$ at the harmonic frequencies of $\omega_k = 2k\pi/T_0$ with $k = 0, 1, 2, \ldots, [N/2]$ up to the Nyquist frequency.
- In the engineering practice, to shape the RC system performance, the synthesis of the RC actually involves the selections of k_{rc}, $G_f(z)$, and $Q(z)$.

2.1.3 Digital RC system and design

Owing to the high-order time delay z^{-N} from its input to its output, the dynamic response of a RC controller is much slower than that of an instantaneous feedback controller. Therefore, RC controllers are usually added on to the closed-loop instantaneous feedback control systems to complementarily achieve optimal performance – the "rough" instantaneous feedback controller ensures good robustness and fast dynamic response, and the add-on "fine" feed-forward RC controller offers high control accuracy [8–21].

Figure 2.3 shows two add-on digital RC systems, where $G_p(z)$ is the plant, $G_c(z)$ is a conventional feedback controller, $G_{rc}(z)$ is a digital modified RC controller with predefined k_{rc}, $Q(z)$, and $G_f(z)$, $r(z)$ is the reference input, $y(z)$ is the output, $e(z) = r(z) - y(z)$ is the tracking error, and $d(z)$ is the disturbance. It is easy to find that the cascaded RC system [16,19] shown in Figure 2.3(b) is equivalent to the plug-in RC system shown in Figure 2.3(a). The plug-in RC is more widely used in practical applications.

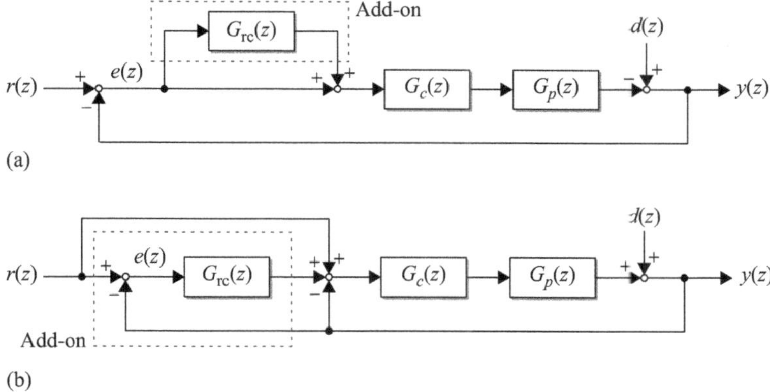

Figure 2.3 Add-on digital RC systems: (a) a plug-in RC system and (b) a cascaded RC system

According to Figure 2.3, without the plug-in RC $G_{rc}(z)$, the transfer function $H(z)$ of the entire feedback control system can be written as:

$$H(z) = \frac{G_c(z)G_p(z)}{1 + G_c(z)G_p(z)} = \frac{z^{-d}B^+(z)B^-(z)}{A(z)} \tag{2.7}$$

where d represents known delay steps; all characteristic roots of $A(z) = 0$ are inside the unit circle; $B^+(z)$ and $B^-(z)$ are the cancellable and un-cancellable parts of $B(z)$, respectively. $B^+(z)$ comprises roots on or outside the unit circle and undesirable roots, which are in the unit circle and $B^+(z)$ comprises roots of $B(z)$ that are not in $B^-(z)$ [4].

Furthermore, when the plug-in RC scheme $G_{rc}(z)$ is absent in Figure 2.3(a), the tracking error $e_0(z)$ can be written as:

$$e_0(z) = r(z) - y(z) = \frac{1}{1 + G_c(z)G_p(z)}[r(z) - d(z)] = S_0(z)[r(z) - d(z)]$$

where the sensitivity function $S_0(z)$ is given as:

$$S_0(z) = \frac{1}{1 + G_c(z)G_p(z)} \tag{2.8}$$

The sensitivity function $S_0(z)$ describes the transfer function from system noise and disturbance to the output of the feedback system without the add-on RC. A lower $|S_0(e^{j\omega})|$ suggests a higher attenuation of the noise and disturbance for the closed-loop control system with the add-on RC.

When considering the plug-in RC scheme $G_{rc}(z)$ as shown in Figure 2.3(a), the output $y(z)$ can be derived as:

$$y(z) = \frac{[1 + G_{rc}(z)]H(z)}{1 + G_{rc}(z)H(z)} r(z) + \frac{[1 + G_c(z)G_p(z)]^{-1}}{1 + G_{rc}(z)H(z)} d(z) \tag{2.9}$$

Substituting $G_{rc}(z)$ of (2.5) into (2.9) yields:

$$y(z) = \frac{[1 - (1 - k_{rc}G_f(z))z^{-N}Q(z)]H(z)}{1 - [(1 - k_{rc}G_f(z)H(z))Q(z)]z^{-N}} r(z) + \frac{(1 + G_c(z)G_p(z))^{-1}(1 - Q(z)z^{-N})}{1 - [(1 - k_{rc}G_f(z)H(z))Q(z)]z^{-N}} d(z) \tag{2.10}$$

And then the tracking error can be written as:

$$e(z) = \frac{1 - Q(z)z^{-N}}{1 - [(1 - k_{rc}G_f(z)H(z))Q(z)]z^{-N}} e_0(z) = S_1(z)S_0(z)(r(z) - d(z))$$

where the added sensitivity function $S_1(z)$ is given as:

$$S_1(z) = \frac{1 - Q(z)z^{-N}}{1 - [(1 - k_{rc}G_f(z)H(z))Q(z)]z^{-N}} \tag{2.11}$$

Similarly, this sensitivity function $S_1(z)$ describes the cascaded transfer function from system noise and disturbance to the output of the feedback system with the add-on RC. Hence, a lower $|S_1(z)|$ means that a further higher attenuation of the measurement noise and disturbance. When $|S_1(e^{j\omega})| \to 0$, i.e., $|1 - Q(e^{j\omega})e^{-jN\omega}| \to 0$ at all harmonic frequencies of $\omega_k = 2k\pi/T_0$ with $k = 0, 1, 2, \ldots$, means that the system is immune to the noise and disturbance at these harmonic frequencies.

From (2.10) and (2.11), the RC system stability can be determined by the characteristic equations of sensitivity functions $S_0(z)$ and $S_1(z)$. The sensitivity functions $S_0(z)$ and $S_1(z)$ can be used to evaluate the performance of the RC system, such as steady-state accuracy, dynamic response, and stability.

2.1.3.1 Steady-state error

From (2.11), at all harmonic frequencies of $\omega_k = 2k\pi/T_0$ with $k = 0, 1, 2, \ldots$, when $1 - Q(z)z^{-N} \to 0$ with $Q(z) \to 1$, the tracking error $e(z)$ will asymptotically converge to zero. That is to say, the RC can achieve zero-error steady-state tracking of any periodic signal $r(z)$ or $d(z)$ at all the harmonic frequencies ω_k below the Nyquist frequency. While, in accordance with (2.11), if $Q(z) \neq 1$, the tracking error $e(z)$ would not approach zero. Obviously, the filter $Q(z)$ can be used to select the harmonics for compensation.

2.1.3.2 Asymptotic convergence rate

According to (2.8), for periodic signals $r(t)$ and $d(t)$ with the same period of $T_0 = NT_s$, the tracking error $e_0(t)$ will be a periodic signal with the same period of T_0 in the steady-state. From (2.11), the transfer function $e(z)$ can be derived as:

$$\begin{aligned}
e(z) &= A(z)z^{-N}e(z) + (1 - Q(z))z^{-N}e_0(z) \\
&= A(z)z^{-N}\left[A(z)z^{-N}e(z) + (1 - Q(z))z^{-N}e_0(z)\right] + (1 - Q(z))z^{-N}e_0(z) \\
&= A^2(z)z^{-2N}e(z) + \left(A(z)z^{-N} + 1\right)(1 - Q(z))z^{-N}e_0(z) \\
&= A^k(z)z^{-kN}e(z) + \left(A^{k-1}(z)z^{-(k-1)N} + A^{k-2}(z)z^{-(k-2)N} + \ldots + 1\right) \\
&\quad \times (1 - Q(z))z^{-N}e_0(z) \\
&= A^k(z)z^{-kN}e(z) + \frac{1 - A^k(z)z^{-kN}}{1 - A(z)z^{-N}}(1 - Q(z))z^{-N}e_0(z) \overset{k\to\infty}{\to} e(z) \\
&= \left(A^\infty(z)z^{-\infty N}\right)e(z) + \frac{1 - A^\infty(z)z^{-\infty N}}{1 - A(z)z^{-N}}(1 - Q(z))z^{-N}e_0(z)
\end{aligned} \tag{2.12}$$

in which $A(z) = \left(1 - k_{rc}G_f(z)H(z)\right)Q(z)$ and $k = 0, 1, 2, \ldots$.

Remark 2.1 *From (2.12), if $|A(z)| < 1$ and $Q(z) = 1$, the tracking error e(z) will asymptotically converge to zero when the time instant number $k \to \infty$. Moreover, the smaller $|A(z)|$ is, the faster the error convergence rate is. Hence, $|A(z)|$ can be defined as a* convergence index.

2.1.3.3 Stability criteria

According to (2.8)–(2.12), it is clear that the overall plug-in RC system is stable if the two following stability conditions are met [4–7], [12–19]:

1. Roots of $1 + G_c(z)G_p(z) = 0$ are inside the unit circle, i.e., without the plug-in RC scheme, and the closed-loop feedback control system $H(z)$ is stable.
2. Roots of $1 - \left(1 - k_{rc}G_f(z)H(z)\right)Q(z)z^{-N} = 0$ are inside the unit circle, i.e., the convergence index $|A(z)| < 1$. That leads to a sufficient stability criterion as:

$$|A(z)| = \left|\left(1 - k_{rc}G_f(z)H(z)\right)Q(z)\right| < 1, \quad \text{for } z = e^{j\omega} \text{ with } \omega < \frac{\pi}{T_s} \tag{2.13}$$

2.1.3.4 Zero-phase compensation design

The compensation filter $G_f(z)$ [4,5] can be chosen as:

$$G_f(z) = \frac{z^d A(z)B^-(z^{-1})}{B^+(z)b} \tag{2.14}$$

where $b \geq \max|B^-(e^{j\omega})|^2$. Since the RC will introduce N-step delay z^{-N} and $N \geq d$, $G_f(z)$ with a noncausal component z^d can be implemented in the RC. Therefore, the product of $G_f(z)$ and $H(z)$ can be written as:

$$G_f(z)H(z) = \frac{B^-(z)B^-(z^{-1})}{b} = \frac{|B^-(e^{j\omega})|^2}{b}\angle 0^\circ \leq 1 \tag{2.15}$$

Equation (2.15) indicates that $G_f(z)$ exactly compensates the system delay from $H(z)$, i.e., zero-phase compensation is achieved. With zero-phase compensation and $|Q(z)| \leq 1$, (2.13) can be rewritten as:

$$|A(z)| = \left|(1 - k_{\text{rc}}G_f(z)H(z))Q(z)\right| < 1 \Rightarrow \left|1 - k_{\text{rc}}\frac{|B^-(z)|^2}{b}\right| < \frac{1}{|Q(z)|} \tag{2.16}$$

Therefore, according to (2.15), (2.16), and $|Q(z)| \leq 1$, a sufficient stability criterion for the RC system can be significantly simplified as:

$$0 < k_{\text{rc}} < 2 \tag{2.17}$$

Remark 2.2 *Equations (2.16) and (2.17) indicate that, an LPF $Q(z)$ with $|Q(z)| \leq 1$ leads to a larger stability range for the RC gain k_{rc}, and then enhances the robustness of RC systems. On the other hand, from (2.16), the introduction of the LPF $Q(z)$ with $|Q(z)| \to 1$ at low frequencies and $|Q(z)| \to 0$ at high frequencies, leads to a perfect tracking of the periodic signals at major low harmonic frequencies but degrades the tracking accuracy at minor high harmonic frequencies. An LPF $Q(z)$ with larger bandwidth will lead to higher tracking accuracy, but a smaller stability region for the RC gain. Therefore, the LPF $Q(z)$ is used to selectively compensate the harmonics of interest, and it can be employed to improve the control system stability at the cost of a minor reduction of the tracking accuracy.*

Remark 2.3 *The phase-lead compensator $G_f(z)$ of (2.14) leads to a very simple stability criterion for the design of the RC gain k_{rc}, and then significantly simplifies the synthesis of the RC systems. From (2.12) and (2.16), the error convergence rate is determined by k_{rc}.*

However, in practice, due to model uncertainties of the control plant $G_p(z)$, it is impossible to obtain an accurate transfer function $H(z)$, especially in the high frequency band. Alternatively, the actual transfer function of $G_f(z)H(z)$ can be written as:

$$G_f(z)H(z) = \frac{B^-(z^{-1})B^-(z)}{b}(1 + \Delta(z)) = \left|G_{fH}(e^{j\omega})\right|e^{j\theta_{fH}(\omega)} \tag{2.18}$$

where $\Delta(z)$ denotes the uncertainties assumed to be bounded by $|\Delta(z)| \leq \varepsilon$ with $\varepsilon > 0$ and $\Delta(z)$ is stable; and $\left|G_{fH}(e^{j\omega})\right| \leq 1 + \varepsilon$. From (2.18), it is clear that the uncertainties $\Delta(z)$ would cause $\theta_{fH}(\omega) \neq 0$, which means that the zero-phase-error compensation may not be achieved due to the uncertainties. In such general cases, the stability criterion is further derived as follows:

Theorem 2.1 *Consider the RC system shown in Figure 2.3 with the constraint of (2.18),*
if $2k\pi - \frac{\pi}{2} < \theta_{fH}(\omega) < 2k\pi + \frac{\pi}{2}, k = 0, 1, 2, \ldots,$ *then*

$$0 < k_{\mathrm{rc}} < \frac{2\cos\left(\theta_{fH}(\omega)\right)}{1+\varepsilon} \tag{2.19}$$

will enable the RC system to be asymptotically stable;
if $2k\pi + \frac{\pi}{2} < \theta_{fH}(\omega) < 2k\pi + \frac{3\pi}{2}, k = 0, 1, 2, \ldots,$ *then*

$$\frac{2\cos\left(\theta_{fH}(\omega)\right)}{1+\varepsilon} < k_{\mathrm{rc}} < 0 \tag{2.20}$$

will enable the closed-loop RC system to be asymptotically stable.

Proof: From the constraint of (2.18), we have:

$$\left|G_f(z)H(z)\right| = \left|\frac{B^-(z^{-1})B^-(z)}{b}(1+\Delta(z))\right| = \left|G_{fH}(z)\right| \le 1+\varepsilon$$

Therefore, the stability criterion of (2.12) can be rewritten as:

$$\left|1 - k_{\mathrm{rc}}G_f(z)H(z)\right| = \left|1 - k_{\mathrm{rc}}\left|G_{fH}(e^{j\omega})\right|\left(\cos\ \theta_{fH}(\omega) + j\sin\ \theta_{fH}(\omega)\right)\right| < \frac{1}{|Q(e^{j\omega})|}$$

And then applying the Euler's identity results in:

$$1 - 2k_{\mathrm{rc}}\left|G_{fH}(e^{j\omega})\right|\cos\ \theta_{fH}(\omega) + \left(k_{\mathrm{rc}}\left|G_{fH}(e^{j\omega})\right|\right)^2 < \frac{1}{|Q(e^{j\omega})|^2}$$

and further

$$k_{\mathrm{rc}}^2 < \frac{1 - |Q(e^{j\omega})|^2}{\left|G_{fH}(e^{j\omega})\right|^2|Q(e^{j\omega})|^2} + \frac{2k_{\mathrm{rc}}\cos\ \theta_{fH}(\omega)}{\left|G_{fH}(e^{j\omega})\right|}$$

Since $|Q(e^{j\omega})| \le 1$ and $\left|G_{fH}(e^{j\omega})\right| \le 1+\varepsilon$, if $2k\pi - \frac{\pi}{2} < \theta_{fH}(\omega) < 2k\pi + \frac{\pi}{2}, k = 0, 1, 2, \ldots,$ then

$$0 < k_{\mathrm{rc}} < \frac{2\cos\left(\theta_{fH}(\omega)\right)}{1+\varepsilon}$$

will sufficiently enable the RC system to be asymptotically stable;

if $2k\pi + \frac{\pi}{2} < \theta_{fH}(\omega) < 2k\pi + \frac{3\pi}{2}, k = 0, 1, 2, \ldots$, then

$$\frac{2\cos(\theta_{fH}(\omega))}{1+\varepsilon} < k_{rc} < 0$$

will enable the closed-loop RC system to be asymptotically stable.

Remark 2.4 *Theorem 2.1 offers a general stability criterion for practical RC systems. It indicates that uncertainties may reduce the stability range for the RC gain* k_{rc}. *When* $\theta_{fH}(\omega) = 0$, *meaning that the zero phase compensation is achieved, it will not only lead to a larger stability range for* k_{rc} *and thus a faster dynamic response, but also simplify the design of RC systems.*

2.1.3.5 Linear phase-lead compensation design

As aforementioned, it is difficult to make $\theta_{fH}(\omega) = 0$, especially at the high-frequency band. In practice, to simplify the design of the RC system, a simple but effective linear phase-lead compensator $G_f(z)$ is usually employed as [18]:

$$G_f(z) = z^p \tag{2.21}$$

where p is the phase-lead steps and $N \gg p$. The linear phase-lead compensator z^p is employed to compensate the delays of the feedback control system H(z). Therefore, this phase-lead compensation RC can be given as:

$$G_{\mathrm{rc}}(z) = k_{\mathrm{rc}} \frac{Q(z)z^{-N+p}}{1 - Q(z)z^{-N}} \tag{2.22}$$

Obviously, the RC in (2.22) is a direct digital form of the continuous-time RC in (2.3). The linear phase-lead compensator z^p in (2.22) is corresponding to e^{sT_c} in (2.3). As shown in Figure 2.4, the linear phase-lead compensator can be easily implemented in the digital RC systems with minor modifications.

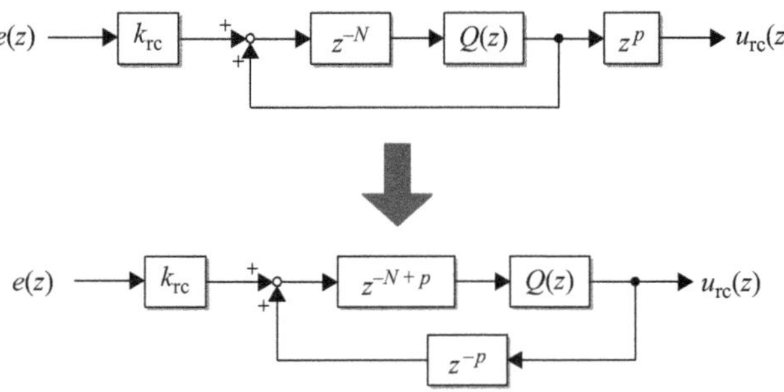

Figure 2.4 General linear phase-lead compensation for an RC system

Correspondingly, the stability criterion of (2.12) can be rewritten as:

$$|(1 - k_{\rm rc} z^p H(z))Q(z)| < 1, \quad \text{for } z = e^{j\omega} \text{ with } \omega < \frac{\pi}{T_s} \tag{2.23}$$

Following the frequency-domain analysis approach, the closed-loop transfer function *H*(*z*) can be expressed as $H(e^{j\omega}) = |H(e^{j\omega})|e^{j\theta_H(\omega)}$, and then (2.23) becomes:

$$k_{\rm rc}^2 < \frac{1 - |Q(e^{j\omega})|^2}{|H(e^{j\omega})|^2 |Q(e^{j\omega})|^2} + \frac{2k_{\rm rc}\cos(\theta_H + p\omega)}{|H(e^{j\omega})|}$$

If $2k\pi - \frac{\pi}{2} < \theta_H + p\omega < 2k\pi + \frac{\pi}{2}, k = 0, 1, 2, \ldots$, then

$$0 < k_{\rm rc} < \frac{2\cos(\theta_H + p\omega)}{|H(e^{j\omega})|} \tag{2.23a}$$

will sufficiently enable the RC system to be asymptotically stable;
if $2k\pi + \frac{\pi}{2} < \theta_H + p\omega < 2k\pi + \frac{3\pi}{2}, k = 0, 1, 2, \ldots$, then

$$\frac{2\cos(\theta_H + p\omega)}{|H(e^{j\omega})|} < k_{\rm rc} < 0 \tag{2.23b}$$

will enable the closed-loop RC system to be asymptotically stable.

For example, considering a given digital feedback control system with the sampling frequency being $f_s = 10$ kHz, the transfer function is given as:

$$H(z) = \frac{0.5z + 0.432}{z^2 - 0.487z + 0.429} \tag{2.23c}$$

The responses of the feedback system *H*(*z*) with the plug-in RC linear phase-lead compensation can be obtained as shown in Figure 2.5. It can be seen that within the stability phase range of $\pm\pi/2$ in (2.23a), the phase-lead compensation with $p = 2$ yields both the widest stable frequency bandwidth of 4.5 kHz and the largest stability range of [0, 1.1] for RC gains. Compared with the Nyquist frequency of 5 kHz and the upper bound value of 2 in (2.17) for an idea zero-phase compensation RC system, it is a very feasible way to employ linear phase-lead compensation method in the design of RC systems.

In practice, the linear phase-lead compensator z^p offers the simplest and most effective solution to the compensation filter $G_f(z)$ for RC controllers.

2.1.4 Two alternative RC schemes

There are other ways to realize RC schemes in practice, and they are shown in Figure 2.6.

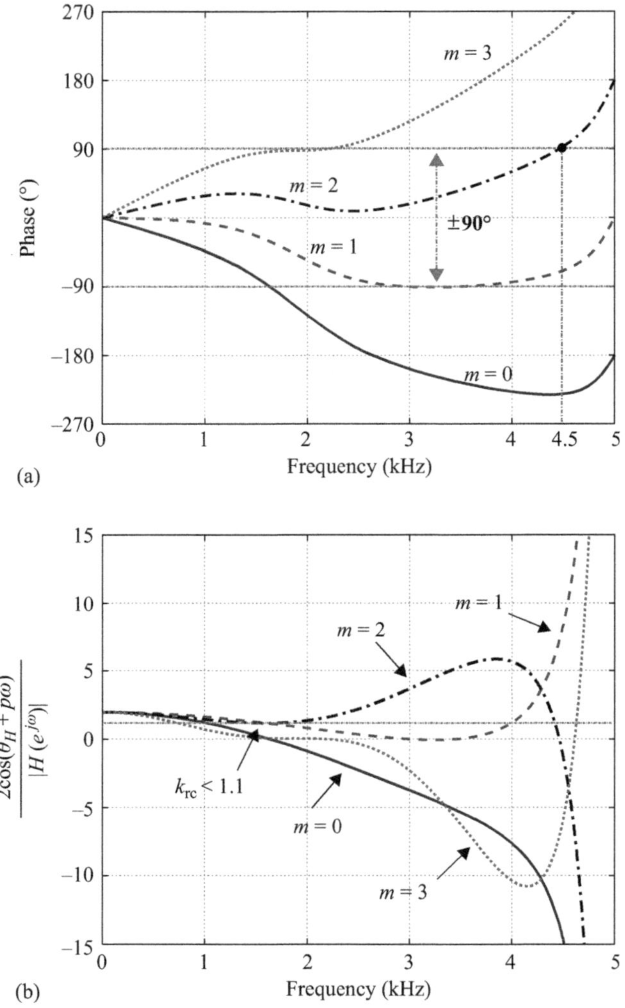

Figure 2.5 Responses of the phase-lead-compensated system under different compensation steps p $= 0, 1, 2, 3$*, where* $f_s = 10$ *kHz: (a) phase of the compensated system* $G_f(z)H(z)$ *(i.e.,* $\theta_H + p\omega$*) and (b) corresponding boundary of the RC gains in (2.23a)*

As shown in Figure 2.6, the two other variant RC schemes can be expressed as:

$$G_1(z) = k_{\text{rc}} \frac{1}{1 - Q(z)z^{-N}} G_{f1}(z), \text{ Variant 1} \tag{2.24}$$

$$G_2(z) = k_{\text{rc}} \frac{z^{-N}}{1 - Q(z)z^{-N}} G_{f2}(z), \text{ Variant 2} \tag{2.25}$$

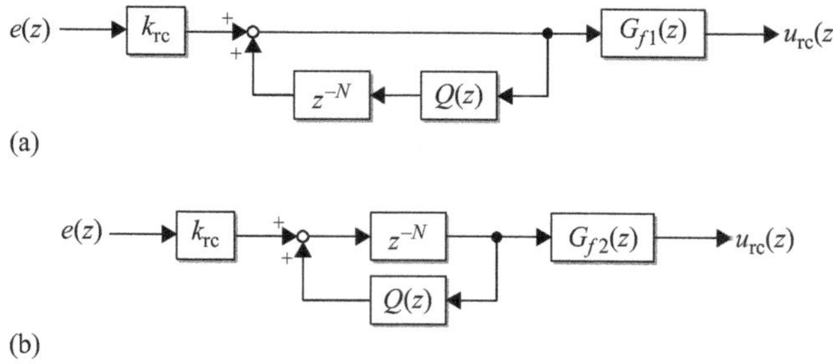

Figure 2.6 Other RC implementations: (a) Variant 1 and (b) Variant 2, where $G_{f1}(z)$ and $G_{f2}(z)$ are phase-lead compensators

When Variant 1 RC scheme $G_1(z)$ is plugged into the closed-loop system to replace the classic RC scheme $G_{rc}(z)$ shown in Figure 2.3, the stability criterion can be written as:

$$\left|Q(z) - k_{rc}G_{f1}(z)H(z)z^N\right| < 1 \tag{2.26}$$

Compared with the classic RC systems, if $G_{f1}(z) = z^{-N}Q(z)G_f(z)$ is adopted to decouple $Q(z)$ and k_{rc}, (2.26) will be equivalent to (2.13), and the synthesis of the RC system will be simplified. However, the implementation of Variant 1 RC controller with decoupling $G_{f1}(z)$ requires two delay elements z^{-N} and two LPFs $Q(z)$. If $G_{f1}(z) \neq z^{-N}Q(z)G_f(z)$, the synthesis of the RC system becomes very complicated due to the coupling between $Q(z)$ and k_{rc}.

When Variant 2 RC scheme $G_2(z)$ is plugged into the closed-loop system to replace the classic RC scheme $G_{rc}(z)$ shown in Figure 2.3, the stability criterion can be written as:

$$\left|Q(z) - k_{rc}G_{f2}(z)H(z)\right| < 1 \tag{2.27}$$

Compared with the classic RC scheme, if $G_{f2}(z) = Q(z)G_f(z)$ is adopted to decouple $Q(z)$ and k_{rc}, (2.27) will be equivalent to (2.13), and the synthesis of the RC system will be simplified. However, the LPF $Q(z)$ is needed by both $G_{f2}(z) = Q(z)G_f(z)$ and the feedback loop in the implementation of Variant 2 RC controller with decoupling $G_{f2}(z)$. If $G_{f2}(z) \neq Q(z)G_f(z)$, the synthesis of the RC system becomes complicated.

In contrast to the above variant RC systems, the classic RC offers a better solution in terms of design and implementation.

2.2 Multiple resonant control (MRSC)

2.2.1 Internal models of harmonics

Any periodic signal can be decomposed into the sum of a (possibly an infinite) set of harmonics (i.e., sines and cosines) plus its DC mean value (i.e., DC component) in the form of Fourier series. Correspondingly, the internal model of a periodic signal is equivalent to the sum of the internal models of its harmonics and DC component, as shown in (2.2). The internal models (i.e., signal generator) of any (in-quadrature) harmonic can be written as [32,33]:

$$\begin{aligned} \hat{G}_{h1}(s) &= \mathcal{L}\{\cos(\omega_h t)\} = \frac{s}{s^2+\omega_h^2} = \frac{1}{2}\left(\frac{1}{s+j\omega_h} + \frac{1}{s-j\omega_h}\right) \\ \hat{G}_{h2}(s) &= \mathcal{L}\{\sin(\omega_h t)\} = \frac{\omega_h}{s^2+\omega_h^2} = \frac{1}{2}\left(\frac{j}{s+j\omega_h} - \frac{j}{s-j\omega_h}\right) \end{aligned} \tag{2.28}$$

in which $\omega_h = h\omega_0$ is the signal angular frequency with ω_0 being the signal fundamental angular frequency and $h \in \mathbb{N}$ is the harmonic order. It is noted that both $\hat{G}_{h1}(s)$ and $\hat{G}_{h2}(s)$ have poles at $s = \pm jh\omega_0$. Hence, they exhibit infinite gains at harmonic frequencies, i.e., $\pm h\omega_0$. Therefore, zero steady-state error tracking of any harmonic with the resonant frequency at $\pm h\omega_0$ can be achieved, if either $\hat{G}_{h1}(s)$ or $\hat{G}_{h2}(s)$ is included into the closed-loop system. Both $\hat{G}_{h1}(s)$ and $\hat{G}_{h2}(s)$ are the *internal models* for harmonics with the resonant frequencies at $\pm\omega_h$.

2.2.2 MRSC scheme

Based on the internal models in (2.28), a general implementation form of the IMP-based resonant control (RSC) at the harmonic frequency of ω_h is shown in Figure 2.7.

The corresponding transfer function for the general RSC is [34]:

$$G_h(s) = \frac{u_{rsc}(s)}{e(s)} = \{k_h \cos(\omega_h t + \phi_h)\} = k_h \frac{s\cos\phi_h - \omega_n \sin\phi_h}{s^2+\omega_h^2} \tag{2.29}$$

where k_h is the control gain for tuning the error convergence rate, and ϕ_h is the phase-lead angle for system delay compensation at the resonant frequency $\pm\omega_h$. In practice, a simplified RSC in (2.28) with $\phi_h = 0$ can be found in many applications, i.e., $\hat{G}_{h1}(s)$ is commonly employed [23].

In order to achieve high control accuracy, multiple resonant controllers (MRSC) are often connected in parallel to compensate multiple selected harmonic

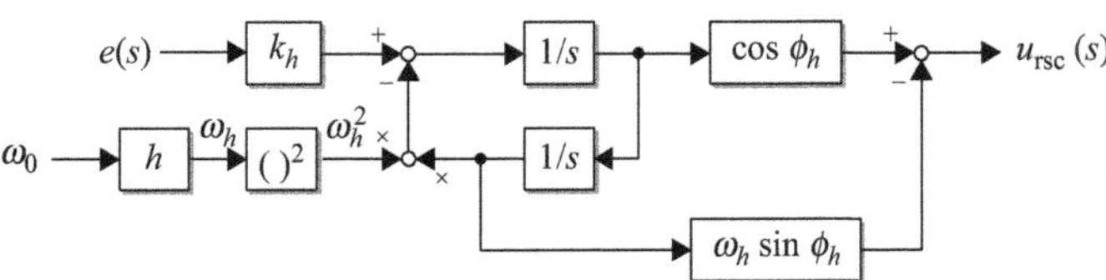

Figure 2.7 General structure of a resonant controller with the resonant frequency at $\pm\omega_h$

frequencies simultaneously, especially to eliminate major harmonic distortions. The MRSC can be simply written as:

$$G_M(s) = \sum_{h \in N_h} G_h(s) = \sum_{h \in N_h} \left(k_h \frac{s \cos \phi_h - \omega_h \sin \phi_h}{s^2 + \omega_h^2} \right) \tag{2.30}$$

where N_h represents the set of the orders of selected harmonics.

Note that each RSC component $G_h(s)$ in the MRSC $G_M(s)$ can independently choose its gain k_h and the compensation phase ϕ_h at each selected harmonic frequency. Therefore, it allows the MRSC to achieve satisfactory tracking performance: high tracking accuracy, fast dynamic response, and good robustness. Nevertheless, if a large number of RSC components are embedded, the resultant MRSC would be associated with heavy parallel computation burden and increased parameter tuning difficulty.

Among different discrete-time implementations of the RSC controllers, the Tustin transform can assure the resonant peaks of discretized RSC components would not significantly differ from the expected resonant frequencies of the corresponding continuous-time ones. The MRSC scheme of (2.30) digitized by using pre-warping Tustin transform can be written as [34]:

$$\begin{aligned} G_M(z) &= \sum_{h \in N_h} G_h(z) \\ &= \sum_{h \in N_h} \left\{ k_h \frac{\frac{1}{2}(1 - z^{-2}) \cos \phi_h \sin(\omega_h T_s) - (1 + 2z^{-1} + z^{-2}) \sin \phi_h \sin^2\left(\frac{\omega_h T_s}{2}\right)}{\omega_h (1 - 2z^{-1} \cos(\omega_h T_s) + z^{-2})} \right\} \end{aligned} \tag{2.31}$$

where T_s denotes the system sampling period.

2.2.3 Digital MRSC system and design

RSC components are also commonly added on to the closed-loop instantaneous feedback control systems to complementarily achieve optimal performance, e.g., proportional-resonant (PR) control systems. Figure 2.8 presents an add-on MRSC system, where $e(z)$ is the control error, $r(z)$ denotes the reference input, $y(z)$ represents the control output, $d(z)$ is the disturbance, $G_c(z)$ is the feedback controller, $G_p(z)$ is the control plant, and $G_M(z)$ denotes the digital MRSC scheme.

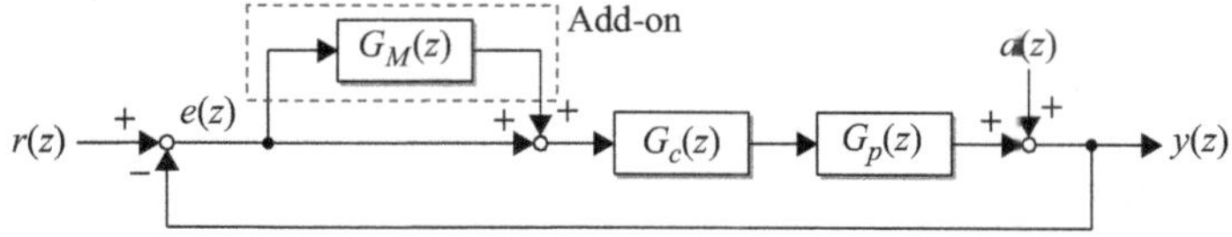

Figure 2.8 Add-on digital MRSC system

The transfer function from the reference input $r(z)$ and the disturbance $d(z)$ to the error $e(z)$ can be expressed as follows:

$$e(z) = \frac{1}{1 + G_c(z)G_p(z)} \cdot \frac{1}{1 + G_M(z)H(z)}(r(z) - d(z)) \tag{2.32}$$

where $H(z) = G_c(z)G_p(z)/\left(1 + G_c(z)G_p(z)\right)$ denotes the closed-loop transfer function without the MRSC unit $G_M(z)$.

2.2.3.1 Steady-state error

Since $G_M(z) \to \infty$ at $\omega = \omega_h$:

$$\lim_{\omega=\omega_h} |e(z)| \to 0 \quad \text{for} \quad z = e^{j\omega} \tag{2.33}$$

which indicates that the MRSC can achieve zero steady-state tracking errors at harmonic frequencies ω_h.

2.2.3.2 Stability criteria

From (2.32), it can be derived that the overall closed-loop MRSC system shown in Figure 2.7 is stable if the following conditions hold [34]:

1. Roots of $1 + G_c(z)G_p(z) = 0$ are inside the unit circle, i.e., $H(z)$ is stable.
2. Roots of $1 + G_M(z)H(z) = 0$ are inside the unit circle.

Unlike the classic RC systems, the stability range of the control gains k_h for the MRSC systems cannot directly be derived from the above conditions. Empirical evidences show that the less the selected harmonics in the MRSC are, the larger the stability range of the control gains k_h is.

2.2.3.3 Zero-phase compensation design

Similar to the synthesis of the RC systems, the phase angle ϕ_h for $G_h(z)$ will be chosen to exactly compensate the delay of the closed-loop system $H(z)$ at the corresponding harmonic frequency ω_h as:

$$\phi_h = -\angle H(e^{j\omega_h}) = -\theta_h \tag{2.34}$$

This means that zero-phase compensation is obtained at $\omega = \omega_h$ [32,34]. It will allow a large stability range for its gain k_h at the harmonic frequency ω_h.

For an MRSC, each RSC component $G_h(s)$ can independently choose its gain k_h and its phase compensation angle ϕ_h at its corresponding frequency ω_h. Since the dynamic response at any harmonic frequency is proportional to the gain for the corresponding RSC component, a larger gain will yield faster dynamic response. Moreover, the dynamic response of an RSC component is proportional to its corresponding resonant frequency ω_h. Therefore, in order to optimize its overall performance in response to a periodic signal (disturbance) with an unevenly distributed harmonic spectrum, the following rules-of-thumb can be applied to the synthesis of the MRSC scheme:

- The gain for each RSC component would be proportional to its portion in the total harmonic spectrum.
- To enforce all harmonic frequencies at the same convergence rate, the gain for each RSC component would be inversely proportional to its resonant frequency.

2.2.4 RSC – Generalized integrator for sinusoidal signals

Balanced three-phase sinusoidal signals can be transformed into two orthogonal signals in the stationary frame (SF), and then into two DC signals in the synchronous reference frame (SRF). Furthermore, any set of unbalanced three-phase sinusoidal signals can be decomposed into the sum of one set of balanced three-phase positive-sequence sinusoidal signals, one set of balanced three-phase negative-sequence sinusoidal signals, and one set of balanced zero-sequence sinusoidal signals. Therefore, for three-phase power plants, zero steady-state tracking of sinusoidal signals can be achieved by using proportional–integral (PI) controllers in the SRF [23,37–40] due to its infinite gain at the DC frequency.

For a three-phase system, the reference frame transformation ($abc \rightarrow \alpha\beta$) can be expressed as:

$$\begin{pmatrix} x_\alpha \\ x_\beta \end{pmatrix} = \underbrace{\frac{2}{3}\begin{pmatrix} 1 & -\dfrac{1}{2} & -\dfrac{1}{2} \\ 0 & \dfrac{\sqrt{3}}{2} & -\dfrac{\sqrt{3}}{2} \end{pmatrix}}_{\mathbf{C}} \begin{pmatrix} x_a \\ x_b \\ x_c \end{pmatrix} \tag{2.35}$$

where $\mathbf{C}$ is the Clark transform matrix, x_α and x_β are variables in the stationary orthogonal $\alpha\beta$ frame, x_a, x_b, x_c are variables in the stationary symmetric abc frame and $x_a + x_b + x_c = 0$. Following, the transformation from the stationary orthogonal $\alpha\beta$ frame to the synchronous rotating orthogonal dq frame can be expressed as:

$$\begin{pmatrix} x_d \\ x_q \end{pmatrix} = \begin{pmatrix} \cos\theta & \sin\theta \\ -\sin\theta & \cos\theta \end{pmatrix} \begin{pmatrix} x_\alpha \\ x_\beta \end{pmatrix} \tag{2.36}$$

where x_d and x_q are variables in the synchronous rotating orthogonal dq frame and $\theta = \omega t$ with t being the time instant and ω being the rotating speed of the dq frame, is the angle between the dq and $\alpha\beta$ frames.

In the vector form, the transformations between the synchronous dq frame and the stationary $\alpha\beta$ frame can be written as:

$$\begin{aligned} \vec{x}_{dq} &= \vec{x}_{\alpha\beta} e^{-j\omega_0 t} = \vec{x}_{\alpha\beta}(\cos(\omega t) - j\sin(\omega t)) \\ \vec{x}_{\alpha\beta} &= \vec{x}_{dq} e^{j\omega_0 t} = \vec{x}_{dq}(\cos(\omega t) + j\sin(\omega t)) \end{aligned} \tag{2.37}$$

where $\vec{x}_{dq} = x_d + jx_q$ and $\vec{x}_{\alpha\beta} = x_\alpha + jx_\beta$. The transformation is also illustrated in Figure 2.9.

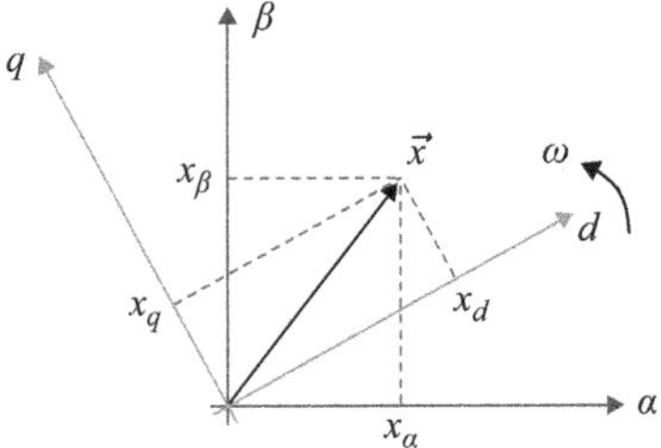

Figure 2.9 Relationship between the synchronous dq frame and the stationary αβ frame

In a three-phase system, $G_{dq}(s)$ in both positive-sequence and negative-sequence SRF synchronous dq frame are integral regulators as:

$$G_{dq}^{+}(s) = G_{dq}^{-}(s) = G_{dq}(s) = \begin{pmatrix} \frac{k_i}{s} & 0 \\ 0 & \frac{k_i}{s} \end{pmatrix} \tag{2.38}$$

where k_i is the coefficient.

As shown in Figure 2.10, the equivalent controller in the stationary $\alpha\beta$ frame for the combination of the integral controllers $G_{dq}(s)$ in both positive- and negative-sequence SRFs can then be written as:

$$G_{\alpha\beta}(s) = \frac{1}{2}\begin{pmatrix} G_{dq}(s+j\omega)+G_{dq}(s-j\omega) & jG_{dq}(s+j\omega)-G_{dq}(s-j\omega) \\ -jG_{dq}(s+j\omega)+jG_{dq}(s-j\omega) & G_{dq}(s+j\omega)+G_{dq}(s-j\omega) \end{pmatrix}$$

$$= \overbrace{\frac{1}{2}\begin{pmatrix} \frac{2k_is}{s^2+\omega^2} & \frac{2k_i\omega}{s^2+\omega^2} \\ -\frac{2k_i\omega}{s^2+\omega^2} & \frac{2k_is}{s^2+\omega^2} \end{pmatrix}}^{G_{\alpha\beta}^{+}(s)} + \overbrace{\frac{1}{2}\begin{pmatrix} \frac{2k_is}{s^2+\omega^2} & -\frac{2k_i\omega}{s^2+\omega^2} \\ \frac{2k_i\omega}{s^2+\omega^2} & \frac{2k_is}{s^2+\omega^2} \end{pmatrix}}^{G_{\alpha\beta}^{-}(s)} = \begin{pmatrix} \frac{2k_is}{s^2+\omega^2} & 0 \\ 0 & \frac{2k_is}{s^2+\omega^2} \end{pmatrix} \tag{2.39}$$

where ω is the rotating velocity of the SRF [33].

Furthermore, the generic RSC of (2.29) can be rewritten as [32]:

$$G_h(s) = k_h\frac{s\cos\phi_h - \omega_h\sin\phi_h}{s^2+\omega_h^2} = \frac{k_h}{2}\left(\frac{1}{s-j\omega_h}e^{j\phi_h} + \frac{1}{s+j\omega_h}e^{j-\phi_h}\right) \tag{2.40}$$

From (2.39) and (2.40), it is clear that the RSC in the stationary frame is equivalent to the combination of two integrators in both positive- and negative-sequence SRFs. Therefore, RSC components are taken as *generalized integrators* [23].

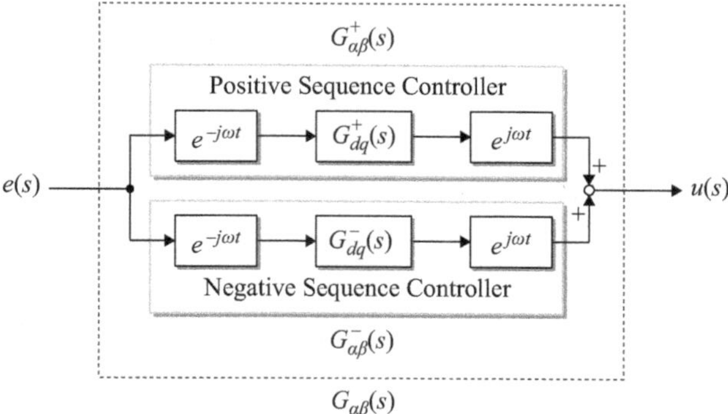

Figure 2.10 Equivalent RSC scheme in the stationary frame using integrators in the synchronous rotating frame

Furthermore, a PR controller in the stationary frame is equivalent to the combination of two PI controllers in both positive- and negative-sequence SRFs.

Nevertheless, compared with the PR controller in the SF, the equivalent PI controllers in the SRFs take much more computation and complexity due to reference frame transformations, and they are more sensitive to noise and error appearing in the synchronization. Moreover, the PR controller can also be directly applied to single-phase systems.

2.3 Discrete Fourier transform (DFT)-based RC

Using the Fourier series, any periodic signal can be decomposed into the sum of a simple fundamental-frequency component and its harmonics. According to the superposition principle, the steady-state control system response to a periodic signal can be calculated by adding up the steady-state frequency response to each of these frequencies. To deal with periodic signals with selective harmonics, the DFT [35,36] can then be adopted to flexibly construct the internal models of the harmonics of interest.

2.3.1 DFT-based internal model of interested harmonics

A DFT-based FIR filter with a window equal to one fundamental period N (more specifically, a discrete cosine transform) is given as [35,36]:

$$F_{\mathrm{dh}}(z) = \frac{2}{N}\sum_{i=1}^{N}\cos\left(\frac{2\pi}{N}h(i+N_a)\right)z^{-i} = \left(\sum_{i=1}^{N}a_h(i)z^{-i}\right)z^{N_a} = Q_{\mathrm{dh}}(z)z^{N_a} \tag{2.41}$$

where i, N, h, and N_a represent the i-th sample point, the number of samples per fundamental period, the harmonic order, and the phase-lead compensation steps.

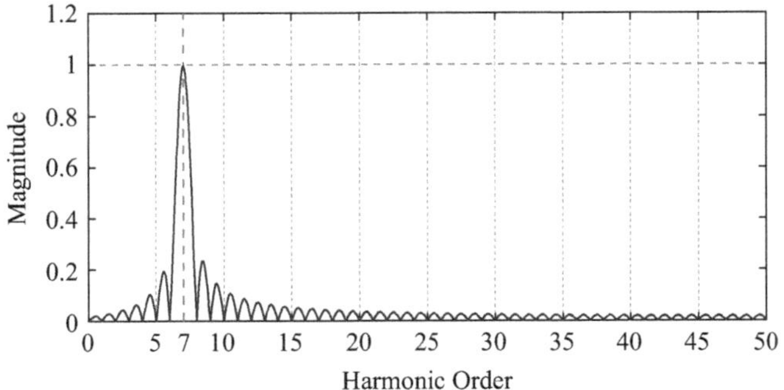

Figure 2.11 Magnitude response of $F_{dh}(z)$ with $N = 100$, $N_a = 0$ and $h = 7$

Equation (2.41) actually produces a cosine signal with the amplitude being $2/N$. This means that the DFT-based FIR filter creates a band-pass filter $Q_{\mathrm{dh}}(z)$ with a unity gain at the selected h-th order harmonic frequency and a zero gain at the unselected harmonic frequencies. For example, if $N = 100$, $N_a = 0$, and $h = 7$, the corresponding magnitude response of $F_{\mathrm{dh}}(z)$ is given in Figure 2.11, which only allows the seventh-order harmonic to pass through.

Therefore, in order to create a comb filter for selected harmonics (giving the appearance of a comb), a DFT-based FIR filter $F_{\mathrm{DFT}}(z)$ can be given as [35,36]:

$$F_{\mathrm{DFT}}(z) = \sum_{h \in N_h} F_{\mathrm{dh}}(z) = \frac{2}{N} \sum_{i=1}^{N} \left(\sum_{h \in N_h}^{N_h} \cos\left[\frac{2\pi}{N} h(i + N_a) \right] \right) z^{-i} \tag{2.42}$$

where $F_{\mathrm{dh}}(z)$ denotes the band-pass filter only for the h-th order harmonic and N_h is the set of the orders of the selected harmonic frequencies, respectively. Similarly, the comb filter $F_{\mathrm{DFT}}(z)$ of (2.42) can be rewritten as:

$$F_{\mathrm{DFT}}(z) = \left(\sum_{i=1}^{N} b_h(i) z^{-i} \right) z^{N_a} = Q_D(z) z^{N_a} \tag{2.43}$$

in which $Q_D(z)$ is actually a band-pass comb filter with $|Q_D(e^{j\omega_h})| = 1$ at the selected harmonic frequencies and $|Q_D(e^{j\omega_h})| = 0$ at the unselected harmonic frequencies. Figure 2.12 gives the implementation of such a comb filter $F_{\mathrm{DFT}}(z)$, which might involve a large amount of calculation efforts and thus it is suitable for fixed-point digital signal processor (DSP) implementation.

Like the classic RC, the digital signal generator for the selected harmonics can be formed by the band-pass filter $F_{\mathrm{DFT}}(z)$ with a positive feedback loop. The internal model of the selected harmonics can be written as [35,36]:

$$\hat{G}_{\mathrm{DFT}}(z) = \frac{F_{\mathrm{DFT}}(z)}{1 - F_{\mathrm{DFT}}(z)} \tag{2.44}$$

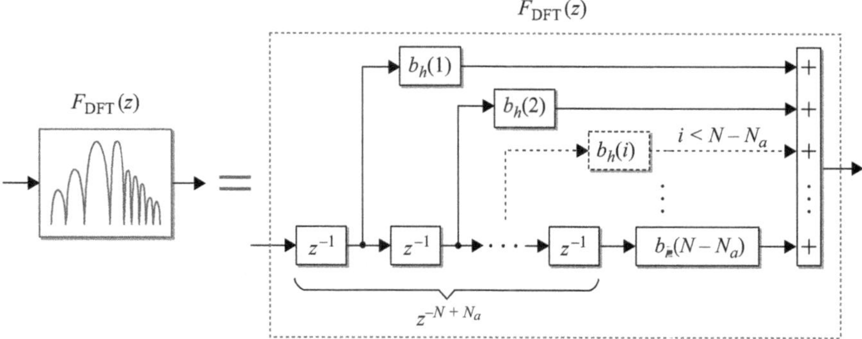

Figure 2.12 Implementation of a comb filter $F_{DFT}(z)$

where $\hat{G}_{\rm DFT}(z)$ is an DFT-based *internal model* of the selected harmonics, offering a flexible way to selectively compensate the harmonics of interest by changing the coefficients of the FIR filter $F_{\rm DFT}(z)$. For example, if $N = 100$, $N_a = 0$, and $h = 0, 1, 2, \ldots, 49$ (i.e., $N_h = 49$ and all harmonics below the Nyqusit frequency are selected), (2.43) becomes an all-pass filter as follows:

$$F_{\rm DFT}(z) = \sum_{h \in N_h} F_{\rm dh}(z) = \frac{2}{100} \sum_{i=1}^{100} \left(\sum_{h=0}^{49} \cos\left(h \frac{2\pi i}{100} \right) \right) z^{-i} = z^{-100}$$

Therefore, the corresponding $\hat{G}_{\rm DFT}(z)$ of (2.44) becomes:

$$\hat{G}_{\rm DFT}(z) = \frac{z^{-100}}{1 - z^{-100}}$$

which is a typical internal model for the classic RC.

2.3.2 DFT-based RC scheme

Based on the internal model $\hat{G}_{\rm DFT}(s)$ of (2.44), a DFT-based RC shown in Figure 2.13 can be derived as:

$$G_{\rm DFT}(z) = \frac{u_{\rm rc}(z)}{e(z)} = \frac{k_F F_{\rm DFT}(z)}{1 - F_{\rm DFT}(z) z^{-N_a}} = \frac{k_F Q_D(z) z^{N_a}}{1 - Q_D(z)} \tag{2.45}$$

in which k_F is the control gain for tuning the dynamic response, and N_a represents the number of leading steps for phase-lead compensation.

Remark 2.5 *To achieve selective harmonics compensation, the DFT-based RC system $G_{\rm DFT}(z)$ of (2.45) is virtually equivalent to the classic RC of (2.5) with a linear phase-lead compensator $G_f(z) = z^{N_a}$ and a selective notch filter $|Q(e^{j\omega})| \leq 1$. It is also equivalent to an MRSC scheme of (2 30) with identical coefficients and phase-lead compensation at all selected harmonic frequencies.*

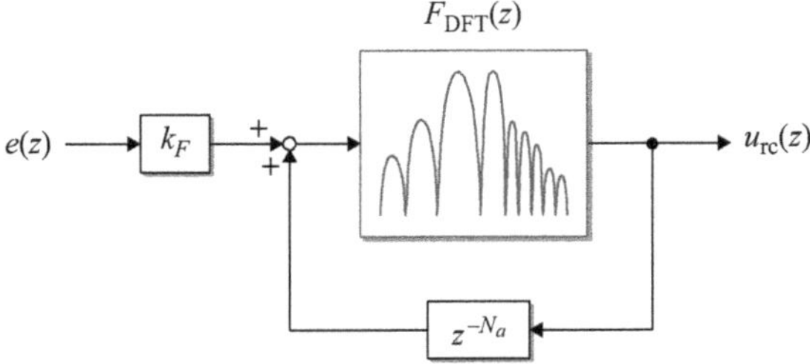

Figure 2.13 A DFT-based RC scheme $G_{DFT}(z)$

Remark 2.6 *The DFT-based RC $G_{\text{DFT}}(z)$ provides a flexible selective harmonic compensator. Like the classic RC, its computational complexity is independent of the number of selected harmonics to be compensated. However, the implementation of $G_{\text{DFT}}(z)$ would involve a large amount of parallel computation, which is proportional to the fundamental period N for the FIR filter $F_{\text{FDT}}(z)$ of (2.42). Therefore, $G_{\text{DFT}}(z)$ is more suitable for high-performance fixed-point DSP implementation.*

2.3.3 DFT-based RC system and design

Figure 2.14 presents a plug-in DFT-based RC system, where $e(z)$ is the control error, $r(z)$ denotes the reference input, $y(z)$ denotes the control output, $d(z)$ is the disturbance, $G_c(z)$ is the feedback controller, $G_p(z)$ is the control plant, and $G_{\text{DFT}}(z)$ denotes the DFT-based RC. The transfer function from the reference input $r(z)$ and the disturbance $d(z)$ to the error $e(z)$ can be expressed as:

$$e(z) = \frac{1}{1 + G_c(z)G_p(z)} \cdot \frac{1}{1 + G_{\text{DFT}}(z)H(z)}(r(z) - d(z)) \tag{2.46}$$

where $H(z) = G_c(z)G_p(z)/\left(1 + G_c(z)G_p(z)\right)$ denotes the closed-loop transfer function without the plug-in RC $G_{\text{DFT}}(z)$.

From (2.46), it can also be derived that the overall closed-loop DFT-based RC system shown in Figure 2.14 is stable if the following conditions hold:

1. Roots of $1 + G_c(z)G_p(z) = 0$ are inside the unit circle, i.e., $H(z)$ is stable.
2. Roots of $1 + G_{\text{DFT}}(z)H(z) = 0$ are inside the unit circle.

The above stability conditions for the plug-in DFT-based RC system are actually compatible to other plug-in PC systems. Similar to MRSC systems, the stability range of the control gain k_F for the DFT-based RC systems cannot directly be derived from the above conditions. However, empirical evidences show that the less the selected harmonics in DFT-based RC are, the larger the stability range of the control gain k_F is.

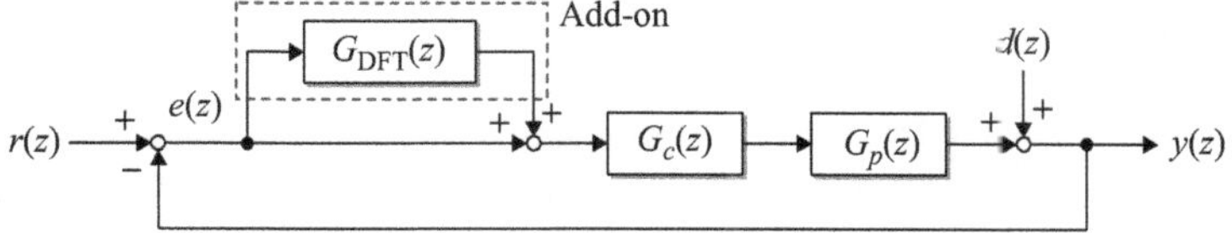

Figure 2.14 A plug-in (add-on) DFT-based RC system

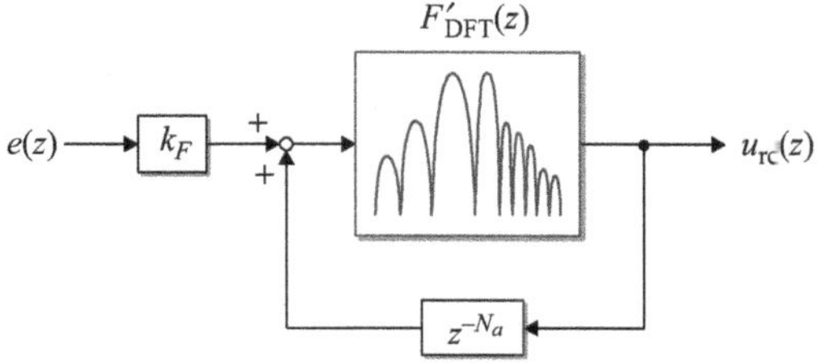

Figure 2.15 A modified DFT-based RC scheme $G'_{DFT}(z)$

2.3.4 Modified DFT-based RC scheme

In order to enable independent tuning of the coefficient at each selected harmonic frequency, the DFT-based RC is modified, as shown in Fig. 2.15. It can be seen that a modified band-pass comb filter is adopted, which can be expressed as:

$$F'_{\text{DFT}}(z) = \sum_{h \in N_h} F'_{\text{dh}}(z) = \sum_{h \in N_h} \left\{ \frac{2}{N} \sum_{i=1}^{N} \left(a_h \cos\left[\frac{2\pi}{N} h(i + N_a) \right] \right) z^{-i} \right\} \tag{2.47}$$

where a_h denotes the weighted coefficient for the selected h-th order harmonic and $\sum_{h \in N_h} a_h = 1$. The corresponding modified DFT-based RC can thus be written as:

$$G'_{\text{DFT}}(z) = \frac{u_{\text{rc}}(z)}{e(z)} = \frac{F'_{\text{DFT}}(z)}{1 - F'_{\text{DFT}}(z) z^{-N_a}} = \frac{k_F \sum_{h \in N_h} F'_{\text{dh}}(z)}{1 - z^{-N_a} \sum_{h \in N_h} F'_{\text{dh}}(z)} \tag{2.48}$$

Remark 2.7 *The modified DFT-based RC scheme of (2.48) is equivalent to an MRSC controller with an identical phase compensation gain for all the RSC components. However, it still allows us to optimize its dynamic response by tuning the coefficient for each selected RSC component (i.e., tuning the weighted coefficient a_h).*

2.4 Basis function

In mathematics, a basis function is an element of a particular basis for a function space. Every continuous function in the function space can be represented as a linear combination of the basis functions [41]. For example, any periodic signal or function can be decomposed into the sum of a (possibly infinite) set of simple oscillating functions, namely sines and cosines (or, equivalently, complex exponentials). A periodic function $f(t)$ with a period of T_0, which is integrable within an interval $[t_0, t_0+T_0]$, can be decomposed into [42]:

$$f(t) = \frac{a_0}{2} + \sum_{n=1}^{\infty}\left(a_n \cos\frac{2n\pi t}{T_0} + b_n \sin\frac{2n\pi t}{T_0}\right) = \sum_{n=0}^{\infty} c_n e^{j\frac{2n\pi t}{T_0}} \tag{2.49}$$

where

$$a_0 = \frac{1}{T}\int_{t_0}^{t_0+T_0} f(t)dt$$

$$a_n = \frac{1}{T}\int_{t_0}^{t_0+T_0} f(t)\cos\left(\frac{2n\pi t}{T_0}\right)dt$$

$$b_n = \frac{1}{T}\int_{t_0}^{t_0+T_0} f(t)\sin\left(\frac{2n\pi t}{T_0}\right)dt$$

$$c_n = \frac{1}{T}\int_{t_0}^{t_0+T_0} f(t)e^{-j\frac{2n\pi t}{T_0}}dt$$

In (2.49) the complex exponential functions $e^{-jn\omega_0 t}$, or the sine functions $\sin(n\omega_0 t)$ and the cosine functions $\cos(n\omega_0 t)$ with $\omega_0 = 2\pi/T_0$ construct the basis functions for the periodic function $f(t)$. Consequently, taking the *Laplace* transform, the internal model for the periodic function $f(t)$ can be correspondingly represented as a linear combination of resonant controllers – the internal models for sine and cosine functions, i.e., basis functions. Therefore, (2.49) can be written as:

$$F(s) = \frac{a_0}{2} + \sum_{n=1}^{\infty}\left(\frac{a_n s}{s^2 + (n\omega_0)^2} + \frac{b_n(n\omega_0)}{s^2 + (n\omega_0)^2}\right) = \sum_{n=-\infty}^{\infty}\left(\frac{c_n}{s - j(n\omega_0)}\right) \tag{2.50}$$

where the resonant controllers are basis functions for the periodic function $f(t)$. Compared with (2.2), (2.50) can be treated a basis function-based RC. Furthermore, (2.50) can be discretized into its discrete-time form by means of the Tustin transformation with frequency pre-warping as:

$$s = \frac{(n\omega_0)}{\tan(n\omega_0 T/2)}\frac{z-1}{z+1} \tag{2.51}$$

Similarly, a new set of basis functions for the RC of periodic signals was developed in [43,44]. Assuming that a periodic reference signal with a period of T_0 is uniformly sampled with the sampling period of T_s, the corresponding discrete sequence within one period is given as $r(k)$, where $k = 0, 1, 2, \ldots, N-1$ and $N = T_0/T_s$ is assumed to be an odd integer. From the Fourier analysis [8], this discrete periodic signal $r(k)$ can be uniquely represented by the inverse Fourier transform as:

$$r(k) = \frac{1}{N}\sum_{m=0}^{N-1} R\left(e^{j2\pi m/N}\right)e^{j2\pi mk/N} \tag{2.52}$$

where N is the number of samples within a period and $R(e^{j2\pi m/N})$ ($m = 0, 1, 2, \ldots, N-1$) are the frequency components in the periodic signal. Note that the discrete frequencies are at $0, 2\pi/N, \ldots, 2\pi(N-1)/N$. For simplicity, the fundamental frequency is denoted as $\omega_0 = 2\pi/N$.

The z-transform of the signal $r(k)$ is defined as:

$$R(z) = \sum_{k=0}^{M-1} r(k)z^{-k} \tag{2.53}$$

Substituting (2.52) into (2.53) yields [44]:

$$R(z) = \sum_{m=-(N-1)/2}^{(N-1)/2} R(e^{jm\omega_0})H_m(z) \tag{2.54}$$

where $H_m(z)$ is termed the m-th order harmonic filter and has the form:

$$H_m(z) = \frac{1}{N}\frac{1 - z^{-N}}{1 - e^{jm\omega_0}z^{-1}} = \frac{1}{N}\left(1 + e^{jm\omega_0}z^{-1} + \ldots + e^{j(N-1)m\omega_0}z^{-(N-1)}\right) \tag{2.55}$$

where the odd number N is assumed to include the zero frequency.

The m-th order harmonic filter $H_m(z)$ in (2.55) can be treated as two-cascaded filters as follows:

$$H_m(z) = \overbrace{\left(\frac{1}{N}\frac{1}{1 - e^{jm\omega_0}z^{-1}}\right)}^{\text{generalized integrator at } m\omega_0} \cdot \overbrace{\left(1 - z^{-N}\right)}^{\substack{\text{comb filter with notches} \\ \text{at all harmonic frequencies}}} \tag{2.56}$$

Remark 2.8 *From (2.56), it can be found that $H_m(z)$ is actually a band-pass filter at the harmonic frequency $m\omega_0$, which is similar to the DFT-based FIR filter $F_{\mathrm{dh}}(z)$ in (2.41). $H_m(z)$ in (2.56) offers a general selective harmonic band-pass filter.*

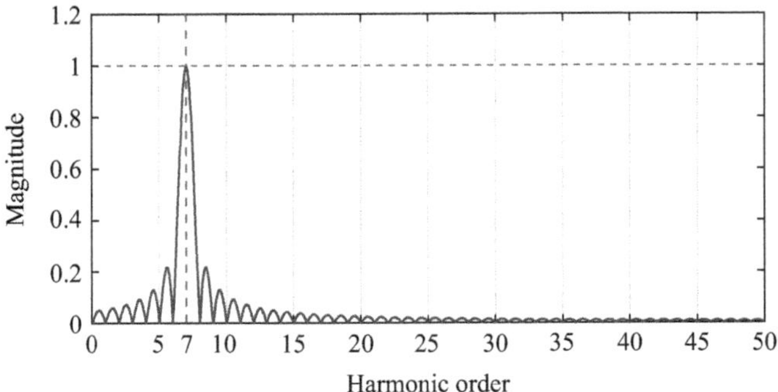

Figure 2.16 Magnitude response of the seventh-order harmonic band-pass filter $H_7(z)$ with $N = 200$ and $h = 7$

For instance, Figure 2.16 gives the magnitude response of the seventh-order harmonic band-pass filter:

$$H_7(z) = \left(\frac{1}{N}\frac{1}{1 - e^{j7\omega_0}z^{-1}}\right) \cdot (1 - z^{-N})$$

Furthermore, $R(z)$ in (2.54) can also be written in terms of real (denoted as **Re**) and imaginary (denoted as **Im**) parts of the frequency component $R(e^{jm\omega_0})$ as:

$$R(z) = \frac{1}{N}\frac{1 - z^{-N}}{1 - z^{-1}}R(e^{j0}) + \sum_{m=1}^{(N-1)/2}\left[\mathbf{Re}\left(R(e^{jm\omega_0})F_{Rm}(z)\right) + \mathbf{Im}\left(R(e^{jm\omega_0})F_{Im}(z)\right)\right] \tag{2.57}$$

where $F_{Rm}(z)$ and $F_{Im}(z)$ are the mth-order harmonic filters given by:

$$F_{Rm}(z) = \frac{1}{N}\frac{2(1 - \cos(m\omega_0)z^{-1})(1 - z^{-N})}{1 - 2\cos(m\omega_0)z^{-1} + z^{-2}} \tag{2.58}$$

and

$$F_{Im}(z) = \frac{1}{N}\frac{2\sin(m\omega_0)z^{-1}(1 - z^{-N})}{1 - 2\cos(m\omega_0)z^{-1} + z^{-2}} \tag{2.59}$$

being two second-order filters. Note that the denominators of the filters in (2.58) and (2.59) are actually the same as that of the resonant controller in (2.31).

Remark 2.9 *The classic RC, RSC, and DFT-based RC are further bridged by the basis functions. All the basis function sets for these three fundamental periodic controllers are based on the IMP for regulating the periodic signals of interest. An individual basis function for a periodic signal is corresponding to a resonant controller at the targeted harmonic frequency [43–45].*

2.5 Summary

By including the internal models of interested harmonics in three different ways, fundamental IMP-based PC strategies – the classic RC, the MRSC, and the DFT-based RC, provide very simple but effective control solutions to the compensation of periodic signals. However, fundamental periodic controllers fail to offer optimal compensation of interested harmonics with high accuracy, fast dynamic response, good robustness, and easy implementation. These fundamental PC strategies face important challenges as follows:

1. Classic repetitive control –
   - Consumes light computation and it is easy for implementation due to its compact form.
   - It can achieve high accuracy due to its included internal models of all harmonics.
   - It is impossible for the classic RC to optimize its dynamic response by independently tuning the coefficient and phase compensation at each selected harmonic frequency. Thus, the classic RC often yields slow dynamic response.
   - The stability range of the control gain k_{rc} can be derived from the entire RC system stability conditions.
   - It assumes that the fundamental period N is an integer and it is known.
   - The linear phase-lead step p is an integer.
2. Multiple resonant control –
   - Offers very fast dynamic response and it has good system stability. Both control gains k_h and phase compensation ϕ_h at each selected harmonic frequency can be independently and finely tuned.
   - It provides flexible selective harmonics compensation. It can improve its control accuracy by flexibly adding parallel RSC components.
   - Empirical evidences show that the less the selected harmonics in the MRSC are, the larger the stability range of the control gains k_h is.
   - A large amount of RSC components are needed to achieve high control accuracy, but will increase the parallel computation burden and it has tuning difficulties in implementation.
3. DFT-based repetitive control –
   - It is equivalent to the classic RC with a phase-lead compensator.
   - A modified DFT-based RC allows optimizing its dynamic response by tuning the control gain at each selected harmonic frequency.
   - Empirical evidences show that the less the selected harmonics in the DFT-based RC are, the larger the stability range of the control gains is.
   - It provides flexible selective harmonics compensation. Its implementation would involve a large amount of parallel computation, which is proportional to the fundamental period N. It is thus more suitable for high-performance fixed-point DSP implementation.
   - The linear phase-lead step N_a is an integer.

Obviously, internal models of harmonics act as generalized "integrators" for periodic signals. The plug-in control strategy that includes feedback control plus PC can actually be regarded as *general proportional–integral–derivative* (*GPID*) control strategy for periodic signals. The classic RC, the MRSC, and the DFT-based RC form the fundamental control schemes of a unified IMP-based PC. In order to achieve optimal selective harmonics compensation, advanced IMP-based PC strategies will be investigated in the next chapter.

References

[1] Francis, B.A. and Wonham, W.M., "The internal model principle of control theory," *Automatica*, vol. 12, no. 5, pp. 457–465, 1976.

[2] Goodwin, G. C., Graebe, S. F., and Salgado, M. E., *Control System Design*, Prentice Hall, 2000.

[3] Hara, S., Yamamoto, Y., Omata, T., and Nakano, M., "Repetitive control system: A new type servo system for periodic exogenous signals," *IEEE Trans. Automatic Control*, vol. 33, no. 7, pp. 659–668, 1988.

[4] Tomizuka, M., Tsao, T.-C., and Chew, K.-K., "Analysis and synthesis of discrete-time repetitive controllers," *ASME: J. Dyn. Syst. Meas. Control*, vol. 111, pp. 353–358, 1989.

[5] Broberg, H. L. and Molyet, R. G., "A new approach to phase cancelation in repetitive control," *IEEE Ind. Appl. Society Annual Meeting*, pp. 1766–1770, 2–6 Oct. 1994.

[6] Longman, R. W., "Iterative learning control and repetitive control for engineering practice," *Int. J. Control*, vol. 73, no. 10, pp. 930–954, 2000.

[7] Longman, R., "On the theory and design of linear repetitive control systems," *European J. Control,"* vol. 16, no. 5, pp. 447–496, 2010.

[8] Hornik, T. and Zhong, Q.-C., "A current-control strategy for voltage-source inverters in microgrids based on H∞ and repetitive control," *IEEE Trans. Power Electron.*, vol. 26, no. 3, pp. 943–952, 2011.

[9] Zhou, K., Qiu, Z. and Yang, Y., "Current Harmonics Suppression of Single-Phase PWM Rectifiers," in *Proc. of 2012 3rd IEEE International Symposium on Power Electronics for Distributed Generation Systems (PEDG)*, pp. 54–57, June 25–28, 2012.

[10] Haneyoshi, T., Kawamura, A., and Hoft, R. G., "Waveform compensation of PWM inverter with cyclic fluctuating loads," *IEEE Trans. Ind. Appl.*, vol. 24, no. 4, pp. 582–589, 1988.

[11] Tzou, Y.-Y., Ou, R.-S., Jung, S.-L., and Chang, M.-Y., "High-performance programmable AC power source with low harmonic distortion using DSP based repetitive control technique," *IEEE Trans. Power Electron.*, vol. 12, no. 4, pp. 715–725, 1997.

[12] Zhou, K. and Wang, D. "Zero tracking error controller for three-phase CVCF PWM inverter," *Electronics Letters*, vol. 36, no. 10, pp. 864–865, 2000.

[13] Zhou, K., Wang, D., and Low, K.S., "Periodic errors elimination in CVCF PWM DC/AC converter systems: A repetitive control approach," *IEE Proc. – Control Theory and Appl.*, vol. 147, no. 6, pp. 694–700, 2000.

[14] Zhou, K. and Wang, D., "Digital repetitive learning controller for three-phase CVCF PWM inverter," *IEEE Trans. Ind. Electron.*, vol. 48, no. 4, pp. 820–830, 2001.

[15] Zhou, K. and Wang, D., "Unified robust zero error tracking control of CVCF PWM converters," *IEEE Trans. Circuits and Systems (I)* vol. 49, no. 4, pp. 492–501, 2002.

[16] Zhou, K. and Wang, D., "Digital repetitive controlled 3-phase PWM rectifier," *IEEE Trans. Power Electron.*, vol. 18, no. 1, pp. 309–316, 2003.

[17] Zhang, K., Kang, Y., Xiong, J., and Chen, J., "Direct repetitive control of SPWM inverter for UPS purpose," *IEEE Trans. Power Electron.*, vol. 18, no. 3, pp. 784–792, 2003.

[18] Zhang, B., Wang, D., Zhou, K., Wang, Y., "Linear phase-lead compensated repetitive control of PWM inverter," *IEEE Trans. Industrial Electronics*, vol. 55, no. 4, pp. 1595–1602, 2008.

[19] Ye, Y., Zhang, B., Zhou, K., Wang, D., and Wang, Y., "High-performance cascade type repetitive controller for CVCF PWM inverter: analysis and design," *IET Electrical Power Appl.*, vol. 1, no. 1, pp. 112–118, 2007.

[20] Ye, Y., Zhou, K., Zhang, B., Wang, D., and Wang, J., "High-performance repetitive control of PWM DC-AC converters with real-time phase-lead FIR filter. *IEEE Trans. Circuits and Systems (II)*, vol. 53, no. 8, pp. 768–772, 2006.

[21] Yang, Y., Zhou, K. and Lu, W. "Robust repetitive control scheme for three-phase CVCF PWM inverters," *IET Power Electron.*, vol. 5, no. 6, pp. 669–677, 2012.

[22] Zmood, D.N., Holmes, D.G., and Bode, G.H., "Frequency-domain analysis of three-phase linear current regulators," *IEEE Trans. Ind. Appl.*, vol. 37, no. 2, pp. 601–610, 2001.

[23] Yuan, X., Merk, W., Stemmler, H., and Allmeling, J., "Stationary frame generalized integrators for current control of active power filters with zero steady-state error for current harmonics of concern under unbalanced and distorted operating conditions," *IEEE Trans. Ind. Appl.*, vol. 38, no. 2, pp. 523–532, 2002.

[24] Sato, Y., Ishizuka, T., Nezu, K., and Kataoka, T., "A new control strategy for voltage-type PWM rectifiers to realize zero steady-state control error in input current," *IEEE Trans. Ind. Appl.*, vol. 34, no. 3, pp. 480–486, 1998.

[25] Fukuda, S. and Yoda, T., "A novel current-tracking method for active filters based on a sinusoidal internal model," *IEEE Trans. Ind. Appl.*, vol. 37, no. 3, pp. 888–895, 2001.

[26] Liserre, M., Teodorescu, R., and Blaabjerg, F., "Multiple harmonics control for three-phase grid converter systems with the use of PI-RES current controller in a rotating frame," *IEEE Trans. Power Electron.*, vol. 21, no. 3, pp. 836–841, 2006.

[27] Teodorescu, R., Blaabjerg, F., Liserre, M., and Loh, P.C., "Proportional-resonant controllers and filters for grid-connected voltage-source converters," *IEE Proc. – Electric Power Appl.*, vol. 153, no. 5, pp. 750–762, 2006.

[28] Liu, C., Blaabjerg, F., Chen, W. and Xu, D., "Stator Current Harmonic Control with Resonant Controller for Doubly Fed Induction Generator," *IEEE Trans. on Power Electron.*, vol. 27, no. 7, pp. 3207–3220, 2012.

[29] Rodriguez, P., Candela, J.I., Luna, A., Asiminoaei, L., Teodorescu, R., and Blaabjerg, F., "Current harmonics cancellation in three-phase four-wire systems by using a four-branch star filtering topology," *IEEE Trans. Power Electron.*, vol. 24, no. 8, pp. 1939–1950, 2009.

[30] Lascu, C., Asiminoaei, L., Boldea, I., and Blaabjerg, F., "High-performance current controller for selective harmonic compensation in active power filters," *IEEE Trans. Power Electron.*, vol. 22, no. 5, pp. 1826–1835, 2007.

[31] Lascu, C., Asiminoaei, L., Boldea, I., and Blaabjerg, F., "Frequency response analysis of current controllers for selective harmonic compensation in active power filters," *IEEE Trans. Ind. Electron.*, vol. 56, no. 2, pp. 337–347, 2009.

[32] Yepes, G., Freijedo, F.D., Lopez, O., and Doval-Gandoy, J., "High-performance digital resonant controllers implemented with two integrators," *IEEE Trans. Power Electron.*, vol. 6, no. 2, pp. 563–576, 2011.

[33] Teodorescu, R., Liserre, M., and Rodríguez, P., *Grid Converters for Photovoltaic and Wind Power Systems*, Wiley-IEEE, 2011.

[34] Yang, Y., Zhou, K., and Cheng, M., "Phase compensation multi-resonant control of CVCF PWM converters," *IEEE Trans. Power Electron.*, vol. 28, no. 8, pp. 3923–3930, 2013.

[35] Mattavelli, P., "A closed-loop selective harmonic compensation for active filters," *IEEE Trans. Ind. Appl.*, vol. 37, no. 1, pp. 81–89, 2001.

[36] Mattavelli, P., Marafão, F. P., "Repetitive-based control for selective harmonic compensation in active power filters," *IEEE Trans. Ind. Electron.*, vol. 51, no. 5, pp. 1018–1024, 2004.

[37] Rowan, T.M. and Kerkman, R.J., "A new synchronous current regulator and an analysis of current-regulated PWM inverters," *IEEE Trans. Ind. Appl.*, vol. IA-22, pp. 678–690, Mar./Apr. 1986.

[38] Blaabjerg, F., Teodorescu, R., Liserre, M., and Timbus, A. V., "Overview of control and grid synchronization for distributed power generation systems," *IEEE Trans. Ind. Electron.*, vol. 53, no. 5, pp. 1398–1409, Oct. 2006.

[39] Timbus, A.V., Liserre, M., Teodorescu, R., Rodriguez, P., and Blaabjerg, F., "Evaluation of current controllers for distributed power generation systems," *IEEE Trans. Power Electron.*, vol. 24, no. 3, pp. 654–664, Mar. 2009.

[40] Kulkarni, A. and John, V., "Mitigation of lower order harmonics in a grid-connected single-phase PV inverter," *IEEE Trans. Power Electron.*, vol. 28, no. 11, pp. 5024–5037, Nov. 2013.

[41] https://en.wikipedia.org/wiki/Basis_function

[42] https://en.wikipedia.org/wiki/Fourier_series
[43] Wang, L., Chai, S., and Rogers, E., "Predictive repetitive control based on frequency decomposition," in Proc. of 2010 American Control Conference, Baltimore, MD, USA, pp. 4277–4282, Jun. 30–Jul. 02, 2010.
[44] Wang, L. and Cluett, W. R., *From Plant Data to Process Control: Ideas for Process Identification and PID Design*. 1st ed. Taylor and Francis, 2000.
[45] Shi, Y., Longman, R. W., and Nagashimac, M., "Small gain stability theory for matched basis function repetitive control," *Acta Astronautica*, vol. 95, pp. 260–271, Feb.–Mar. 2014.

Chapter 3
Advanced periodic control for power harmonics mitigation

Abstract

The rapidly growing amount of power harmonics injected by power electronic converter interfaced loads and distributed generator systems may cause serious power quality problems in the electrical power systems, such as harmonic losses, resonances, device malfunction, and even entire system instability. Power harmonics induced by power electronics converters usually concentrate on some specific frequencies. For instance, single-phase H-bridge converters mainly produce $(4k \pm 1)$ $(k = 1, 2, \ldots)$-order power harmonics, while n-pulse $(n = 6, 12, \ldots)$ converters, such as in high-voltage direct current (HVDC) transmission systems, mainly produce $(nk \pm 1)$ $(k = 1, 2, \ldots)$-order power harmonics [1–7], and also diode rectifiers loads used in many applications disturb the grid. As discussed in Chapter 2, the IMP-based fundamental periodic controllers fail to optimally compensate power harmonics with high control accuracy, while maintaining fast dynamic response, guaranteeing robustness, and being feasible for implementation.

To address these issues, this chapter will explore advanced IMP-based PC technologies [8] for optimal power harmonics compensation. The general design methodology also comprises an internal model for selective power harmonic signals and synthesis methods for plug-in PC systems. These advanced PC schemes not only exploit the periodicity of reference/disturbance, but also account for its harmonic frequency distribution. The relationship between fundamental periodic controllers and advanced ones will be demonstrated.

3.1 Parallel structure repetitive control (PSRC)

Parallel RSC controllers enable the multiple RSC (MRSC) controller to independently and finely tune the control gain and the phase compensation angle for every resonant controller (RSC) at each selected harmonic frequency, and then to offer both fast dynamic response and good system stability. However, the MRSC may need a large amount of RSC components to achieve high control accuracy, but at the cost of heavy parallel computational burden and tuning difficulties in practical implementation. Classic repetitive controller (CRC) consumes light computation and is easy for implementation due to its compact delay-based infinite impulse response (IIR) filter form. The CRC can achieve high accuracy due to its included internal models of all harmonics, but cannot optimize its dynamic response by independently tuning control gains and phase compensation at each selected harmonic frequency.

Considering the featured power harmonics distribution, a parallel structure repetitive control (PSRC) RC which includes complex internal models for featured power harmonics, is developed to take advantages of both the MRSC and the CRC for better power harmonics mitigation in this section.

3.1.1 Complex internal model of selective harmonics

Given that $(nk \pm 1)$-order harmonics dominate the power harmonics, as shown in Figure 3.1, in order to compensate selected $(nk \pm m)$ $(k = 0, 1, 2, \ldots$ and $m < n)$-order harmonics, like the internal model in (2.2) for the CRC, a compact delay-based selective harmonics generator, i.e., the internal model of $(nk \pm m)$-order harmonics, is formulated in [9,10] as:

$$\hat{G}_{\pm m}(s) = \frac{u(s)}{e(s)} = \frac{e^{-sT_0/n \pm j(2\pi m/n)}}{1 - e^{-sT_0/n \pm j(2\pi m/n)}} = \frac{e^{-2\pi[s/(n\omega_0) \mp j(m/n)]}}{1 - e^{-2\pi[s/(n\omega_0) \mp j(m/n)]}}$$
$$= \frac{e^{\pm j(2\pi m/n)}}{e^{sT_0/n} - e^{\pm j(2\pi m/n)}} \tag{3.1}$$

in which T_0 has been defined previously, n and m are integers with $n > m \geq 0$. Figure 3.2 exemplifies the internal model of $nk + m$, where $n = 6$ and $m = 1$. Since

$$\hat{G}_{\pm m}(s) = \frac{e^{-2\pi[s/(n\omega_0) \mp j(m/n)]}}{1 - e^{-2\pi[s/(n\omega_0) \mp j(m/n)]}}$$
$$= -\frac{1}{2} + \frac{n}{T_0} \cdot \frac{1}{s \mp jm\omega_0} + \frac{2n}{T_0} \sum_{k=1}^{+\infty} \frac{s \mp jm\omega_0}{(s \mp jm\omega_0)^2 + n^2k^2\omega_0^2} \tag{3.2}$$

where $\omega_0 = 2\pi f_0 = 2\pi/T_0$, $k = 0, 1, 2, \ldots$ and $m = 0, 1, 2, \ldots, n-1$. From (3.2), it is clear that the complex internal model of (3.1) is equivalent to a parallel combination of RSC components at $\pm(nk \pm m)$-order harmonic frequencies with the coefficient $2n/T_0$. Notice that the signal generator $\hat{G}_m(s)$ has poles at $s = j(\pm nk \pm m)\omega_0$, and then exhibits infinite gains at these harmonic frequencies. The infinite gains will lead to zero tracking errors exclusively at these frequencies if $\hat{G}_m(s)$ is included into the closed-loop system.

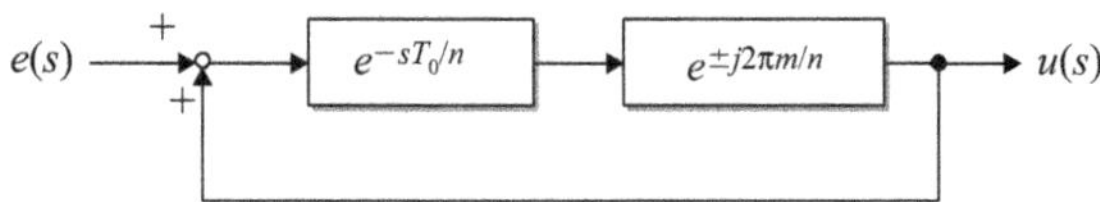

Figure 3.1 Complex internal model for selective harmonics in (3.1)

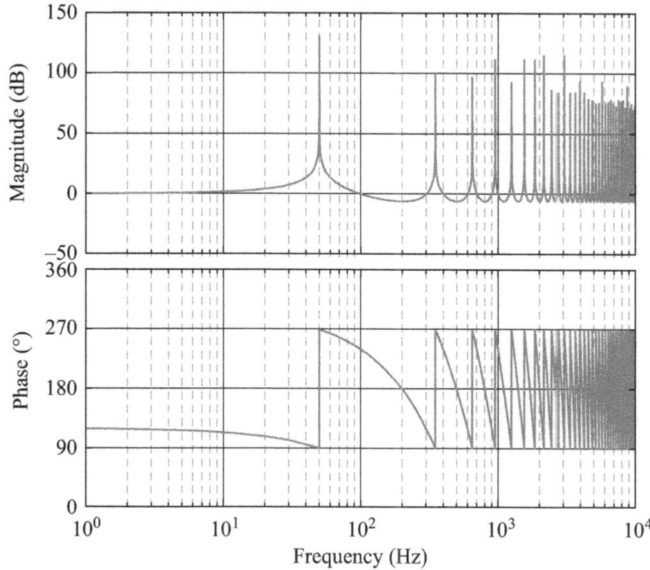

Figure 3.2 Bode plots of the internal model of 6k + 1 harmonics in (3.1), where $T_0 = 0.02$ s

Remark 3.1 *To compensate the featured ($nk \pm m$)-order harmonics, (3.1) offers a compact tailor-made complex internal model of $\pm(nk \pm m)$-order harmonic frequencies.*

Consequently, in order to cover all harmonics for compensation, a general parallel-structure internal model can be given as:

$$\hat{G}_{\text{prc}}(s) = \sum_{m=0}^{n-1} \left[\hat{G}_m(s)\right] = \sum_{m=0}^{n-1} \left[\frac{e^{j(2\pi m/n)}}{e^{sT_0/n} - e^{j(2\pi m/n)}}\right] \tag{3.3}$$

where all the internal models of $\pm nk + m$ order harmonics are connected in parallel.

3.1.2 Parallel structure RC

As mentioned in Chapter 2, for multiple harmonics compensation, the MRSC [11–25] offers much faster dynamic response than what the CRC [25–35] does because all RSC components of the MRSC are connected in parallel and allow independent tuning of the control gain k_h and phase compensation ϕ_h for each RSC component. Likewise, in order to achieve much faster dynamic response for the RC, based on the internal model of (3.3), a prototype parallel structure repetitive controller (PSRC) can be created as:

$$G_{\text{psrc}}(z) = \sum_{m=0}^{n-1} \left[k_{\text{pm}}\hat{G}_m(s)\right] = \sum_{m=0}^{n-1} \left[k_{\text{pm}} \frac{e^{j(2\pi m/n)}}{e^{sT_0/n} - e^{j(2\pi m/n)}}\right] \tag{3.4}$$

where k_{pm} with the subscript $m = 0, 1, \ldots, n-1$ are control gains for tuning the convergence rate.

Lemma 3.1 *Let* $k_{pm} = k_{rc}/n$, $m = 0, 1, 2, \ldots, n-1$, *the prototype PSRC scheme of (3.4) will be equivalent to a prototype CRC [9,10]. Then, the following equation is tenable:*

$$G_{psrc}(s) = \sum_{m=0}^{n-1}\left[k_{pm}\hat{G}_m(s)\right] = \sum_{i=0}^{n-1}\left[\left(\frac{k_{rc}}{n}\right)\frac{e^{j(2\pi m/n)}}{e^{sT_0/n} - e^{j(2\pi m/n)}}\right] = k_{rc}\frac{e^{-sT_0}}{1 - e^{-sT_0}} \tag{3.5}$$

Proof: From (2.3), (3.2), and (3.4), we know that the prototype PSRC can be written as:

$$\begin{aligned} G_{psrc}(s) &= \sum_{m=0}^{n-1}\left[\frac{k_{rc}}{n}\hat{G}_m(s)\right] \\ &= \sum_{m=0}^{n-1}\frac{k_{rc}}{n}\left[-\frac{1}{2} + \frac{n}{T_0}\cdot\frac{1}{s - jm\omega_0} + \frac{2n}{T_0}\sum_{k=1}^{+\infty}\frac{s - jm\omega_0}{(s - jm\omega_0)^2 + n^2k^2\omega_0^2}\right] \\ &= k_{rc}\left[-\frac{1}{2} + \frac{1}{T_0 s} + \frac{2}{T_0}\sum_{n=1}^{\infty}\frac{s}{s^2 + (n\omega_0)^2}\right] = k_{rc}\frac{e^{-sT_0}}{1 - e^{-sT_0}} = G_{rc}(s). \end{aligned} \tag{3.6}$$

In order to compensate multiple harmonics, the MRSC requires the same amount of paralleled RSC components; as shown in Figure 3.3, while the number of paralleled compact delay-based internal models for selected $(nk \pm m)$-order harmonics in the PSRC is only up to n. A much less number of paralleled compact delay-based internal models with weighted coefficients enables the PSRC to achieve a good trade-off among the fast dynamic response, the light computation, and the easy implementation. The PSRC naturally bridges the CRC and the MRSC.

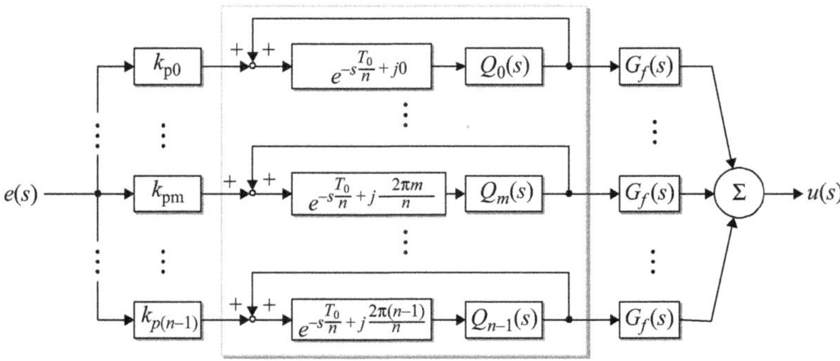

Figure 3.3 General parallel-structure RC scheme with n units

In practical PSRC applications, a low-pass or band-pass filter $Q_m(z)$ is usually used to select the interested harmonics for compensation and enhance the system robustness, as shown in Figure 3.3. Furthermore, a phase-lead compensator $G_f(z)$ is also incorporated in Figure 3.3 to improve the system stability and control performance [36–41]. Straightforwardly, considering practical implementation issues, the PSRC scheme is given as [9,10]:

$$G_{\mathrm{psrc}}(s) = \sum_{m=0}^{n-1}\left[k_{\mathrm{pm}}\hat{G}'_m(s)G_f(s)\right] = \sum_{m=0}^{n-1}\left[k_{\mathrm{pm}}\frac{e^{j(2\pi m/n)}Q_m(s)}{e^{sT_0/n} - e^{j(2\pi m/n)}Q_m(s)}G_f(s)\right] \tag{3.7}$$

where $\hat{G}'_m(s)$ is the modified $\hat{G}_m(s)$ with $m = 0, 1, \ldots, n-1$.

From (3.2) and (3.7), it is known that:

- When $Q_m(s) \to 1$, the PSRC $G_{\mathrm{psrc}}(s) \to \infty$ at poles $s = j(\pm nk + m)\omega_0$, and thus offering exact tracking control of periodic signals at selected frequencies $(\pm nk + m)f_0$. The PSRC offers an accurate control solution to the compensation of selected harmonics.
- The coefficients for $\pm nk + m$ order harmonic frequencies in (3.4) are n times as large as those of the CRC of (2.2). Hence, when $Q_m(s) \to 1$, the error convergence rate at $\pm nk + m$ order harmonic frequencies employing the PSRC $G_{\mathrm{psrc}}(s)$ of (3.7) will be n times faster than what the CRC can achieve.

Lemma 3.2 *Let $k_{\mathrm{pm}} = k_{\mathrm{rc}}/n$ and $Q_m(s) = Q(s)$ with $m = 0, 1, 2, \ldots, n-1$, the PSRC scheme of (3.7) is equivalent to the CRC scheme with the control gain being k_{rc} and low-pass or band-pass filter $Q^n(s)$ (as shown in Figure 3.4), i.e.:*

$$G_{\mathrm{psrc}}(s) = \frac{k_{\mathrm{rc}}}{n}\sum_{m=0}^{n-1}\frac{e^{j(2\pi m/n)}Q(s)}{e^{sT_0/n} - e^{j(2\pi m/n)}Q(s)}G_f(s) = k_{\mathrm{rc}}\frac{Q^n(s)}{e^{sT_0} - Q^n(s)}G_f(s) \tag{3.8}$$

Proof: When $n = 1$, we can easily obtain that (3.8) is tenable. When $n > 1$, let $x = e^{sT_0/n}$. Thus, if the following equation can be proved, (3.8) is then tenable:

$$\frac{1}{n}\sum_{m=0}^{n-1}\frac{e^{j(2\pi m/n)}}{x - e^{j(2\pi m/n)}} = \frac{\frac{1}{n}\left\{\sum_{m=0}^{n-1}\left[e^{j(2\pi m/n)}\prod_{\substack{i=0\\ i\neq m}}^{n-1}\left(x - e^{j(2\pi i/n)}\right)\right]\right\}}{\prod_{m=0}^{n-1}\left(x - e^{j(2\pi m/n)}\right)} = \frac{1}{x^n - 1} \tag{3.9}$$

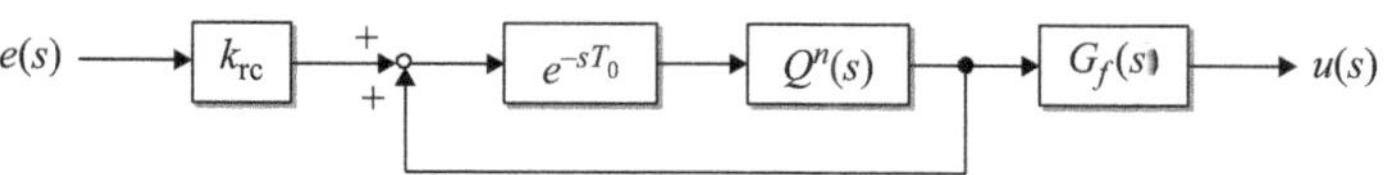

Figure 3.4 Equivalent structure of the PSRC scheme to the CRC scheme

Because the roots of $x^n - 1 = 0$ are $e^{j2\pi k/n}$ with $k = 0, 1, \ldots, n - 1$, we can obtain:

$$x^n - 1 = \prod_{k=0}^{n-1} \left(x - e^{j(2\pi k/n)}\right)$$

Therefore, if the following equation can be proved, (3.9) is tenable:

$$\sum_{m=0}^{n-1} \left[e^{j(2\pi m/n)} \prod_{\substack{i=0 \\ i \neq m}}^{n-1} \left(x - e^{j(2\pi i/n)}\right) \right] = n \tag{3.10}$$

Let

$$q(x) = \sum_{m=0}^{n-1} \left[e^{j(2\pi m/n)} \prod_{\substack{i=0 \\ i \neq m}}^{n-1} \left(x - e^{j(2\pi i/n)}\right) \right] - np(x) = x^n - 1$$

$$= \prod_{m=0}^{n-1} \left(x - e^{j(2\pi m/n)}\right) a_m = e^{j(2\pi m/n)}, m = 0, 1, \ldots, n - 1$$

We have to prove that

$$q(x) = 0$$

Notice that

$$q(a_m) = a_m \cdot p'(a_m) - n = a_m \cdot n \cdot a_m^{n-1} - n = 0, m = 0, 1, \ldots, n - 1 \tag{3.11}$$

Thus, (3.9) and (3.10) are proved, and then (3.8) is proved. Lemma 3.2 is proved.

In practical applications, the PSRC is usually implemented in its digital form. As shown in Figure 3.5, the digital PSRC can be written as:

$$G_{\text{psrc}}(z) = \sum_{m=0}^{n-1} \left[k_{\text{pm}} \hat{G}_m'(z) G_f(z) \right] = \sum_{m=0}^{n-1} \left[k_{\text{pm}} \frac{e^{j(2\pi m/n)} \cdot z^{-N/n} Q_m(z)}{1 - e^{j(2\pi m/n)} \cdot z^{-N/n} Q_m(z)} G_f(z) \right] \tag{3.12}$$

where $N = T_0/T_s \in \mathbb{N}$ with T_s being the sampling period, $\hat{G}_m'(z)$ is the digital form of the modified $\hat{G}_m'(s)$ with $m = 0, 1, \ldots, n - 1$, $G_f(z)$ is a phase-lead compensator, such as a linear phase-lead compensator $G_f(z) = z^p$ with $p \in \mathbb{N}$, and the filter $Q_m(z)$ often employs a finite impulse response (FIR) low-pass filter as follows:

$$Q_m(z) = \left(\sum_{i=1}^{n} a_i z^i + a_0 + \sum_{i=1}^{n} a_i z^{-i} \right) \bigg/ \left(2 \sum_{i=1}^{n} a_i + a_0 \right) \tag{3.13}$$

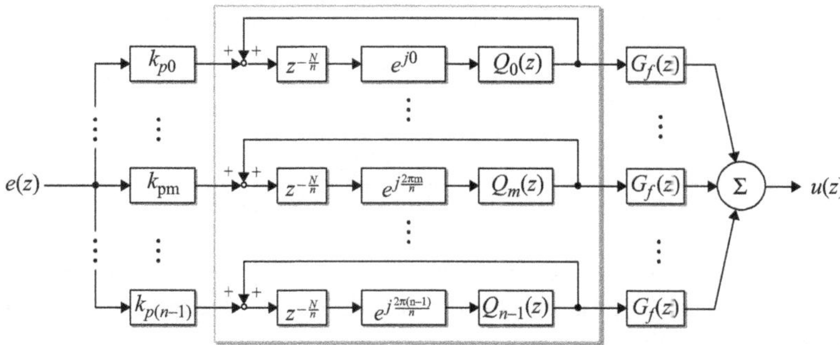

Figure 3.5 General digital parallel-structure RC (PSRC) scheme

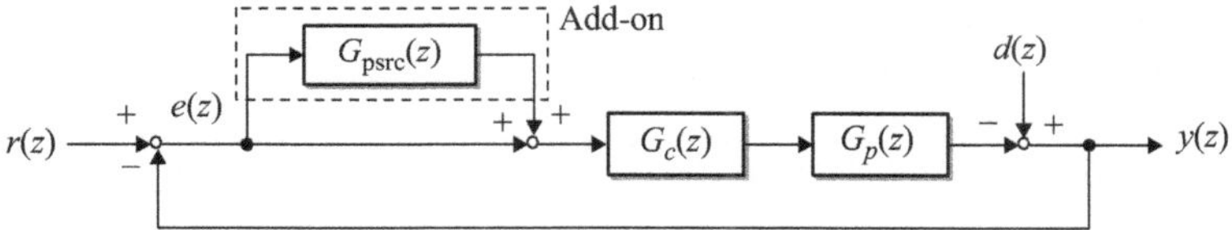

Figure 3.6 Digital add-on parallel-structure repetitive control system

where $a_i > 0$ and $n \in \mathbb{N}$ with $n \leq (N-1)/2$. It is clear that $|Q_m(e^{j\omega})| \leq 1$ and $\angle Q_m(e^{j\omega}) = 0$.

Obviously, when $|Q_m(e^{j\omega})| \to 1$, the poles of (3.12) are located at $(\pm nk + m)\omega_0$, with $|\pm nk + m| \leq N/2$, $m = 0, 1, 2, \ldots, n-1$, $k \in \mathbb{N}$. Likewise, let $k_{pm} = k_{rc}/n$, $m = 0, 1, 2, \ldots, n-1$, the digital PSRC becomes the digital CRC scheme.

Remark 3.2 *The total number of memory cells of the PSRC is N, which is equal to that of the CRC. If any group of harmonics is insignificant, its corresponding internal model can be removed from (3.4). This implies that the digital PSRC occupies the same or less amount of data memory as what the digital CRC does.*

3.1.3 Digital PSRC system and design

Figure 3.6 shows a typical closed-loop control system with a plug-in digital PSRC system, where $G_p(z)$ is the plant, $G_c(z)$ is the conventional feedback controller, $G_{psrc}(z)$ is the PSRC scheme, $r(z)$ is the reference input, $y(z)$ is the output, $e(z) = r(z) - y(z)$ is the tracking error and the input of $G_{psrc}(z)$, and $d(z)$ is the disturbance [40,41]. $Q_m(z)$ with $|Q_m(e^{j\omega})| \leq 1$ with $m = 0, 1, \ldots, n-1$ is usually employed in the PSRC scheme to select the harmonics of interest for compensation and make a good trade-off between the tracking accuracy and the system robustness [36–38].

In Figure 3.6, the output $y(z)$ of the entire system can be expressed as:

$$y(z) = G_r(z) \cdot r(z) + G_d(z) \cdot d(z)$$
$$= \frac{[1 + G_{\text{psrc}}(z)]H(z)}{1 + G_{\text{psrc}}(z)H(z)} r(z) + \frac{[1 + G_c(z)G_p(z)]^{-1}}{1 + G_{\text{psrc}}(z)H(z)} d(z) \tag{3.14}$$

where $G_r(z)$ is the transfer function from $r(z)$ to $y(z)$, $G_d(z)$ is the transfer function from $d(z)$ to $y(z)$, and $H(z)$ is the transfer function of the feedback control system without the add-on PSRC system. And:

$$H(z) = \frac{y(z)}{r(z)} = \frac{G_c(z)G_p(z)}{1 + G_c(z)G_p(z)} \tag{3.15}$$

is assumed to be asymptotically stable.

Without loss of generality, $H(z)$ can be described by:

$$H(z) = \frac{B(z)}{A(z)} = \frac{z^{-c}B^+(z)B^-(z)}{A(z)} \tag{3.16}$$

where c represents known delay steps with $c \in [0, N/n]$, all characteristic roots of $A(z) = 0$ are inside the unit circle, $B^+(z)$ and $B^-(z)$ are the cancelable and uncancelable parts of $B(z)$, respectively [36–38], $B^-(z)$ comprises roots on or outside the unit circle and undesirable roots, which are in the unit circle and $B^+(z)$ comprises roots of $B(z)$, which are not in $B^-(z)$.

The compensation filter $G_f(z)$ can be chosen as [9,10], [36–38]:

$$G_f(z) = \frac{z^c A(z)B^-(z^{-1})}{B^+(z)b} \tag{3.17}$$

where $b \geq \max|B^-(e^{j\omega})|^2$. Since the PSRC will introduce N/n-step delay $z^{-N/n}$ (usually $n \ll N$) and $N/n \geq c$, $G_f(z)$ with a noncausal component z^c can be implemented in the PSRC system [9,10].

Therefore, there is an inverse function $G_f(z)$ of $H(z)$ such that:

$$G_{fH}(z) = G_f(z)H(z) = \frac{B^-(z^{-1})B^-(z)}{b} \tag{3.18}$$

Thus, let $G_f(z)H(z) = N_{GH}(\omega)e^{j\theta_{GH}}$ with $z = e^{j\omega}$, and we have:

$$N_{GH}(\omega) = |G_f(e^{j\omega})H(e^{j\omega})| = \frac{B^-(z^{-1})B^-(z)}{b} \leq 1 \text{ and } \theta_{GH} = 0 \tag{3.19}$$

Equation (3.19) implies that zero-phase compensation is achieved. Zero-phase compensation will significantly simplify the design of the PSRC [36–38].

Theorem 3.1 *For the closed-loop PSRC system shown in Figure 3.6 with the constraints (3.14)–(3.19), if the control gains $k_m \geq 0$ (the subscript $m = 0, 1, 2, \ldots, n-1$) satisfy the following inequalities [9,10]:*

$$0 < \sum_{i=0}^{n-1} k_m < 2 \tag{3.20}$$

then the closed-loop PSRC system is asymptotically stable.

Proof: From (3.14) to (3.19), and Figure 3.6, we can obtain:

$$\begin{aligned} y(z) &= G_r(z)r(z) + G_d(z)d(z) \\ &= \frac{\left[1 + \sum_{m=0}^{n-1}\left[k_{\text{pm}}\hat{G}'_m(z)\right]G_f(z)\right]H(z)}{1 + G_{fH}(z)\sum_{m=0}^{n-1}\left[k_{\text{pm}}\hat{G}'_m(z)\right]}r(z) + \frac{\left[1 + G_c(z)G_p(z)\right]^{-1}}{1 + G_{fH}(z)\sum_{m=0}^{n-1}\left[k_{\text{pm}}\hat{G}'_m(z)\right]}d(z) \end{aligned} \tag{3.21}$$

Let $z = |z|e^{j\omega}$ with $|z| = a$, and

$$Q_m(z) = N_{\text{qm}}(\omega)e^{j\theta_{\text{qm}}(\omega)}, \quad m = 0, 1, 2, \ldots, n-1$$

where

$$N_{\text{qm}}(\omega) = \left|Q_m(e^{j\omega})\right| \leq 1, \quad m = 0, 1, 2, \ldots, n-1$$

then we have:

$$\begin{aligned} \text{Re}\left[\hat{G}'_m(z)\right] &= \text{Re}\left[\frac{e^{j(2\pi m/n)}z^{-N/n}\cdot Q_m(z)}{1 - e^{j(2\pi m/n)}z^{-N/n}Q_m(z)}\right] = \text{Re}\left[\frac{1}{e^{-j(2\pi m/n)}z^{N/n}Q_m^{-1}(z) - 1}\right] \\ &= \text{Re}\left[\left(\frac{a^{N/n}}{N_{\text{qm}}(\omega)}e^{j\left(\frac{N\omega}{n} - \theta_{\text{qm}}(\omega) - \frac{2\pi}{n}m\right)} - 1\right)^{-1}\right], \quad m = 0, 1, 2, \ldots, n-1 \end{aligned} \tag{3.22}$$

Let $b_m = a^{N/n}/N_{\text{qm}}(\omega)$ and $\theta_m = N\omega/n - \theta_{\text{qm}}(\omega) - 2\pi m/n$, with $m = 0, 1, \ldots, n-1$. If $|z| = a \geq 1$, we have $b_m \geq a^{N/n} \geq a \geq 1$ due to usually $N \gg n$ in practice. Equation (3.22) can be rewritten as:

$$\begin{aligned} \text{Re}\left[\hat{G}'_m(z)\right] &= \text{Re}\left[\frac{1}{b_m e^{j\theta_m} - 1}\right] = \text{Re}\left[\frac{(b_m\cos\theta_m - 1) - jb_m\sin\theta_m}{(b_m\cos\theta_m - 1)^2 + (b_m\sin\theta_m)^2}\right] \\ &= \frac{b_m\cos\theta_m - 1}{b_m^2 + 1 - 2b_m\cos\theta_i}, \quad m = 0, 1, 2, \ldots, n-1 \end{aligned} \tag{3.23}$$

If $b_m \cos\theta_m - 1 \geq 0$ $(m = 0, 1, 2, \ldots, n-1)$, we have:

$$\mathrm{Re}\left[\hat{G}'_m(z)\right] = \frac{b_m \cos\theta_m - 1}{b_m^2 + 1 - 2b_m \cos\theta_m} \geq 0, \quad m = 0, 1, 2, \ldots, n-1$$

If $b_m \cos\theta_m - 1 < 0$ $(m = 0, 1, 2, \ldots, n-1)$, we have:

$$\frac{b_m^2 + 1 - 2b_m \cos\theta_m}{b_m \cos\theta_m - 1} = \frac{b_m^2 - 1}{b_m \cos\theta_m - 1} - 2 \leq \frac{a^2 - 1}{b_m \cos\theta_m - 1} - 2 \leq -2$$

$$\forall |z| = a \geq 1, \quad m = 0, 1, 2, \ldots, n-1$$

As a consequence:

$$\mathrm{Re}\left[\hat{G}'_m(z)\right] \geq -\frac{1}{2}, \quad \forall |z| \geq 1, \quad m = 0, 1, 2, \ldots, n-1 \tag{3.24}$$

Since

$$b \geq \max\left(\left|B^-(e^{j\omega})\right|^2\right) = \max\left(B^-(e^{-j\omega})B^-(e^{j\omega})\right) > 0$$

Then

$$G_{fH}(e^{j\omega}) = \frac{B^-(e^{j\omega})B^-(e^{-j\omega})}{b} \leq 1 \Rightarrow \frac{1}{G_{fH}(e^{j\omega})} \geq 1 \Rightarrow -\frac{1}{G_{fH}(e^{j\omega})} \leq -1$$

From (3.14) to (3.19) and (3.24), we have:

$$\begin{aligned} &\min_{|z|\geq 1} \mathrm{Re}\left[\sum_{m=0}^{n-1}\left[k_m \hat{G}'_m(z)\right]\right] = \min_{|z|\geq 1}\left(\sum_{m=0}^{n-1}\left(k_m \mathrm{Re}\left[\hat{G}'_m(z)\right]\right)\right) \\ &\geq \min_{|z|\geq 1}\left(\sum_{m=0}^{n-1}\left(k_m\left(-\frac{1}{2}\right)\right)\right) = \min_{|z|\geq 1}\left(\left(-\frac{1}{2}\right)\sum_{m=0}^{n-1} k_m\right) > -\frac{1}{2}\cdot 2 \\ &= -1 \geq -\frac{1}{G_{fH}(e^{j\omega})}, \quad \forall |z| \geq 1 \end{aligned}$$

This implies that:

$$1 + G_{fH}(z)\sum_{m=0}^{n-1}\left[k_m \cdot \hat{G}'_m(z)\right] \neq 0, \quad \forall |z| \geq 1 \tag{3.25}$$

Thus, all the poles p_i with the subscript $i = 0, 1, \ldots N-1$ of the transfer functions $G_r(z)$ and $G_d(z)$ in (3.21) are inside the unit circle $|z| = 1$, i.e., $|p_i| < 1$. Finally, the asymptotical stability of $H(z)$ implies the asymptotical stability of the closed-loop digital PSRC system in Figure 3.6.

Remark 3.3 *Theorem 3.1 indicates that both the sum of the control gains k_m of the PSRC of (3.12) and the control gain k_{rc} of the CRC of (2.5) has the same stability range. It can be easily seen that the stability criteria for the CRC system [28–32] is compatible to Theorem 3.1. Since the PSRC provides a universal framework for housing various RC schemes [9,10], Theorem 3.1 offers a universal stability criterion for these RC systems.*

Remark 3.4 *Equation (3.2) indicates the coefficients for $\pm nk + m$ order harmonic frequencies in (3.2) are n times as large as that of the CRC of (2.5). Moreover, Theorem 3.1 shows that both the sum of control gains k_m of the PSRC and the control gain k_{rc} of the CRC has the same stability range. Within the stability range, large k_m for the major harmonics and small k_m (or even zero) for the minor harmonics will enable the PSRC systems to achieve much faster (up to n times) error convergence rate than what the CRC systems do. Hence, the error convergence rate of the PSRC systems can be optimized by independently tuning the control gains k_m according to the harmonics distribution.*

Corollary 3.1: If the PSRC system in Figure 3.6 with the control gains $k_m = k_{rc}/n$, and filters $Q_m(z) = Q(z)$, where $m = 0, 1, 2, \ldots, n-1$, fulfils the following condition [9,10]:

$$\left|Q^n(z)\left(1 - k_{rc}G_f(z)H(z)\right)\right| < 1 \tag{3.26}$$

then it is asymptotically stable, and the control gain k_{rc} satisfies:

$$0 < k_{rc} < 2 \tag{3.27}$$

Proof: If $k_m = k_{rc}/n$, and $Q_m(z) = Q(z)$, where $m = 0, 1, 2, \ldots, n-1$, according to Lemma 3.2, the PSRC system in Figure 3.6 becomes the CRC system with the control gain k_{rc} and filter $Q^n(z)$. From Figure 3.6 and (3.14), $G_r(z)$ and $G_d(z)$ can be derived as:

$$\begin{aligned} y(z) = G_r(z)r(z) + G_d(z)d(z) &= \frac{\left\{1 - z^{-N}Q^n(z)\left[1 - k_{rc}G_f(z)\right]\right\}H(z)}{1 + z^{-N}Q^n(z)\cdot\left[1 - k_{rc}G_f(z)\cdot H(z)\right]}r(z) \\ &+ \frac{\left[1 - z^{-N}Q^n(z)\right]\left[1 + G_c(z)G_p(z)\right]^{-1}}{1 + z^{-N}Q^n(z)\left[1 - k_{rc}G_f(z)H(z)\right]}d(z) \end{aligned} \tag{3.28}$$

Obviously, if $|Q^n(z)\,(1 - k_{rc}\,G_f(z)\,H(z))| < 1$, then all the poles p_i $(i = 0, 1, \ldots, N-1)$ of $G_r(z)$ and $G_d(z)$ in (3.28) are inside the unit circle $|z| = 1$. Therefore, the asymptotical stability of $H(z)$ implies that the closed-loop system transfer functions $G_r(z)$ and $G_d(z)$ in (3.28) are asymptotically stable. Corresponding criteria can be found in the CRC system [32].

Let

$$Q(z) = N_q(\omega)\exp\left(j\theta_q(\omega)\right)$$

where $N_q(\omega) = |Q(e^{j\omega})| \leq 1$.

Then, the formula in (3.21) can be expressed as:

$$N_q^n(\omega)\left|1 - k_{rc}N_{GH}(\omega)e^{j\theta_{GH}}\right| < 1$$

Or equivalently,

$$|1 - k_{rc}N_{GH}(\omega)\cos\,\theta_{GH} - jk_{rc}N_{GH}(\omega)\sin\,\theta_{GH}| < \frac{1}{N_q^n(\omega)} \tag{3.29}$$

Using the norm definition [40,41], we have:

$$0 < k_{rc} < \frac{1 - N_q^{2n}(\omega)}{N_q^{2n}(\omega)\cdot k_{rc}\cdot N_{GH}^2(\omega)} + \frac{2\cos\,\theta_{GH}}{N_{GH}(\omega)}$$

Since $(1 - N_q^{2n}(\omega))/(k_{rc}N_{GH}^2(\omega)N_q^{2n}(\omega)) \geq 0$, to ensure the system stability conservatively, we can choose:

$$0 < k_{rc} < \frac{2\cos\,\theta_{GH}}{N_{GH}(\omega)} \leq \frac{2}{N_{GH}(\omega)}$$

Similarly, the conservative stability range of k_{rc} can be obtained as:

$$0 < k_{rc} < 2 \tag{3.30}$$

Theorem 3.2 *If the closed-loop PSRC system in Figure 3.6 with $Q_m(z) = 1$, $m = 0, 1, 2, \ldots, n-1$ is asymptotically stable, then the error e(k) in Figure 3.6 converges asymptotically to zero when its spectral content corresponds to the frequencies of the roots of $z^N = 1$ [9,10].*

Proof: The error transfer function $T(z)$ of the closed-loop PSRC system is given by:

$$T(z) = \frac{e(z)}{r(z) - d(z)} = \frac{1}{1 + G_c(z)G_p(z)}\cdot\frac{G_x(z)}{G_x(z) + G_y(z)}$$

where

$$G_x(z) = \prod_{m=0}^{n-1}\left(1 - e^{j(2\pi m/n)}\cdot z^{-N/n}\cdot Q_m(z)\right)$$

$$G_y(z) = \sum_{p=0}^{n-1}\left[k_p e^{j\frac{p\cdot 2\pi}{n}} z^{-\frac{N}{n}} \prod_{\substack{m=0\\ m\neq p}}^{n-1}\left(1 - e^{j\frac{2\pi m}{n}} z^{-\frac{N}{n}} Q_m(z)\right)\right] \tag{3.31}$$

Since $H(z)$ and $G_r(z)$ are asymptotically stable, it is clear that if $Q_m(z)=1$, $m=1,2,\ldots,n-1$, then:

$$\left|T(e^{j\omega})\right|=0,\quad \forall z=e^{j\omega}\quad \text{such that } z^N=1 \tag{3.32}$$

Therefore, the error $e(k)$ in Figure. 3.6 converges asymptotically to zero.

Remark 3.5 *Theorem 3.2 indicates that all periodic errors, whose frequencies are integer multiples of the fundamental frequency, can be completely eliminated by the PSRC with* $Q_m(z)=1$, $m=0, 1, 2, \ldots, n-1$.

In practice, to simplify the design of the PSRC, a linear phase-lead compensator $G_f(z)=z^c$ is usually employed in the PSRC of (3.12) instead of the inverse transfer function of $H(z)$ [36–38], and an FIR low-pass filter $Q_m(z)=\alpha_1 z+\alpha_0+\alpha_1 z^{-1}$ with $2\alpha_1+\alpha_0=1$, $\alpha_0\geq 0$ and $\alpha_1\geq 0$ is also employed.

3.2 Selective harmonic control (SHC)

To facilitate the implementation of the PSRC in (3.4), a real delay-based internal model of selective $(nk\pm m)$-order harmonics is developed to replace the previous complex one in this section. As a result, the selective harmonic control (SHC) scheme is established.

3.2.1 Real internal model of selective harmonics

It should be pointed out that it is not easy to implement the complex internal model. In practical applications, as shown in Figure 3.7, a real internal model of selected $(nk\pm m)$-order harmonics can be obtained by combining two conjugate complex internal models as follows [42,43]:

$$\begin{aligned}\hat{G}_{\mathrm{sm}}(s)&=\frac{1}{2}\left(\hat{G}_m(s)+\hat{G}_{-m}(s)\right)=\frac{1}{2}\left(\frac{e^{\frac{-sT_0}{n}+j\frac{2\pi m}{n}}}{1-e^{\frac{-sT_0}{n}+j\frac{2\pi m}{n}}}+\frac{e^{\frac{-sT_0}{n}-j\frac{2\pi m}{n}}}{1-e^{\frac{-sT_0}{n}-j\frac{2\pi m}{n}}}\right)\\&=\frac{\cos(2\pi m/n)e^{\frac{sT_0}{n}}-1}{e^{\frac{2sT_0}{n}}-2\cos(2\pi m/n)e^{\frac{sT_0}{n}}+1}\end{aligned} \tag{3.33}$$

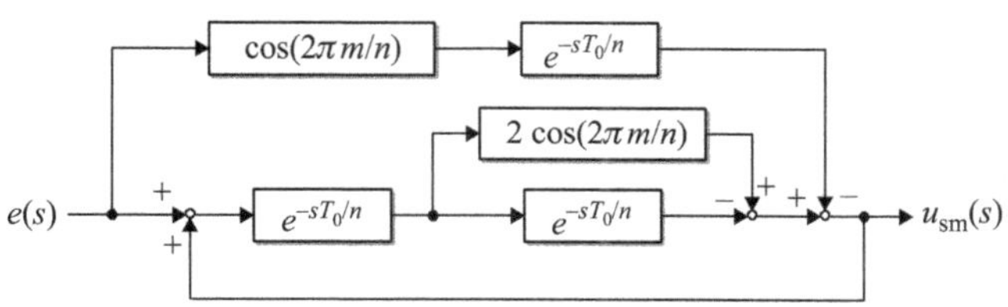

Figure 3.7 Real internal model for selected harmonics $\hat{G}_{sm}(s)$

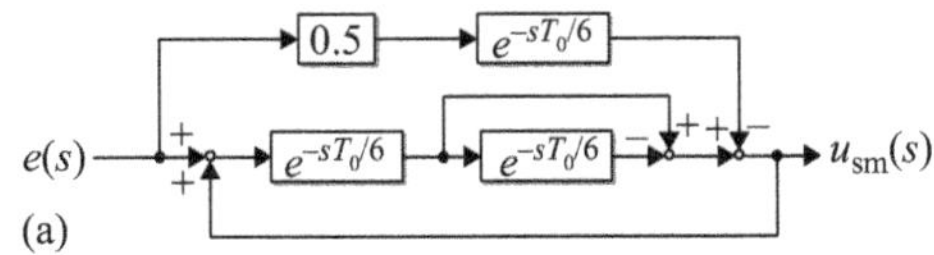

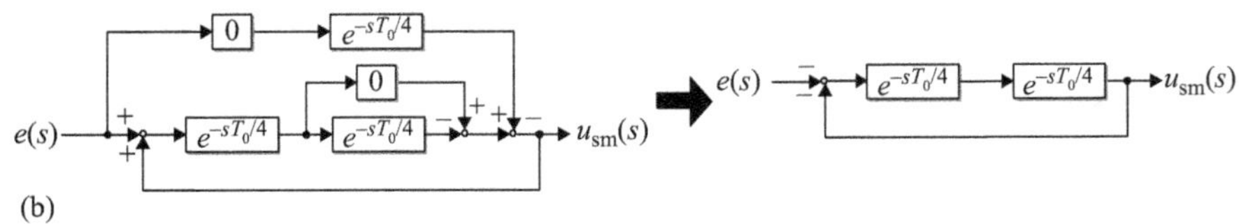

Figure 3.8 Internal models for selective harmonics: (a) $\hat{G}_{6k\pm1}(s)$ of (3.34) to compensate h-order ($h = 6k \pm 1$) harmonics and (b) $\hat{G}_{4k\pm1}(s)$ of (3.35) to compensate h-order ($h = 4k \pm 1$) harmonics, where $h > 0$ and $k = 0, 1, 2, \ldots$

Remark 3.6 *The real internal model $\hat{G}_{sm}(s)$ of (3.33) provides a universal formula for the internal model for various selective harmonics.*

For example, let $n = 6$ and $m = 1$, an internal model of $(6k \pm 1)$-order harmonics with $\cos(2\pi/6) = 0.5$ (as shown in Figure 3.8(a)) will be obtained as [7]:

$$\hat{G}_{6k\pm1}(s) = \frac{u_{\rm sm}(s)}{e(s)} = \frac{0.5e^{s\frac{T_0}{6}} - 1}{e^{s\frac{T_0}{3}} - e^{s\frac{T_0}{6}} + 1} \tag{3.34}$$

whose Bode plots are presented in Figure 3.9. It can be seen in Figure 3.9 that the internal model for $(6k \pm 1)$-order harmonics $\hat{G}_{6k\pm1}(s)$ has infinite gains at the harmonics of $6k \pm 1$ times of the fundamental frequency, meaning that it can selectively cancel out the $(6k \pm 1)$-order harmonics when plugged into a stable closed-loop system.

Let $n = 4$ and $m = 1$, an internal model for $(4k \pm 1)$-order harmonics (i.e., odd harmonics [44,45]) with $\cos(2\pi/4) = 0$ (as shown in Figure 3.8(b)) can be obtained as [44,45]:

$$\hat{G}_{4k\pm1}(s) = \frac{u_{\rm sm}(s)}{e(s)} = \frac{-1}{e^{s\frac{T_0}{2}} + 1} = \frac{-e^{-s\frac{T_0}{2}}}{1 + e^{-s\frac{T_0}{2}}} \tag{3.35}$$

Moreover, let $n = 1$ and $m = 0$, an internal model for the CRC will be obtained as:

$$\hat{G}_{\rm rc}(s) = \frac{u_{\rm sm}(s)}{e(s)} = \frac{1}{e^{sT_0} - 1} = \frac{e^{-sT_0}}{1 - e^{-sT_0}} \tag{3.36}$$

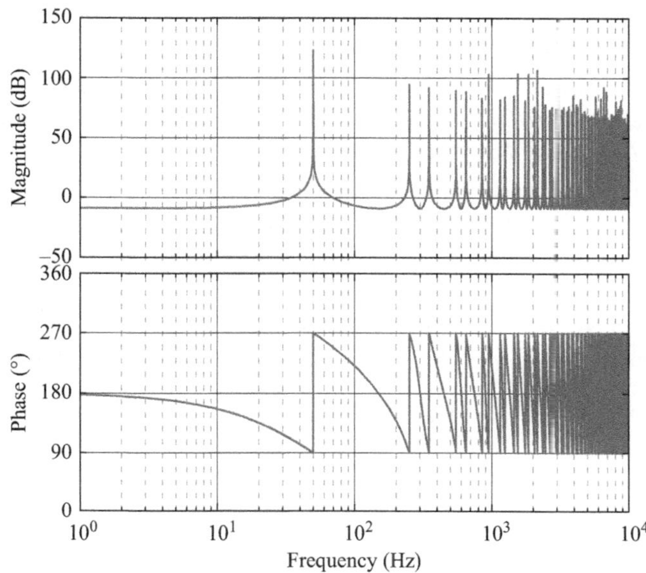

Figure 3.9 Bode plots of the internal model for (6k ± 1)-order harmonics $\hat{G}_{6k\pm1}(s)$, where $T_0 = 0.02$ s

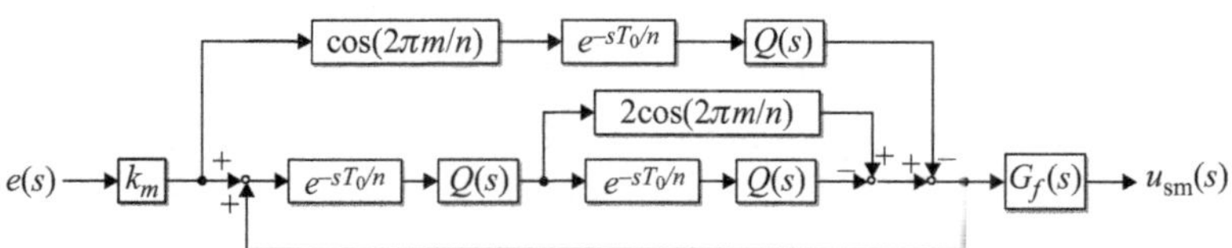

Figure 3.10 Selective harmonic controller $G_{sm}(s)$ in practical applications

3.2.2 Selective harmonic control

As shown in Figure 3.10, based on the real internal model of (3.33), an SHC, i.e., $(nk \pm m)$-order harmonic controller, $G_{sm}(s)$ can be derived as [43]:

$$G_{sm}(s) = \frac{k_m G_f(s)}{2}\left(\frac{e^{\frac{-sT_0}{n}+j\frac{2\pi m}{n}}Q(s)}{1 - e^{\frac{-sT_0}{n}+j\frac{2\pi m}{n}}Q(s)} + \frac{e^{\frac{-sT_0}{n}-j\frac{2\pi m}{n}}Q(s)}{1 - e^{\frac{-sT_0}{n}-j\frac{2\pi m}{n}}Q(s)}\right)$$

$$= k_m G_f(s)\frac{\cos(2\pi m/n)e^{\frac{sT_0}{n}}Q(s) - Q^2(s)}{e^{\frac{2sT_0}{n}} - 2\cos(2\pi m/n)e^{\frac{sT_0}{n}}Q(s) + Q^2(s)} \tag{3.37}$$

where $G_f(s)$ is a phase-lead compensator to stabilize the overall system, and the filter $Q(s)$ is employed to further select the interested frequencies and make a good trade-off between the tracking accuracy and the system robustness as discussed in the previous chapters.

Remark 3.7 *From (3.2) and (3.33), when $n > 2$ and $k_m = k_{rc}$, it is known that the equivalent control gains at ($\pm nk \pm m$)-order harmonic frequencies of the SHC in (3.37) are n/2 times as large as those of the CRC in (2.3). Hence, the error convergence rate of the SHC of (3.37) at ($\pm nk \pm m$)-order harmonic frequencies can be n/2 times faster than that of the CRC in (2.3).*

The SHC of (3.37) provides a universal delay-based IMP-based controller, which is tailored for ($nk \pm m$)-order harmonics compensation. For example, let $n = 1$ and $m = 0$, (3.37) becomes a CRC, let $n = 4$ and $m = 1$, (3.37) becomes an odd harmonic RC [44,45], and let $n = 6$ and $m = 1$, (3.37) becomes a $6k \pm 1$ RC [42,46]. It is thus also named "*$nk \pm m$ order RC*" [42].

As shown in Figure 3.11, the corresponding digital SHC can be written as:

$$G_{\rm sm}(z) = k_m G_f(z) \cdot \frac{\left(\cos(2\pi m/n)Q(z)z^{N/n} - Q^2(z)\right)}{z^{2N/n} - 2\cos(2\pi m/n)Q(z)z^{N/n} + Q^2(z)} \tag{3.38}$$

where $N = T_0/T_s \in \mathbb{N}$ with T_s being the sampling period.

Remark 3.8 *As shown in Figure 3.11, the maximum time delay in the forward channels is 2N/n multiples of the sampling period, and thus the total number of memory cells is $(3N/n) < N$ (if $n > 3$). To compensate ($nk \pm m$)-order harmonics, the SHC occupies less memory than what the CRC does.*

3.2.3 Digital SHC system and design

Figure 3.12 shows a typical plug-in digital SHC system, where $G_p(z)$ is the transfer function of the plant, $G_c(z)$ is the conventional feedback controller, $G_{\rm sm}(z)$ is a complete SHC scheme, $r(z)$ is the reference input, $y(z)$ is the output, $e(z) = r(z) - y(z)$ is the tracking error and the input of $G_{\rm sm}(z)$, and $d(z)$ is the disturbance.

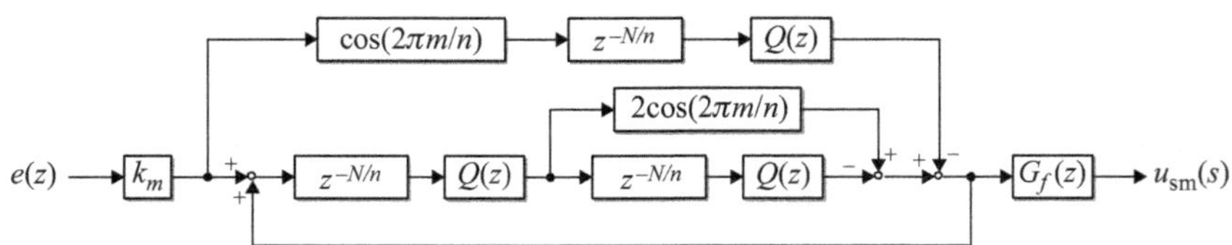

Figure 3.11 Digital selective harmonic controller $G_{sm}(z)$ for compensating the ($nk \pm m$)-order harmonics

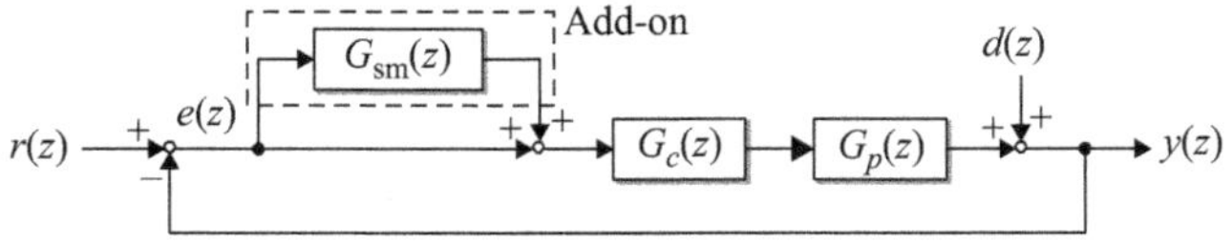

Figure 3.12 Add-on (plug-in) digital selective harmonic control system

The output $y(z)$ of the plug-in digital SHC system can be expressed as:

$$
\begin{aligned}
y(z) &= G_r(z) \cdot r(z) + G_d(z) \cdot d(z) \\
&= \frac{[1 + G_{\mathrm{sm}}(z)]H(z)}{1 + G_{\mathrm{sm}}(z)H(z)} r(z) + \frac{[1 + G_c(z)G_p(z)]^{-1}}{1 + G_{\mathrm{sm}}(z)H(z)} d(z)
\end{aligned} \tag{3.39}
$$

where $G_r(z)$ is the transfer function from $r(z)$ to $y(z)$, $G_d(z)$ is the transfer function from $d(z)$ to $y(z)$, and $H(z)$ is the transfer function of the conventional feedback control system without the plug-in SHC $G_{\mathrm{sm}}(z)$ of (3.38). $H(z)$ is assumed to be asymptotically stable, and it is given as:

$$
H(z) = \frac{G_c(z)G_p(z)}{1 + G_c(z)G_p(z)} \tag{3.40}
$$

Without loss of generality, $H(z)$ can be described by:

$$
H(z) = \frac{B(z)}{A(z)} = \frac{z^{-c}B^{+}(z)B^{-}(z)}{A(z)} \tag{3.41}
$$

where c represents known delay steps with $c \in [0, N/n]$, all characteristic roots of $A(z) = 0$ are inside the unit circle, $B^{+}(z)$ and $B^{-}(z)$ are the cancelable and un-cancelable parts of $B(z)$, respectively [42]. $B^{-}(z)$ comprises roots on or outside the unit circle and undesirable roots, which are in the unit circle and $B^{+}(z)$ comprises of roots of $B(z)$, which are not in $B^{-}(z)$.

The compensation filter $G_f(z)$ can be chosen as [9]:

$$
G_f(z) = \frac{z^{c}A(z)B^{-}(z^{-1})}{B^{+}(z)b} \tag{3.42}
$$

where $b \geq \max|B^{-}(e^{j\omega})|^2$. Since the periodic signal generator in an SHC will introduce N/n-step delay $z^{-N/n}$ (usually $n \ll N$) and $N/n \geq d$, $G_f(z)$ with noncausal component z^c can be implemented in the SHC scheme [42].

In practice, due to model uncertainties and load variations, it is impossible to obtain the exact transfer function $H(z)$. That is, the practical transfer function of $G_f(z)H(z)$ can be written as:

$$
\begin{aligned}
G_{fH}(z) &= G_f(z)H(z) = \frac{B^{-}(z^{-1})B^{-}(z)}{b}(1 + \Delta(z)) = |G_{fH}(e^{j\omega})| \angle\theta_{fH}(\omega) \\
&= |G_{fH}(e^{j\omega})| e^{j\theta_{fH}(\omega)}
\end{aligned} \tag{3.43}
$$

where $\Delta(z)$ denotes the uncertainties assumed to be bounded by $|\Delta(e^{j\omega})| \leq \varepsilon$ with ε being a positive constant, and $\Delta(z)$ is stable. From (3.43), it is clear that the uncertainties $\Delta(z)$ can cause $\theta_{fH}(\omega) \neq 0$, which means that zero-phase error compensation may not be obtained due to the uncertainties.

Theorem 3.3 *Considering the closed-loop SHC system shown in Figure 3.12 with the constraints (3.39) to (3.43), if*

$$2p\pi - \frac{\pi}{2} < \theta_{fH}(\omega) < 2p\pi + \frac{\pi}{2},\ p = 0, 1, 2, \ldots,$$

then

$$0 < k_m < \frac{2\cos(\theta_{fH}(\omega))}{1+\varepsilon}$$

which will enable the closed-loop SHC system to be asymptotically stable; if

$$2p\pi + \frac{\pi}{2} < \theta_{fH}(\omega) < 2p\pi + \frac{3\pi}{2}, p = 0, 1, 2, \ldots,$$

then

$$\frac{2\cos(\theta_{fH}(\omega))}{1+\varepsilon} < k_m < 0$$

which will enable the closed-loop SHC system to be asymptotically stable [43].

Proof: From (3.40), since $H(z)$ is assumed to be asymptotically stable, the poles of $(1 + G_c(z)G_p(z))^{-1}$ are then located inside the unit circle. Therefore, from (3.39), if all the poles of $(1 + G_{\text{sm}}(z)H(z))^{-1}$ are also located inside the unit circle, the overall closed-loop SHC system shown in Figure 3.12 will be asymptotically stable.

The denominator of $(1 + G_{\text{sm}}(z)H(z))^{-1}$ is:

$$z^{2N/n} + (k_m G_{fH}(z) - 2)Q(z)\cos(2\pi m/n)z^{N/n} + (1 - k_m G_{fH}(z))Q^2(z)$$

which can be factorized as:

$$\left(z^{N/n} - \alpha e^{\beta j}\right)\left(z^{N/n} - \alpha e^{-\beta j}\right) \tag{3.44}$$

where

$$\alpha = \pm(1 - k_m G_{fH}(z))^{1/2} Q(z)$$

$$\cos\beta = \pm\frac{k_m G_{fH}(z) - 2}{2(1 - k_m G_{fH}(z))^{1/2}}\cos(2\pi m/n)$$

Therefore, the poles of $(1 + G_{\text{rc}}(z)H(z))^{-1}$ are located inside the unit circle if $|\alpha| < 1$, i.e.:

$$|Q(z)|\left|(1 - k_m G_{fH}(z))^{1/2}\right| < 1$$

And then

$$\left|Q(e^{j\omega})\right|^4\left|1 - k_m\left|G_{fH}(e^{j\omega})\right|e^{j\theta_{fH}(\omega)}\right|^2 < 1 \tag{3.45}$$

Since $|Q(e^{j\omega})| \le 1$, thus from (3.45), if

$$1 - 2k_m\left|G_{fH}(e^{j\omega})\right|\cos\left(\theta_{fH}(\omega)\right) + k_m^2\left|G_{fH}(e^{j\omega})\right|^2 < 1$$

holds, the poles of $(1 + G_{\rm sm}(z)H(z))^{-1}$ are located inside the unit circle.

Since

$$b \ge \max\left(\left|B^-(e^{j\omega})\right|^2\right) = \max\left(B^-(e^{-j\omega})B^-(e^{j\omega})\right) > 0$$

from (3.43), we have:

$$\left|G_{fH}(e^{j\omega})\right| = \left|\frac{B^-(e^{j\omega})B^-(e^{-j\omega})}{b}\left(1 + \Delta(e^{j\omega})\right)\right| \le 1 + \varepsilon \tag{3.46}$$

Therefore, from (3.45) and (3.46), it can be obtained that, if

$$2p\pi - \frac{\pi}{2} < \theta_{fH}(\omega) < 2p\pi + \frac{\pi}{2}, \quad p = 0, 1, 2, \ldots,$$

then

$$0 < k_m < \frac{2\cos\left(\theta_{fH}(\omega)\right)}{1 + \varepsilon}$$

which will enable the closed-loop SHC system shown in Figure 3.12 to be asymptotically stable; if

$$2p\pi + \frac{\pi}{2} < \theta_{fH}(\omega) < 2p\pi + \frac{3\pi}{2}, \quad p = 0, 1, 2, \ldots,$$

then

$$\frac{2\cos\left(\theta_{fH}(\omega)\right)}{1 + \varepsilon} < k_m < 0 \tag{3.47}$$

which will enable the closed-loop SHC system to be asymptotically stable.

Remark 3.9 *Theorem 3.3 provides a stability criterion for the closed-loop SHC system. It can be seen that, if $\theta_{fH}(\omega) = 0$ (i.e., zero-phase error compensation is achieved), the CRC systems [31–33], the odd-harmonic RC systems [44,45], and the $(nk \pm m)$-order RC systems [42,43] have an identical stability range for the RC gain (i.e., the stability criteria for these RC systems are compatible to Theorem 3.1). Theorem 3.1 implies that uncertainties will reduce the stability range for the RC gain k_m. Since the SHC scheme provides a universal framework for housing these RC schemes, Theorem 3.1 offers a universal stability criterion for these RC systems.*

Remark 3.10 *Theorem 3.3 indicates that the control gain k_{rc} for the CRC systems and the control gain k_m for the SHC systems have an identical stability range. Therefore, from Remark 3.6, it can be concluded that the SHC can offer n/2 times faster error converge rate than what the CRC can do.*

For example, with the identical gain, the odd harmonic RC offers up to two times faster error convergence rate in eliminating the odd harmonics than what the CRC does [45]. However, since one SHC only can compensate one group of harmonics (e.g., $(4k \pm 1)$-order harmonics in single-phase systems), the control accuracy of the SHC would be usually lower than what can be achieved by the CRC scheme.

Theorem 3.4 *If the closed-loop SHC system with $Q(z) = 1$ in Figure 3.12 is asymptotically stable, then the error $e(z)$ in Figure 3.12 converges asymptotically to zero at the frequencies of $\pm(nk \pm m)\omega_0$, where $|nk \pm m| \leq N/2$, $k \in \mathbb{N}$.*

Proof: Almost it is the same as that of Theorem 3.2.

Remark 3.11 *Theorem 3.2 indicates that the harmonic components of the error at the frequencies of $\pm(nk \pm m)\omega_0$ with $k \in \mathbb{N}$ can be completely eliminated by the $(nk \pm m)$-order RC controller with $Q(z) = 1$, even under model uncertainties.*

Remark 3.12 *The proof of Theorem 3.3 indicates that the introduction of the low-pass filter $Q(z)$ with $|Q(e^{j\omega})| \leq 1$ will make it easier to ensure all poles $|p_i| < 1$ with $i = 0, 1, \ldots, 2N/n - 1$ of $(1 + G_{sm}(z)H(z))^{-1}$, and then enhance the robustness of the SHC system. On the other hand, the introduction of $Q(z)$ with $|Q(e^{j\omega})| \leq 1$, will cause a significant variation of the poles of the SHC in (3.37), and yield an imperfect cancelation of periodic errors at the harmonic frequencies when $|Q(e^{j\omega})| < 1$. The tracking accuracy will be reduced somewhat. Therefore, the introduction of the filter $Q(z)$ brings a trade-off between tracking accuracy and system robustness in the SHC system.*

In practice, to simplify the design of the SHC scheme, a linear phase-lead compensator $G_f(z) = z^c$ is usually employed in the SHC of (3.38) instead of the inverse transfer function of $H(z)$ [45], and an FIR low-pass filter $Q(z) = \alpha_1 z + \alpha_0 + \alpha_1 z^{-1}$ with $2\alpha_1 + \alpha_0 = 1$, $\alpha_0 \geq 0$ and $\alpha_1 \geq 0$ is also adopted. The entire digital SHC system is shown in Figure 3.13, where the phase-lead compensator is incorporated in the inner control loops, and it is suitable for practical implementations.

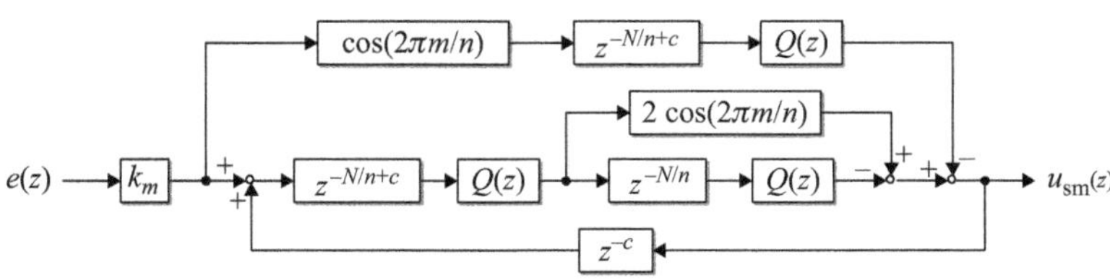

Figure 3.13 Practical Implementation of the SHC scheme with a linear phase-lead compensator

3.3 Optimal harmonic control (OHC)

Replacing the complex internal models of selective harmonics with real ones in (3.33), the PSRC in (3.4) will be transformed into an optimal harmonic control (OHC), which can mitigate power harmonics with high control accuracy, while maintaining fast transient response, guaranteeing robustness, and being feasible for implementation.

3.3.1 Optimal harmonic control

From Theorem 3.4, it is clear that, including the real internal model of (3.33) into a stable closed-loop system will enable the control system to track/reject $(nk \pm m)$-order harmonics. As shown in Figure 3.14, in order to compensate more harmonics for better accuracy while keeping a fast error convergence rate, an optimal harmonic control (OHC) scheme which includes paralleled SHC components of (3.37) tailored for the selected harmonics, is developed as:

$$G_{\mathrm{OHC}}(s) = \sum_{m \in N_m} G_{\mathrm{sm}}(s) \tag{3.48}$$

where m and N_m represent the $(nk \pm m)$ $(k=0,\ 1,\ 2,\ \ldots,$ and $m \leq n/2)$ harmonic order and the set of the selected harmonics, respectively.

Including the paralleled digital SHC modules of (3.38), the corresponding digital OHC can be straightforwardly written as:

$$\begin{aligned} G_{\mathrm{OHC}}(z) &= \sum_{m \in N_m} G_{\mathrm{sm}}(z) \\ &= \sum_{m \in N_m} k_m G_f(z) \cdot \frac{\left[\cos(2\pi m/n) z^{N/n} Q(z) - Q^2(z)\right]}{z^{2N/n} - 2\cos(2\pi m/n) z^{N/n} Q(z) + Q^2(z)} \end{aligned} \tag{3.49}$$

Note that the dual-mode repetitive control (DMRC) [47,48] is a special case of the OHC with $n=4$ and $m=0,\ 1,\ 2$, where $G_{\mathrm{sm}}(z)$ with $m=0,\ 2$ are corresponding to the even harmonic RC and $G_{\mathrm{sm}}(z)$ with $m=1$ corresponds to the odd harmonic RC.

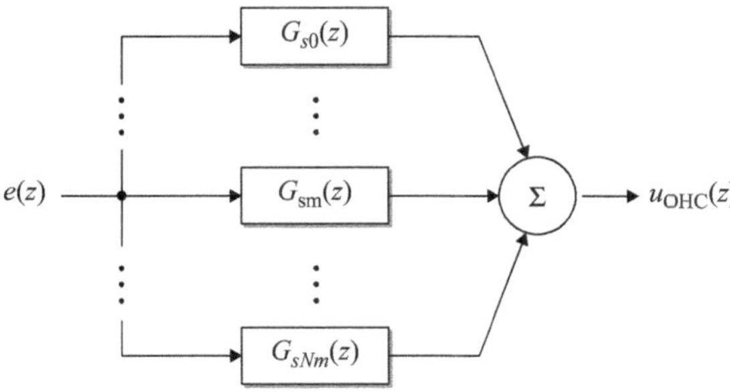

Figure 3.14 Digital optimal harmonic control $G_{OHC}(s)$ where $m \in N_m$

Remark 3.13 *Compared with the complex PSRC of (3.12), the real OHC of (3.49) can at least reduce the number of paralleled SHC modules by half, and thus simplify its implementation.*

3.3.2 Digital OHC system and design

Figure 3.15 shows a typical closed-loop control system with a plug-in OHC $G_{OHC}(z)$ of (3.49) where $G_p(z)$ is the transfer function of the plant, $G_c(z)$ is the feedback controller, $r(z)$ is the reference input, $y(z)$ is the output, $e(z) = r(z) - y(z)$ is the tracking error and the input of $G_{OHC}(z)$, and $d(z)$ is the disturbance.

Theorem 3.5 *With the same assumptions of (3.39)–(3.43), the OHC system with $|Q(e^{j\omega})| \leq 1$ in Figure 3.15 is asymptotically stable if the following two conditions hold [13,14]:*

- $H(z)$ is asymptotically stable;
- The control gains $k_m \geq 0$ satisfy the following inequality:

$$0 < \sum_{m \in N_m} k_m < 2 \tag{3.50}$$

Remark 3.14 *Obviously, the above stability criteria for the OHC systems can be derived from the PSRC systems [9,10], and is compatible to the SHC systems [42,43]. Quite similar to Remark 3.4, the error convergence rate of the OHC systems can be optimized by independently tuning the control gains k_m according to the harmonics distribution. The OHC systems can thus achieve up to n/2 times of the error convergence rate as fast as what the CRC systems can do.*

For example, as a representative of the OHC, the DMRC [47,48] employs a parallel combination of the even-order harmonic RC and the odd-order harmonic RC. The control gains for the even-order RC and the odd-order harmonic RC in the DMRC can be tuned independently. Compared with the CRC, the DMRC with a larger control gain for the odd-order harmonic RC to compensate major odd harmonics can achieve a much faster error convergence rate (up to two times of that achieved by the CRC) without reducing the control accuracy and system robustness [47,48].

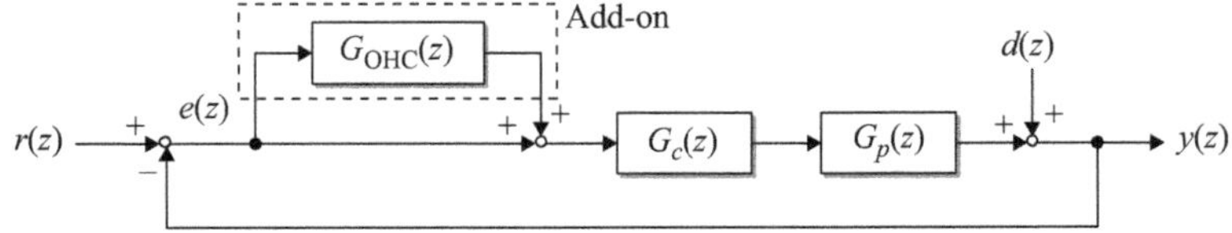

Figure 3.15 Add-on (plug-in) digital OHC system

Moreover, the digital DMRC occupies the same data memory as what the digital CRC does.

In practice, to simplify the design of the OHC, a linear phase-lead compensator $G_f(z) = z^c$ with c being the phase-lead steps is usually employed in the OHC as shown in (3.49), instead of the inverse transfer function of $H(z)$ [46], and an FIR low-pass filter $Q(z) = \alpha_1 z + \alpha_0 + \alpha_1 z^{-1}$ with $2\alpha_1 + \alpha_0 = 1$, $\alpha_0 \geq 0$, and $\alpha_1 \geq 0$ is also used.

Generally speaking, the OHC offers power converters a general optimal IMP-based control solution to the mitigation of the harmonic distortions mainly concentrated at one or multiple harmonics, in particular $(nk \pm m)$ ($k = 1, 2, \ldots,$ and $m = 0, 1, \ldots, n-1$) order harmonic frequencies – high accuracy, fast transient response, cost-effective and easy real-time implementation, and compatible design rules-of-thumb.

3.4 Summary

To efficiently mitigate major $(nk \pm m)$ ($k = 1, 2, \ldots,$ and $m = 0, 1, \ldots, n-1$)-order power harmonics caused by n-pulse power converters, tailor-made delay-based complex and real internal models of interested harmonics are developed. By employing these internal models, advanced periodic controllers are consequently developed, which include the PSRC, the SHC, and the OHC. Among these three advanced periodic controllers, the OHC is actually a combination of the PSRC and the SHC: the parallel structure enables it to take advantages of the PSRC – a fast dynamic response due to the paralleled SHC modules with optimally weighted coefficients; the SHC modules enable it to take cost-effective and easy real-time implementation due to the limited number of delay-based compact real SHC modules. Moreover, including more SHC modules into it, the OHC can further improve control accuracy by removing more harmonics.

The advanced periodic control systems, which employ plug-in structure (i.e., feedback control plus advanced periodic control), are investigated in this chapter. Stability criteria and design method for these advanced periodic control systems have been developed, and are found to be compatible to those of fundamental periodic control systems.

The outstanding features of these three advanced periodic controllers in this chapter for optimal power harmonics mitigation are listed as follows:

1. Parallel structure repetitive control (PSRC) –
   - Fast dynamic response. Control gains k_m for each group of the selected $(\pm nk + m)$-order harmonics can be weighted by the share of harmonics distribution, and be tuned to optimize its dynamic response (up to n times as fast as what the CRC does).
   - High control accuracy. It can flexibly choose harmonics for compensation. The more the selected harmonics in the PSRC are considered, the higher control accuracy the PSRC can achieve.

- Easy for design and tuning. Stability criteria and system design rules-of-thumb for the PSRC systems are compatible to those for the CRC systems.
- Limited parallel computational burden. Compared with the multiple resonant control, a very limited number of paralleled SHC modules in the PSRC lead to less parallel computation.
- Easy for implementation of the PSRC with a limited number of compact delay-based complex SHC modules.

2. Selective harmonic control (SHC) –
   - Fast dynamic response. Equivalent control gains at selected harmonics of the SHC are up to $n/2$ times of those for the CRC systems.
   - Good control accuracy. The SHC is used to only compensate dominant harmonics.
   - Easy for design and tuning. Stability criteria and system design rules-of-thumb for the SHC systems are compatible to those for the CRC systems.
   - Light computation. Compared with the CRC, the compact delay-based SHC consumes less memory elements and consumes similar computation resource.
   - Easy for implementation of the real recursive SHC.

3. Optimal harmonic control (OHC) –
   - Fast dynamic response. Control gains k_m for each group of the selected $(\pm nk \pm m)$-order harmonics can be weighted by the share of harmonics distribution, and be tuned to optimize its dynamic response (up to $n/2$ times as fast as what the CRC does).
   - High control accuracy. It can flexibly select harmonics for compensation. The more the harmonics in the PSRC are selected, the higher control accuracy the OHC can achieve.
   - Easy for design and tuning. Stability criteria and system design rules-of-thumb for the OHC systems are compatible to those for the CRC systems.
   - Very limited parallel computational burden. Compared with the PSRC, the number of the paralleled SHC modules in the OHC is reduced by half.
   - Easy for implementation of the OHC with a very limited number of the compact delay-based real SHC modules.

Generally speaking, in view of dominant $(nk \pm m)$-order harmonics in the power harmonics, the OHC offers power converters a general optimal IMP-based control solution to the power harmonics mitigation with high accuracy, fast dynamic response, cost-effective and easy real-time implementation, and compatible design rules-of-thumb.

References

[1] Bojoi, R.I., Limongi, L.R., Roiu, D., and Tenconi, A., "Enhanced power quality control strategy for single-phase inverters in distributed generation systems," *IEEE Trans. Power Electron.*, vol. 26, no. 3, pp. 798–806, 2011.

[2] Stevens, J., "The issue of harmonic injection from utility integrated photovoltaic systems," *IEEE Trans. Energy Convers.*, vol. 3, no. 3, pp. 507–510, 1988.

[3] Stihi, O. and Ooi, B.T., "A single-phase controlled-current PWM rectifier," *IEEE Trans. Power Electron.*, vol. 3, no. 4, pp. 453–459, 1988.

[4] Teodorescu, R., Liserre, M., and Rodríguez, P., *Grid Converters for Photovoltaic and Wind Power Systems*, Wiley-IEEE, 2011.

[5] Zhou, K., Qiu, Z., Watson, N.R., and Liu, Y., "Mechanism and elimination of harmonic current injection from single-phase grid-connected PWM converters," *IET Power Electron.*, vol. 6, no. 1, pp. 88–95, 2013.

[6] Yang, Y., Zhou, K., and Blaabjerg, F., "Harmonics suppression for single phase grid-connected PV systems in different operation modes," *Long Beach, CA, USA: 28th Annual IEEE Applied Power Electronics Conference and Exposition (APEC)*, pp. 889–896, Mar. 2013.

[7] Escobar, G., Hernandez-Briones, P.G., Martinez, P.R., Hernandez-Gomez, M., and Torres-Olguin, R.E., "A repetitive-based controller for the compensation of $6l \pm 1$ harmonic components," *IEEE Trans. Ind. Electron.*, vol. 55, no. 8, pp. 3150–3158, Aug. 2008.

[8] Francis, B. and Wonham, W.M., "The internal model principle of control theory," *Automatica*, vol. 12, no. 5, pp. 457–465, Sep. 1976.

[9] Lu, W., Zhou, K., Wang, D., and Cheng, M., "A general parallel structure repetitive control scheme for multiphase DC–AC PWM converters," *IEEE Trans. Power Electron.*, vol. 28, no. 8, pp. 3980–3987, 2013.

[10] Lu, W., Zhou, K., and Wang, D., "General parallel structure digital repetitive control," *Int. J. Control*, vol. 86, no. 1, pp. 70–83, Jan. 2013.

[11] Lu, W., Zhou, K., Yang, Y., "A general internal model principle based control scheme for CVCF PWM converters," *Hefei, China: Second IEEE International Symposium on Power Electronics for Distributed Generation Systems (PEDS)*, pp. 485–489, Jun. 2010.

[12] Yang, Y., Zhou, K., Cheng, M., and Zhang, B., "Phase compensation multiresonant control of CVCF PWM converters," *IEEE Trans. Power Electron.*, vol. 28, no. 8, pp. 3923–3930, Aug. 2013.

[13] Yang, Y., Zhou, K., and Cheng, M., "Phase compensation resonant controller for single-phase PWM converters," *IEEE Trans. Ind. Informatics*, vol. 9, no. 2, pp. 957–964, 2013.

[14] Yuan, X., Merk, W., Stemmler, H., and Allmeling, J., "Stationary frame generalized integrators for current control of active power filters with zero steady-state error for current harmonics of concern under unbalanced and

distorted operating conditions," *IEEE Trans. Ind. Appl.*, vol. 38, no. 2, pp. 523–532, 2002.

[15] Liserre, M., Teodorescu, R., and Blaabjerg, F., "Multiple harmonics control for three-phase grid converter systems with the use of PI-RES current controller in a rotating frame," *IEEE Trans. Power Electron.*, vol. 21, no. 3, pp. 836–841, May 2006.

[16] Teodorescu, R., Blaabjerg, F., Liserre, M., and Loh, P.C., "Proportional-resonant controllers and filters for grid-connected voltage-source converters," *IEE Proc. – Electric Power Appl.*, vol. 153, no. 5, pp. 750–762, 2006.

[17] Liu, C., Blaabjerg, F., Chen, W., and Xu, D., "Stator current harmonic control with resonant controller for doubly fed induction generator," *IEEE Trans. Power Electron.*, vol. 27, no. 7, pp. 3207–3220, Jul. 2012.

[18] Rodriguez, P., Candela, J. I., Luna, A., Asiminoaei, L., Teodorescu, R., Blaabjerg, F., "Current harmonics cancellation in three-phase four-wire systems by using a four-branch star filtering topology," *IEEE Trans. Power Electron.*, vol. 24, no. 8, pp. 1939–1950, 2009.

[19] Lascu, C., Asiminoaei, L., Boldea, I., and Blaabjerg, F., "High-performance current controller for selective harmonic compensation in active power filters," *IEEE Trans. Power Electron.*, vol. 22, no. 5, pp. 1826–1835, 2007.

[20] Lascu, C., Asiminoaei, L., Boldea, I., Blaabjerg, F., "Frequency response analysis of current controllers for selective harmonic compensation in active power filters," *IEEE Trans. Ind. Electron.*, vol. 56, no. 2, pp. 337–347, 2009.

[21] Yepes, G., Freijedo, F.D., Lopez, O., and Doval-Gandoy, J., "High-performance digital resonant controllers implemented with two integrators," *IEEE Trans. Power Electron.*, vol. 6, no. 2, pp. 563–576, 2011.

[22] Miret, J., Castilla, M., Matas, J., Guerrero, J.M., and Vasquez, J.C., "Selective harmonic-compensation control for single-phase active power filter with high harmonic rejection," *IEEE Trans. Ind. Electron.*, vol. 56, no. 8, pp. 3117–3127, Aug. 2009.

[23] Castilla, M., Miret, J., Camacho, A., Matas, J., and de Vicuna, L.G., "Reduction of current harmonic distortion in three-phase grid-connected photovoltaic inverters via resonant current control," *IEEE Trans. Ind. Electron.*, vol. 60, no. 4, pp. 1464–1472, Apr. 2013.

[24] Castilla, M., Miret, J., Matas, J., Garcia de Vicuna, L., Guerrero, J., "Control design guidelines for single-phase grid-connected photovoltaic inverters with damped resonant harmonic compensators," *IEEE Trans. Ind. Electron.*, vol. 56, no. 11, pp. 4492–4501, 2009.

[25] Zhou, K., Qiu, Z. and Yang, Y., "Current harmonics suppression of single-phase PWM rectifiers," *Aalborg, Denmark: 3rd IEEE International Symposium on Power Electronics for Distributed Generation Systems (PEDG)*, pp. 54–57, June 25–28, 2012.

[26] Loh, P. C., Tang, Y., Blaabjerg, F., and Wang, P., "Mixed-frame and stationary-frame repetitive control schemes for compensating typical load and grid harmonics," *IET Power Electron.*, vol. 4, no. 2, pp. 218–226, 2011.

[27] Garcia-Cerrada, A., Pinzon-Ardila, O., Feliu-Batlle, V., Roncero-Sanchez, P., and García-González, P., "Application of a repetitive controller for a three-phase active power filter," *IEEE Trans. Power Electron.*, vol. 22, no. 1, pp. 237–246, 2007.

[28] Zhou, K. and Wang, D., "Zero tracking error controller for three-phase CVCF PWM inverter," *Electron. Lett.*, vol. 36, no. 10, pp. 864–865, 2000.

[29] Zhou, K., Wang, D., and Low, K.S., "Periodic errors elimination in CVCF PWM DC/AC converter systems: A repetitive control approach," *IEE Proc. – Control Theory Appl.*, vol. 147, no. 6, pp. 694–700, 2000.

[30] Zhou, K. and Wang, D., "Unified robust zero error tracking control of CVCF PWM converters," *IEEE Trans. Circuits Syst. (I)*, vol. 49, no. 4, pp. 492–501, 2002.

[31] Zhou, K. and Wang, D., "Digital repetitive controlled three-phase PWM rectifier," *IEEE Trans. Power Electron.*, vol. 18, no.1, pp. 309–316, Jan. 2003.

[32] Zhou, K. and Wang, D., "Digital repetitive learning controller for three-phase CVCF PWM inverter," *IEEE Trans. Ind. Electron.*, vol. 48, no. 4, pp. 820–830, 2001.

[33] Ye, Y., Zhang, B., Zhou, K., Wang, D., Wang, Y., "High-performance cascade type repetitive controller for CVCF PWM inverter: Analysis and design," *IET Electric. Power Appl.*, vol. 1, no. 1, pp. 12–118, 2007.

[34] Ye, Y., Zhou, K., Zhang, B., Wang, D., Wang, J., "High-performance repetitive control of PWM DC–AC converters with real-time phase-lead FIR filter," *IEEE Trans. Circuits Syst. (II)*, vol. 53, no. 8, pp. 768–772, 2006.

[35] Yang, Y., Zhou, K., Lu, W., "Robust repetitive control scheme for three-phase CVCF PWM inverters," *IET Power Electron.*, vol. 5, no. 6, pp. 669–677, 2012.

[36] Tomizuka, M., Tsao, T.-C., and Chew, K.-K., "Analysis and synthesis of discrete-time repetitive controllers," *ASME: J. Dyn. Syst. Meas. Control*, vol. 111, pp. 353–358, 1989.

[37] Cosner, C., Anwar, G., and Tomizuka, M., "Plug-in repetitive control for industrial robotic manipulators," *Cincinnati, Ohio, US: IEEE International Conference on Robotics and Automation*, pp. 1970–1975, 1990.

[38] Tomizuka, M., Tsao, T., and Chew, K., "Discrete-time domain analysis and synthesis of repetitive controllers," *Atlanta, GA, United States: American Control Conference*, pp. 860–866, Jun. 1988.

[39] Broberg, H.L., Molyet, R.G., "A new approach to phase cancelation in repetitive control," *Denver, CO, United States: IEEE Ind. Appl. Society Annual Meeting*, 1994.

[40] Zhang, B., Wang, D., Zhou, K., and Wang, Y., "Linear phase-lead compensation repetitive control of a CVCF PWM inverter," *IEEE Trans. Ind. Electron.*, vol. 55, no. 4, pp. 1595–1602, Apr. 2008.

[41] Zhang, B., Zhou, K., Wang, Y., and Wang, D., "Performance improvement of repetitive controlled PWM inverters: A phase-lead compensation solution," *Int. J. Circuit Theory Appl.*, vol. 38, no. 5, pp. 453–469, Jun. 2010.

[42] Lu, W., Zhou, K., Wang, D., and Cheng, M., "A generic digital nk $\pm$ m order harmonic repetitive control scheme for PWM converters," *IEEE Trans. Ind. Electron.*, vol. 61, no. 3, pp. 1516–1527, 2014.

[43] Zhou, K., Lu, W., Yang, Y., and Blaabjerg, F., "Harmonic control: A natural way to bridge resonant control and repetitive control," *Washington, DC, US: American Control Conference*, pp. 3189–3193, 2013.

[44] Costa-Castelló, R., Grinó, R., and Fossas, E., "Odd-harmonic digital repetitive control of a single-phase current active filter," *IEEE Trans. Power Electron.*, vol. 19, no. 4, pp. 1060–1068, Jul. 2004.

[45] Zhou, K., Low, K. S., Wang, D., Luo, F. L., Zhang, B., and Wang, Y., "Zero-phase odd-harmonic repetitive controller for a single-phase PWM inverter," *IEEE Trans. Power Electron.*, vol. 21, no. 1, pp. 193–201, 2006.

[46] Zhou, K., Yang, Y., Blaabjerg, F., and Wang, D., "Optimal selective harmonic control for power harmonics mitigation," *IEEE Trans. Ind. Electron.*, vol. 62, no. 2, pp. 1220–1230, 2015.

[47] Zhou, K., Wang, D., Zhang, B., Wang, Y., Ferreira, J. A., and de Haan, S. W. H., "Dual-mode structure digital repetitive control," *Automatica*, vol. 43, no. 3, pp. 546–554, Mar. 2007.

[48] Zhou, K., Wang, D., Zhang, B., and Wang, Y., "Plug-in dual-mode structure digital repetitive controller for CVCF PWM inverters," *IEEE Tran. Ind. Electron.*, vol. 56, no. 3, pp. 784–791, Mar. 2009.

Chapter 4
Periodic control of power converters

Abstract

This chapter will present six application examples of the periodic control of power converters, which includes the voltage control for constant-voltage-constant-frequency (CVCF) pulse-width modulation (PWM) inverters, CVCF high-frequency link (HFL) single-phase and three-phase inverters, current control for grid-connected single-phase and three-phase PWM inverters, and HFL rectifier. Experiments have been carried out to verify the effectiveness of various fundamental and advanced periodic control schemes for power converters.

4.1 Periodic control (PC) of CVCF single-phase PWM inverters

4.1.1 Background

Periodic errors induced by nonlinear loads (e.g., rectifier loads) are major sources of the total harmonic distortion (THD) for the CVCF PWM inverters that are widely used in various AC power supplies such as the uninterruptable power supply. The internal model principle (IMP)-based PC schemes, such as the repetitive control (RC) and the resonant control (RSC) [1–20], are found to be able to achieve zero-tracking errors, and thus being suitable control solutions to the tracking/eliminating periodic signals/disturbances with a known period. In order to achieve good control performance, the PC schemes are usually added on to a feedback control system in practical applications, such as the plug-in RC schemes [5,6], and the PI-RES (i.e., proportional–integral–resonant) control scheme [19,20]. In those cases, the plug-in PC and the feedback controller are "complementary" – the feedback controller offers fast dynamic response and robustness, and the PC scheme ensures accurate compensation of residual periodic tracking errors. In this section, various PC schemes are developed for power converters to demonstrate their effectiveness. Experiment results are provided to verify the performance of the classic RC (CRC), the odd-order RC, the dual-mode repetitive control (DMRC), and the multiple resonant control (MRSC) schemes.

4.1.2 Modeling and control of single-phase PWM inverters

Figure 4.1 shows a CVCF single-phase PWM inverter with a resistive load or a nonlinear diode rectifier load, where v_c is the output voltage, i_o is the output current, v_{dc} is the DC bus voltage with its nominal value being v_{dcn}, L_f, L_r, C_f, C_r, and R, R_r are the nominal values for the inductor, capacitor, and resistor of the resistive and rectifier loads, correspondingly, and v_{inv} is the output voltage of the PWM inverter. The control objective of such a CVCF PWM inverter is to produce high-quality sinusoidal voltages under various loads.

The dynamics of the PWM inverter with a linear resistive load can be described as [18]:

$$\begin{cases} \dot{\mathbf{X}} = \mathbf{AX} + \mathbf{BU} \\ \mathbf{Y} = \mathbf{CX} + \mathbf{DU} \end{cases}$$

with

$$\mathbf{X} = \begin{bmatrix} v_c(t) \\ \dot{v}_c(t) \end{bmatrix}, \quad \mathbf{Y} = v_c(t), \quad \mathbf{U} = v_{inv}(t).$$

$$\mathbf{A} = \begin{bmatrix} 0 & 1 \\ -\dfrac{1}{L_f C_f} & -\dfrac{1}{C_f R} \end{bmatrix}, \quad \mathbf{B} = \begin{bmatrix} 0 \\ \dfrac{1}{L_f C_f} \end{bmatrix}, \quad \mathbf{C} = [1 \quad 0], \quad \mathbf{D} = 0 \tag{4.1}$$

For a linear system $\dot{\mathbf{X}} = A\mathbf{X} + \mathbf{BU}$, its sampled-data equation can be expressed as $\mathbf{X}(k+1) = e^{\mathrm{A}T_s}\,\mathbf{X}(k) + \int e^{\mathrm{A}(T_s-\tau)}\mathbf{U}(\tau)\mathbf{B}d\tau$ with T_s being the sampling period. Therefore, the sampled-data form of (4.1) can be approximated as [21]:

$$\begin{bmatrix} v_c(k+1) \\ \dot{v}_c(k+1) \end{bmatrix} = \begin{bmatrix} \varphi_{11} & \varphi_{12} \\ \varphi_{21} & \varphi_{22} \end{bmatrix} \begin{bmatrix} v_c(k) \\ \dot{v}_c(k) \end{bmatrix} + \begin{bmatrix} g_1 \\ g_2 \end{bmatrix} v_{inv}(k)$$

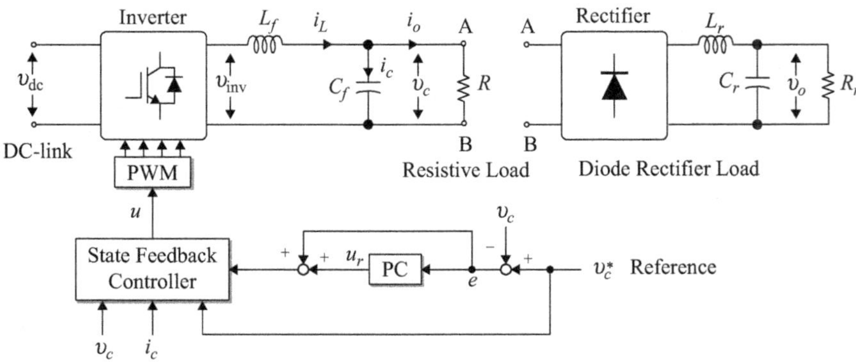

Figure 4.1 Schematic and control structure of a single-phase CVCF PWM inverter with a repetitive controller based periodic control system

where

$$\begin{aligned}
\varphi_{11} &= 1 - T_s^2/(2L_fC_f),\\
\varphi_{22} &= 1 - T_s/(C_fR) - T_s^2/(2L_fC_f) + T_s^2/\left(2C_f^2R^2\right),\\
\varphi_{12} &= T_s - T_s^2/(2C_fR),\\
\varphi_{21} &= -T_s/(L_fC_f) + T_s^2/\left(2L_fC_f^2R\right),\\
g_1 &= T_s^2/(2L_fC_f),\\
g_2 &= T_s/(L_fC_f) - T_s^2/\left(2L_fC_f^2R\right).
\end{aligned} \tag{4.2}$$

A plug-in RC scheme, which comprises a generic state feedback controller (SFC) and a PC, is developed to accurately regulate the output AC voltage to track the reference sinusoidal signal with a low THD and fast dynamic responses.

The generic SFC can be written as [14]:

$$u(k) = -\left(k_1\frac{v_c(k)}{v_{dc}} + k_2\frac{\dot{v}_c(k)}{v_{dc}}\right) + k_{ref}\frac{v_c^*(k)}{v_{dc}} \tag{4.3}$$

where $u(k) = (v_{inv}(k)/v_{dc}) \in [-1,\ 1]$ is the control output, $v_c^*(k)$ is the reference output voltage, and k_1, k_2, and k_{ref} are the feedback control coefficients.

With such an SFC, the state equation of the closed-loop system can be expressed as:

$$\begin{bmatrix} v_c(k+1) \\ \dot{v}_c(k+1) \end{bmatrix} = \begin{bmatrix} \varphi_{11} - g_1k_1 & \varphi_{12} - g_1k_2 \\ \varphi_{21} - g_2k_1 & \varphi_{22} - g_2k_2 \end{bmatrix} \begin{bmatrix} v_c(k) \\ \dot{v}_c(k) \end{bmatrix} + \begin{bmatrix} g_1k_{ref} \\ g_2k_{ref} \end{bmatrix} v_c^*(k) \tag{4.4}$$

The poles of the closed-loop system in (4.4) can be arbitrarily assigned by adjusting the feedback control gain k_1 and k_2. The transfer function from v_c^* to v_c for the closed-loop system with nominal parameter values can then be written as [5,21,22]:

$$H(z) = \frac{m_1z + m_2}{z^2 + p_1z + p_2}$$

where

$$\begin{aligned}
m_1 &= g_1k_{ref},\\
m_2 &= (\varphi_{12} - g_1k_2)g_2k_{ref} - (\varphi_{22} - g_2k_2)g_1k_{ref},\\
p_1 &= -\varphi_{11} + g_1k_1 - \varphi_{22} + g_2k_2,\\
p_2 &= (\varphi_{11} - g_1k_1)(\varphi_{22} - g_2k_2) - (\varphi_{12} - g_1k_2)(\varphi_{21} - g_2k_1).
\end{aligned} \tag{4.5}$$

To deal with the periodic reference and disturbances, a PC scheme $G_{pc}(z)$ is plugged into the SFC-controlled inverter system as shown in Figure 4.2. In the digital control system of Figure 4.2, $r(z)$ is the input reference, $y(z)$ is the output, $e(z)$ is the tracking error, $G_p(z)$ represents the plant model, $G_c(z)$ is the digital form of the SFC scheme, and $d(z)$ is the disturbance.

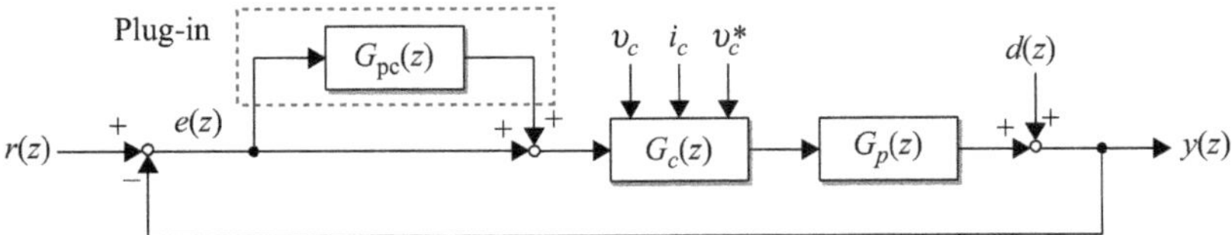

Figure 4.2 Plug-in periodic control system for the single-phase CVCF inverter system

4.1.2.1 Plug-in RC scheme

Since odd-order (i.e., $4k \pm 1$ order with $k = 1, 2, 3, \ldots$) harmonics usually are dominant in the output voltage of the CVCF single-phase PWM inverters, a digital optimal harmonic control (OHC) is then introduced into the control scheme in Figure 4.2 to act as the PC scheme. The OHC $G_{\text{OHC}}(z)$ can be given as [23–27]:

$$\begin{aligned} G_{\text{OHC}}(z) &= \sum_{m \in N_m} G_{\text{sm}}(z) \\ &= \sum_{m \in N_m} k_{\text{rm}} G_f(z) \cdot \frac{\left[\cos(2\pi m/n) z^{N/n} Q(z) - Q^2(z)\right]}{z^{2N/n} - 2\cos(2\pi m/n) z^{N/n} Q(z) + Q^2(z)} \end{aligned} \tag{4.6}$$

where $k_{\text{rm}} \in [k_{\text{r}0}, k_{\text{r}1}, k_{\text{r}2}]$ are the control gains for the corresponding selective harmonic control (SHC) modules $G_{\text{sm}}(z)$ with $m = 0, 1, 2$, $G_f(z)$ is a phase-lead compensator, $Q(z)$ is a low-pass filter, $N = f_s/f_0$ with f_s being the sampling frequency and f_0 being the fundamental frequency of the reference, and N_m is the harmonic set. The SHC modules are exclusively used to compensate the $4k$-order harmonics, $(4k \pm 1)$-order harmonics, and $(4k \pm 2)$-order harmonics, and their corresponding control gains $k_{\text{r}0}$, $k_{\text{r}1}$, and $k_{\text{r}2}$ to avoid any confusion with the previous definitions will be proportional to the ratios of the corresponding harmonics to the total harmonics.

As discussed in Chapter 3, the OHC provides a universal framework for housing various RC schemes such as the CRC and the SHC. For example,

- when $k_{\text{r}0} = k_{\text{r}2} = k_{\text{rc}}/4$ and $k_{\text{r}1} = k_{\text{rc}}/2$, $G_{\text{OHC}}(z)$ in (4.6) will become the CRC as:

$$G_{\text{rc}}(z) = k_{\text{rc}} G_f(z) \frac{Q(z) z^{-N}}{1 - Q(z) z^{-N}} \tag{4.7}$$

- when $k_{\text{r}0} = k_{\text{r}2} = 0$ and $k_{\text{r}1} = k_{\text{o}}$, $G_{\text{OHC}}(z)$ in (4.6) becomes the odd-order harmonic RC [10–13] as:

$$G_{\text{orc}}(z) = -k_{\text{o}} G_f(z) \frac{Q(z) z^{-N/2}}{1 + Q(z) z^{-N/2}} \tag{4.8}$$

- and when $k_{r0} = k_{r2} = k_e/2$ and $k_{r1} = k_o$, $G_{OHC}(z)$ in (4.6) becomes the DMRC [13,14] as

$$G_{drc}(z) = G_{erc}(z) + G_{orc}(z)$$
$$= k_e G_f(z) \frac{z^{-N/2}Q(z)}{1 - z^{-N/2}Q(z)} - k_o G_f(z) \frac{-z^{-N/2}Q(z)}{1 + z^{-N/2}Q(z)} \quad (4.9)$$

where the even-order harmonic RC $G_{erc}(z)$ is equivalent to the combination of $G_{s0}(z)$ and $G_{s2}(z)$. The DMRC is actually an OHC with simplified implementation.

According to the stability criteria (2.23), for the RC systems with a linear phase-lead compensator $G_f(z) = z^p$ [17,18], the stability region for the RC gains for the above RC schemes can be derived as follows:

- For the CRC system if $-\frac{\pi}{2} < \theta_H + p\omega < \frac{\pi}{2}$, then

$$0 < k_{rc} < \frac{2\cos(\theta_H + p\omega)}{|H(e^{j\omega})|}; \quad (4.10)$$

- For the odd-order harmonic RC system, if $-\frac{\pi}{2} < \theta_H + p\omega < \frac{\pi}{2}$, then

$$0 < k_o < \frac{2\cos(\theta_H + p\omega)}{|H(e^{j\omega})|}; \quad (4.11)$$

- For the DMRC system, if $-\frac{\pi}{2} < \theta_H + p\omega < \frac{\pi}{2}$, then

$$0 < k_o + k_e < \frac{2\cos(\theta_H + p\omega)}{|H(e^{j\omega})|}. \quad (4.12)$$

All these three RC schemes are developed according to Figure 4.2. The resultant plug-in PC schemes are then applied to the inverter system shown in Figure 4.1 for experimental validation.

4.1.2.2 Plug-in RSC scheme

To deal with the major harmonic distortion in the low-frequency band, the MRSC at dominant harmonic frequencies can also be added into the SFC system [16,17] as the PC shown in Figure 4.2. The MRSC can be given as:

$$G_M(s) = \sum_{h \in N_h} G_h(s) = \sum_{h \in N_h} \left(k_h \frac{s\cos\theta_h - \omega_h \sin\theta_h}{s^2 + \omega_h^2} \right) \quad (4.13)$$

where $G_h(s)$ is an RSC for the harmonics at the frequency of ω_h with a phase compensation of θ_h and k_h being its control gain, and N_h is the set of interested harmonic orders. Obviously, the control gain k_h for each RSC in the MRSC of $G_M(s)$ can be designed independently with its corresponding phase compensation θ_h. Independent gain and phase compensation for each RSC enable the MRSC to achieve zero-phase compensation at each dominant harmonic frequency (i.e., resonant frequency) with an optimal error convergence rate. Notably, more RSC components can be added into the MRSC to enhance its tracking accuracy at the cost of increased complexity. Compared with the CRC, the MRSC can achieve a higher error convergence rate. Furthermore, since the MRSC is only comprised of RSC components at the dominant harmonic frequencies in the low-frequency band, there is no need to include a low-pass filter $Q(s)$.

The MRSC scheme of (4.13) can be discretized by means of the pre-warping Tustin transform as [16,17]:

$$G_M(z) = \sum_{h \in N_h} G_h(z)$$
$$= \sum_{h \in N_h} \left\{ k_h \frac{\frac{1}{2}(1 - z^{-2})\cos\phi_h \sin(\omega_h T_s) - (1 + 2z^{-1} + z^{-2})\sin\phi_h \sin^2\left(\frac{\omega_h T_s}{2}\right)}{\omega_h (1 - 2z^{-1}\cos(\omega_h T_s) + z^{-2})} \right\} \tag{4.14}$$

where T_s denotes the sampling period.

Since odd-order harmonics in the low-frequency band (e.g., typically up to the 15th-order harmonic) are dominant in the total harmonic distortion in the output voltage of the single-phase PWM inverter, only the low-order RSC components $G_h(s)$ with $h = 3, 5, 7, \ldots, 15$, are included in the MRSC $G_M(s)$. Figure 4.3 shows the entire plug-in MRSC structure.

In order to achieve zero-phase compensation at the harmonic frequency ω_h, the compensation phase θ_h for the hth-order RSC $G_h(z)$ will be chosen as:

$$\theta_h = -\angle H(j\omega_h) \tag{4.15}$$

where $\angle H(j\omega_h)$ is the phase of the closed-loop system $H(z)$ in (4.5) at the harmonic frequency ω_h. The phase compensation will enable the MRSC to tune the RSC gains k_h within a wider range, and then to achieve a faster error convergence rate. Consequently, the control system will become more robust to disturbances.

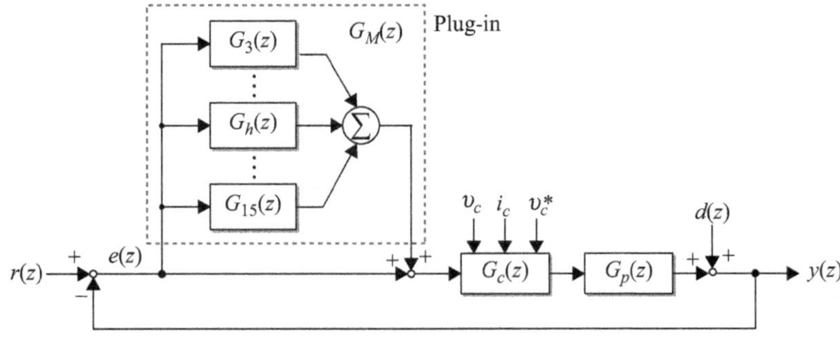

Figure 4.3 Plug-in multiple resonant control system for the single-phase CVCF inverter system

4.1.3 Experimental validation

4.1.3.1 Plug-in RC scheme

To verify the control performance of the RC schemes for the single-phase CVCF inverter system, a dSPACE DS1102-based test rig is built up for experiments. System parameters are listed in Table 4.1. For the SFC of Eq. (4.3), $k_1 = 90$, $k_2 = 8.4 \times 10^{-3}$, $k_{\text{ref}} = 90$ are arbitrarily designed. With the parameters shown in Table 4.1, the closed-loop nominal transfer function $H(z)$ in (4.5) can be derived as:

$$H(z) = \frac{0.5z + 0.432}{z^2 - 0.487z + 0.429} \tag{4.16a}$$

Figure 4.4 shows the trajectories of poles of the system model $H(z)$ in (4.5) when the resistive load R changes from 0.95 Ω to infinity. It is clearly seen that, when $R < 1.1$ Ω, the poles are located outside the unit circle, and the system is unstable. On the other hand, when $R > 1.1$ Ω, the poles are located inside the unit circle, and the system is stable. Hence, the stability range of load for the SFC system $H(z)$ is $[1.1, \infty)$ Ω.

With the SFC system of (4.16a), Figure 4.5 shows the responses of the phase-lead compensated system (i.e., $G_f(z)H(z)$) with different linear phase-lead compensation steps $p = 0, 1, 2, 3$. The corresponding bandwidth and the RC gain range are also listed in Table 4.2. Note that, instead of ±90°, a phase band of ±75° is adopted as the boundary for $(\theta_H + p\omega)$ in (4.10)–(4.12) with a phase margin of 15°. Hereby, the theoretical analysis indicates that the lead step $p = 2$ yields both the widest stable frequency bandwidth and the widest range for RC gains. Since the majority of the harmonic distortions concentrate in the low-frequency band, Figure 4.5 and Table 4.2 demonstrate the feasibility of the RC with a phase-lead compensation.

However, due to various and un-modeled delays, the phase-lead compensator $G_f(z) = z^p$ with an optimal lead step $p = 2$ is usually insufficient for phase compensation in practice. The real value of the optimal phase lead step $p = 3$ for all

Table 4.1 Parameters of the single-phase CVCF inverter system

Parameters	Nominal value	Unit
DC-link voltage V_{dc}	80	V
Inductor filter L_f	20	mH
Capacitor filter C_f	45	μF
Linear resistive load R	15	Ω
Rectifier inductor L_r	1	mH
Rectifier capacitor C_r	500	μF
Rectifier resistor R_r	22	Ω
PWM switching frequency	10	kHz
Sampling frequency	10	kHz
Reference output voltage v_c^*	$50 \sin(100\pi\omega t)$	V

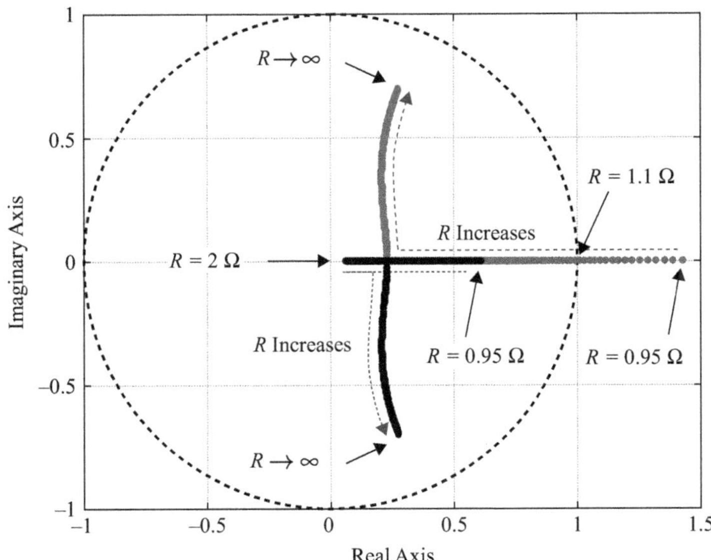

Figure 4.4 Trajectories of poles of the SFC system model with R varying from 0.95 Ω to infinity

three RC schemes is determined by experiments. A low-pass filter $Q(z) = 0.25z + 0.5 + 0.25z^{-1}$ is used in (4.7) to (4.9) in the experiments.

Figure 4.6 shows the steady-state responses of the SFC inverter system with the rectifier load, where the THD of the output voltage v_c is 8.0%. To evaluate the harmonic distribution of the output voltage v_c, a harmonic ratio $h_r(j)$ is defined as:

$$h_r(j) = \frac{\left(\sum_{i=1}^{j} M_i\right)}{\left(\sum_{i=1}^{199} M_i\right)} \times 100\% \tag{4.16b}$$

in which M_i is the magnitude of the ith-order harmonic. Figure 4.7(a) gives the harmonic ratio $h_r(j)$ for the current spectrum shown in Figure 4.6(b). It is indicated in Figure 4.7(a) that over 95% of the harmonics are within the range of [0, 1800] Hz, while it can be seen in Figure 4.7(b) that the bandwidth of the employed $Q(z) = 0.25z + 0.5 + 0.25z^{-1}$ is about 1820 Hz. Therefore, the LPF $Q(z)$ allows the RC schemes to eliminate most of the harmonics in the output voltage v_c. Furthermore, the harmonic ratio for all odd-order harmonics of the output voltage v_c shown in Figure 4.6(b) is about 87%. Obviously, the odd-order harmonics dominate the THD of the output voltage v_c.

Subsequently, the RC schemes are applied to the SFC system to reduce the THD of the output voltage. The same diode rectifier load has been connected to the inverter. Figure 4.8 shows the steady-state performance of the inverter system with different plug-in RC schemes. It can be seen in Figure 4.8 that the THD of the output voltage has

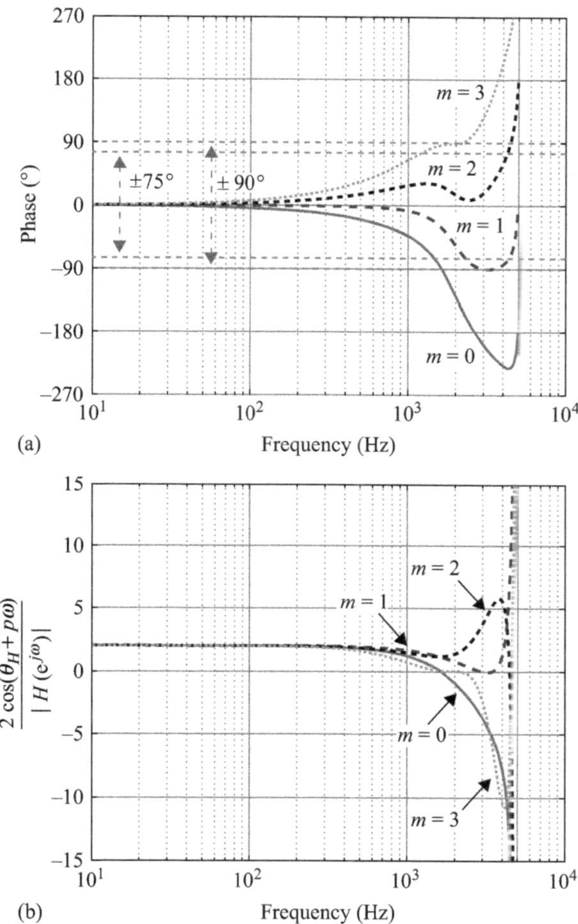

Figure 4.5 Responses of the phase-lead-compensated system under different compensation steps p = 0, 1, 2, 3: (a) phase of the compensated system $G_f(z)H(z)$ (i.e., $\theta_H + p\omega$) and (b) corresponding boundary of the RC gains

Table 4.2 Bandwidth and RC gain range for system stability with different phase-lead step p

Phase-lead step *p*	**0**	**1**	**2**	**3**
Bandwidth, Hz	1500	2300	4400	1150
RC gain range	[0, 0.275]	[0, 0.45]	[0, 1.05]	[0, 0.4]

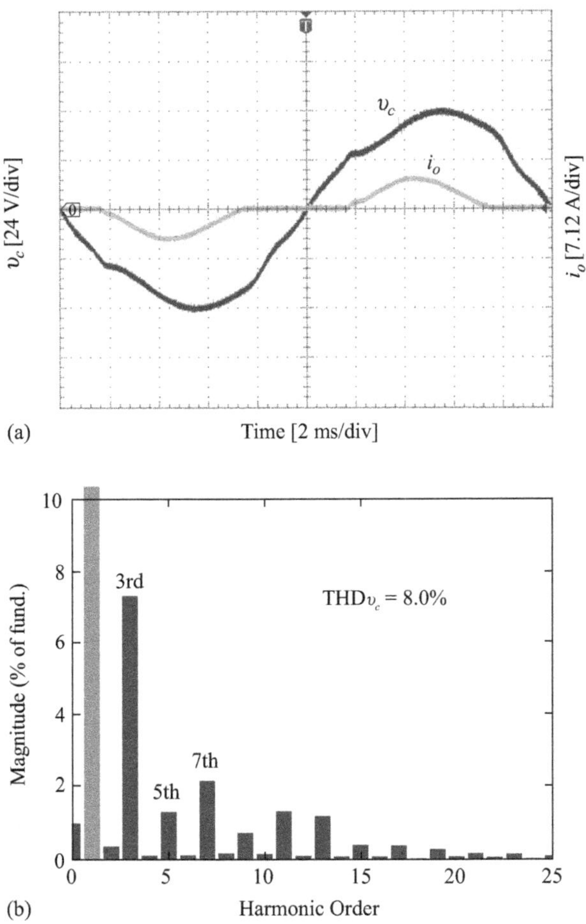

Figure 4.6 Steady-state responses of the state-feedback controlled inverter system shown in Figure 4.1 with a diode rectifier load: (a) output voltage v_c and load current i_o and (b) harmonic distribution of the voltage v_c

been significantly reduced from 8% to around 1%. Specifically, the CRC with the gain $k_{rc} = 1.2$ produces the output voltage v_c with a low THD of 0.8%, while the odd-order harmonic RC with the gain $k_o = 1.2$ gives an output voltage v_c with a slightly increased THD of 1.2%. It is because the odd-order harmonic RC is employed only to mitigate the dominant odd-order harmonics shown in Figure 4.6(a) with a fast dynamic response shown in Figure 4.9. Furthermore, Figure 4.8(c) shows that the DMRC with the gains being $k_o = 0.4$ and $k_e = 0.8$ produces an output voltage v_c with a THD of 0.8%, which is similar to the case when using the CRC compensator. In contrast, when the gains are designed according to the harmonic distribution as $k_o = 0.8$ and $k_e = 0.4$, the DMRC compensated system can produce a high-quality voltage v_c with THD = 0.5% even under the diode rectifier load. The above experimental tests have verified the effectiveness of the RC schemes in terms of harmonic compensation.

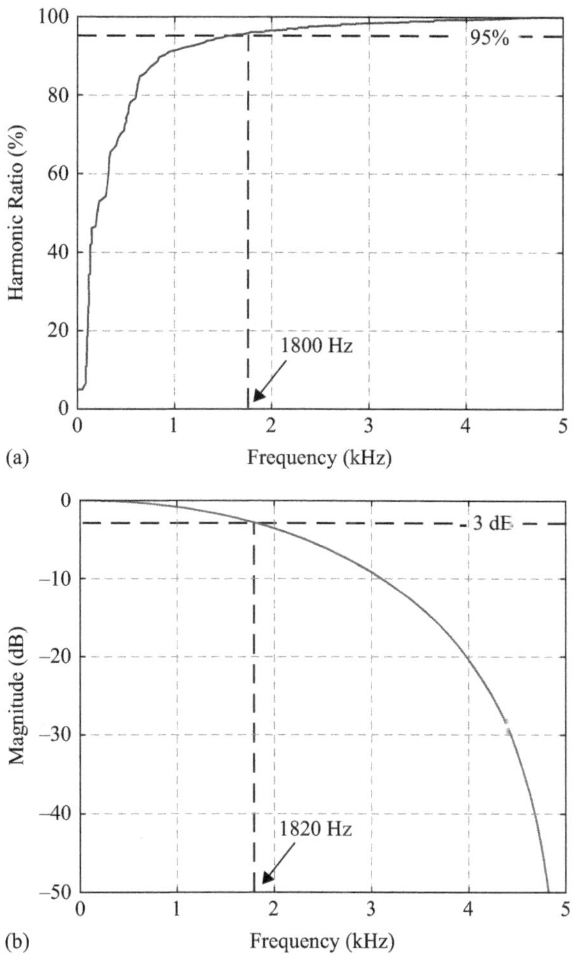

Figure 4.7 Harmonic analysis of the SFC system: (a) harmonic ratio $h_r(j)$ of v_c and (b) magnitude response of $Q(z)$

Figure 4.9 shows the transient tracking error $v_c^* - v_c$ with various RC schemes being plugged into the SFC-controlled inverter system, where the same rectifier load has been used. It can be observed in Figure 4.9 that the odd-order harmonic RC with the gain being $k_o = 1.2$ offers the fastest transient response, as it is also mentioned previously. The DMRC scheme with the gains $k_o = 0.8$ and $k_e = 0.4$ ranks the second, followed by the CRC system with the gain $k_{rc} = 1.2$, as shown in Figure 4.9. Among all the RC schemes, the DMRC with the gains $k_o = 0.4$ and $k_e = 0.8$ presents the slowest convergence rate. According to (4.7)–(4.9), the CRC with $k_{rc} = 1.2$ is equivalent to the DMRC with $k_o = 0.6$ and $k_e = 0.6$. And, an odd-order harmonic controller with $k_o = 1.2$ is equivalent to the DMRC, where the gains should be $k_o = 1.2$ and $k_e = 0$. Moreover, Figure 4.9(c) and (d) imply that, for the DMRC

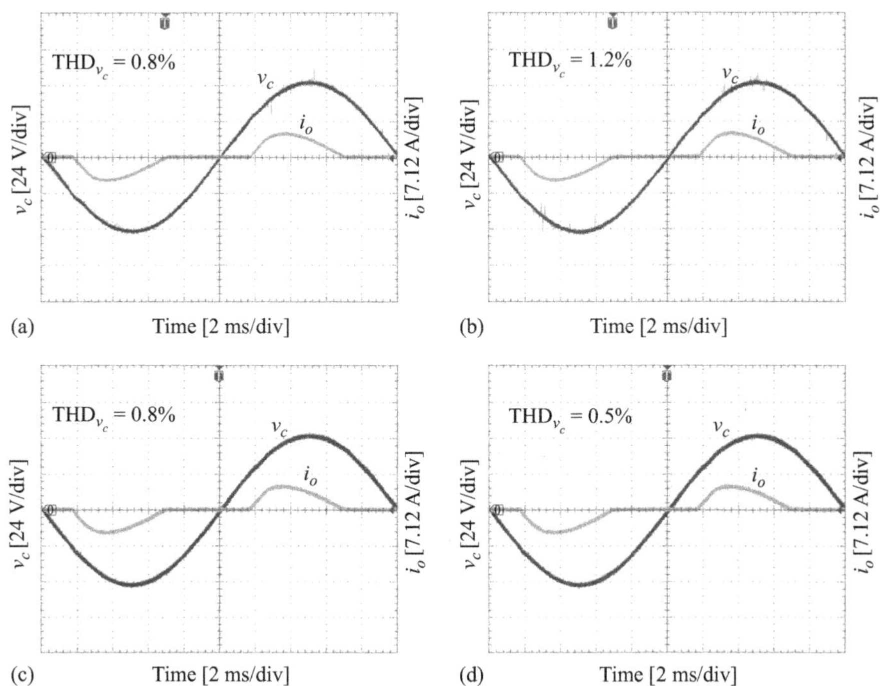

Figure 4.8 Steady-state performance of the single-phase inverter system with the rectifier load using different RC schemes (v_c: output voltage, i_o: load current): (a) the CRC with $k_{rc} = 1.2$, (b) the odd-order harmonic RC with $k_o = 1.2$, (c) the DMRC with $k_o = 0.4$, $k_e = 0.8$, and (d) the DMRC with $k_o = 0.8$, $k_e = 0.4$

scheme, the larger the odd-order harmonic gain k_o is, the faster the tracking error convergence rate is. Additionally, Table 4.3 summarizes the control performance of all plug-in RC schemes. The above results indicate that the DMRC with $k_o = 0.8$ and $k_e = 0.4$ offers a good trade-off between the tracking accuracy and dynamic response. As a special case of the OHC in (4.6), the DMRC can further optimize its performance by finely tuning its control gains according to the harmonic ratio $h_r(j)$.

In addition, Figure 4.10 shows the dynamic performance of the inverter system under load step-changes, where the DMRC scheme has been adopted with $k_o = 0.8$ and $k_e = 0.4$. It can be seen from Figure 4.10 that the DMRC-based inverter system is robust to the sudden load changes between no load and full rectifier load. The settling time for the transient response of the output voltage v_c is about 3~5 cycles (i.e., 60~100 ms).

4.1.3.2 Plug-in RSC scheme

To verify the control performance of the plug-in RSC scheme, a dSPACE DS1102-based test rig is built up referring to Figure 4.1. The system parameters are listed in Table 4.4.

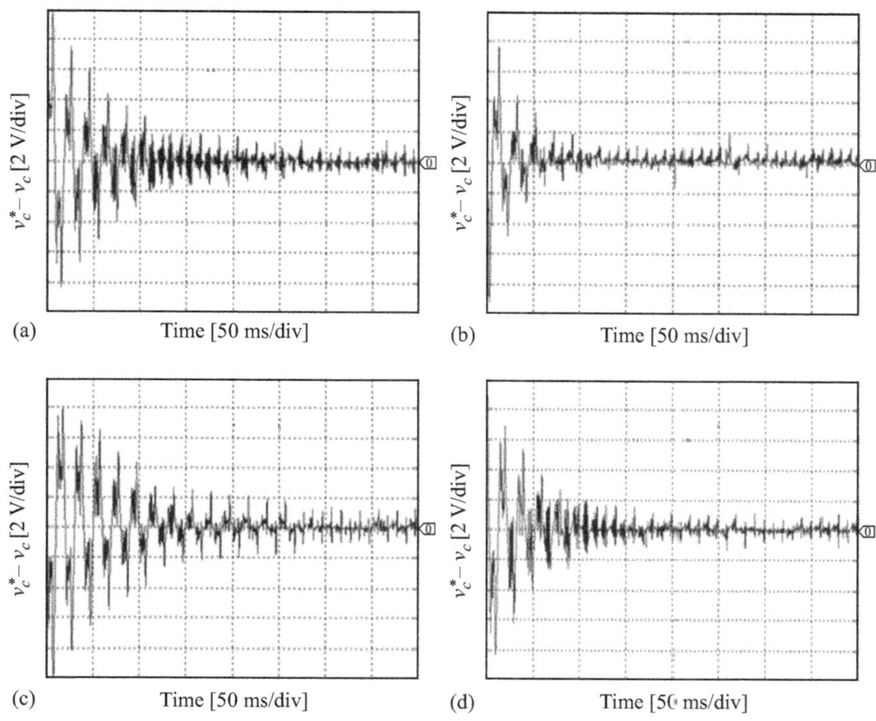

Figure 4.9 *Transient tracking errors $v_c^* - v_c$ for the inverter system using different RC schemes: (a) the CRC with $k_{rc} = 1.2$, (b) the odd-order harmonic RC with $k_o = 1.2$, (c) the DMRC with $k_o = 0.4$, $k_e = 0.8$, and (d) the DMRC with $k_o = 0.8$, $k_e = 0.4$*

Table 4.3 *Summary of the performance of the inverter system with different RC schemes*

RC Type	**THD (%)**	**Convergence time (s)**	**Even harmonics is reduced?**
CRC with $k_{rc} = 1.2$	0.8	0.20	Yes
Odd-harmonic RC with $k_o = 1.2$	1.2	0.10	No
DMRC with $k_o = 0.4$, $k_e = 0.8$	0.8	0.32	Yes
DMRC with $k_o = 0.8$, $k_e = 0.4$	0.5	0.16	Yes

With the nominal load $R = 60\ \Omega$, the SFC in (4.3) is chosen as:

$$u(k) = -\left[27.76 \times \frac{v_c(k)}{v_{dc}} + 4.15 \times 10^{-3} \times \frac{\dot{v}_c(k)}{v_{dc}}\right] + 28.76 \times \frac{v_c^*(k)}{v_{dc}} \qquad (4.17)$$

which arbitrarily assigns the poles of $H(z)$ in (4.5) at $0.135 \pm 0.360j$.

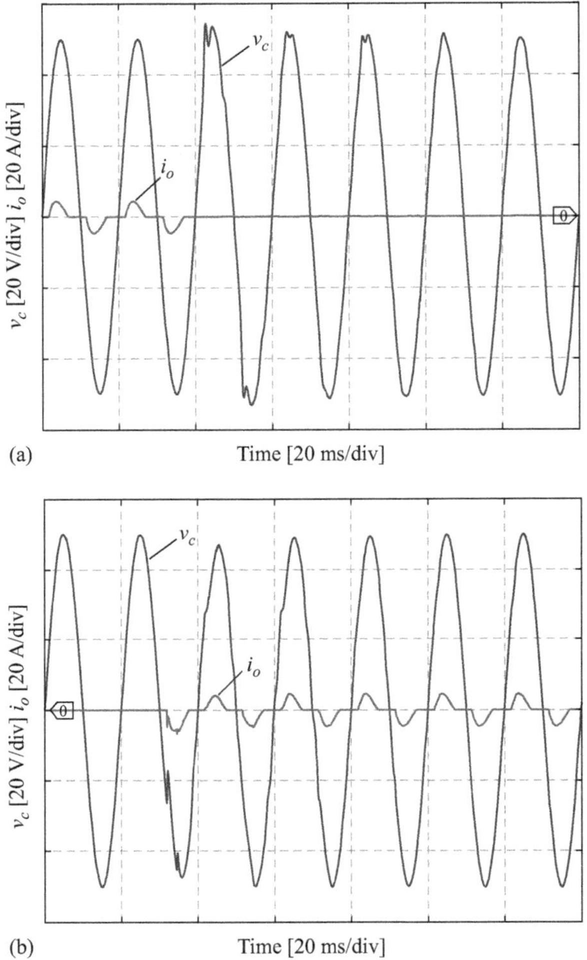

Figure 4.10 Dynamic performance of the inverter system under rectifier load step-changes using the DMRC with $k_o = 0.8$ and $k_e = 0.4$: (a) from full rectifier load to no load and (b) from no load to full rectifier load

Figure 4.11 presents the pole distributions of the closed-loop system $H(z)$ with a resistive load $R \in [0.43, \infty)$ Ω. It can be seen in Figure 4.11 that, when $R \geq 0.5$ Ω, the poles of $H(z)$ in (4.5) are inside the unit circle. Therefore, the closed-loop system $H(z)$ is stable if $R \in [0.5, \infty)$ Ω.

However, due to parameter uncertainties, unknown delays and nonlinear loads, accurate transfer function of the SFC system is different from $H(z)$ shown in (4.5). The actual phase delay of the SFC system at the harmonic frequency ω_h can be measured through experiments, and will be compensated to maximize the stability region of the gain k_h. Moreover, since the error convergence rate at ω_h is proportional to its corresponding gain k_h of $G_h(z)$, a larger k_h yields a faster convergence rate.

Table 4.4 Parameters of a single-phase CVCF PWM inverter system

Parameters	Nominal value	Unit
DC-link voltage v_{dc}	250	V
Inductor filter L_f	3.3	mH
Capacitor filter C_f	100	μF
Linear resistive load R	60	Ω
Rectifier inductor L_r	3.3	mH
Rectifier capacitor C_r	1000	μF
Rectifier resistor R_r	60	Ω
Switching frequency	10	kHz
Sampling frequency	10	kHz
Reference output voltage v_c^*	$155.6\sin(100\pi\omega t)$	V

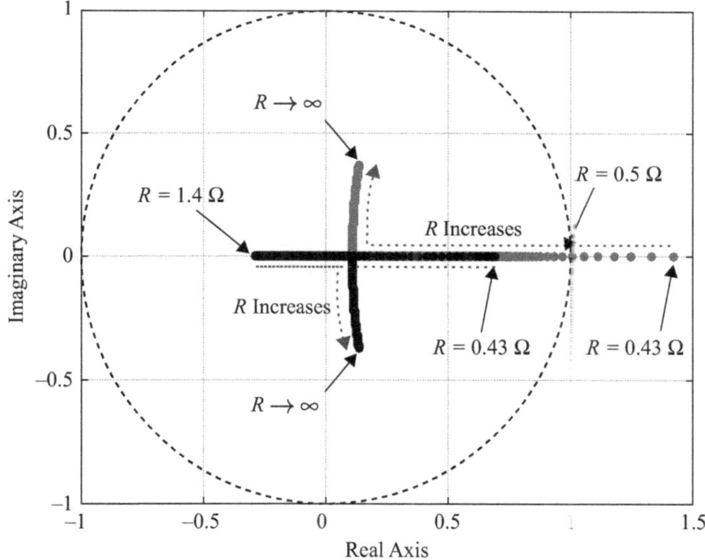

Figure 4.11 Trajectories of poles of the SFC system model with R varying from 0.43 Ω to infinity

However, if the gain k_h is too large to be beyond the stability region, the system may become unstable. Therefore an optimal choice of k_h is needed to enable the MRSC-based inverter system to achieve a satisfactory error convergence rate. Based on the experimental results, the control gains and compensation phase angles of the MRSC $G_M(z)$ are chosen as listed in Table 4.5.

First, the MRSC $G_M(z)$ without any phase compensation (i.e., $\theta_h = 0$) are plugged into the SFC-based inverter system with the rectifier load. When the seventh-order harmonic compensator is plugged into the SFC system, as shown in

Table 4.5 Parameters of the multiple resonant control $G_M(z)$

Gain	Value	Phase compensation	Value
k_1	400	θ_1	7.5°
k_3	60	θ_3	40°
k_5	40	θ_5	50°
k_7	20	θ_7	65°
k_9	10	θ_9	150°
k_{11}	10	θ_{11}	200°
k_{13}	5	θ_{13}	220°
k_{15}	5	θ_{15}	230°

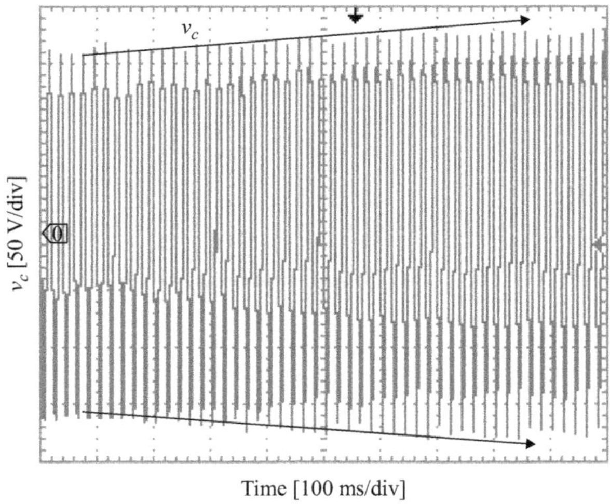

Figure 4.12 Unstable output voltage v_c

Figure 4.12, the inverter system becomes unstable. This confirms the previous discussion, where it is mentioned that the phase compensation should be considered in the MRSC system in order to enhance the system stability.

With the parameters shown in Table 4.5, experiments have been carried out to verify the performance of the MRSC. Figure 4.13 shows the steady-state responses of the inverter system when employing the SFC plus a fundamental-frequency RSC $G_1(z)$. It can be seen in Figure 4.13 that the THD of the output voltage v_c is 5.8% and the third-, fifth-, seventh-, and ninth-order harmonics are dominant distortions in the output voltage v_c. The SFC plus the fundamental-frequency RSC regulator cannot suppress the major low-order harmonics in the output voltage v_c.

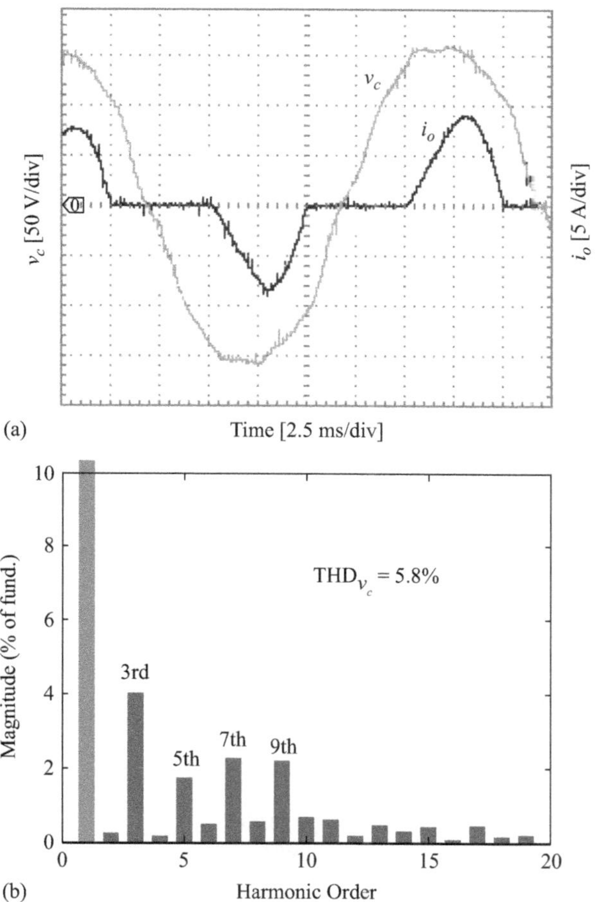

Figure 4.13 Steady-state responses of the inverter system under the rectifier load using the SFC plus a fundamental-frequency RSC: (a) output voltage v_c and load current i_o and (b) harmonic spectrum of the output voltage v_c

In order to further reduce the distortions in the output voltage, the MRSC in (4.13) is then enabled and plugged into the SFC system, as shown in Figure 4.3. Figure 4.14 shows the steady-state responses of the SFC with the plug-in MRSC $G_M(z) = \Sigma G_h(z)$ controlled inverter system under the same load condition, where the dominant odd-order harmonics up to the 15th-order have been compensated by the MRSC $G_M(z)$ with the control parameters listed in Table 4.5. It can be seen in Figure 4.14 that the odd-order harmonics up to the 15th-order are significantly reduced in the harmonic spectrum of the output voltage v_c, and the THD of the output voltage v_c is reduced to 1%, which is much lower than that shown in Figure 4.13.

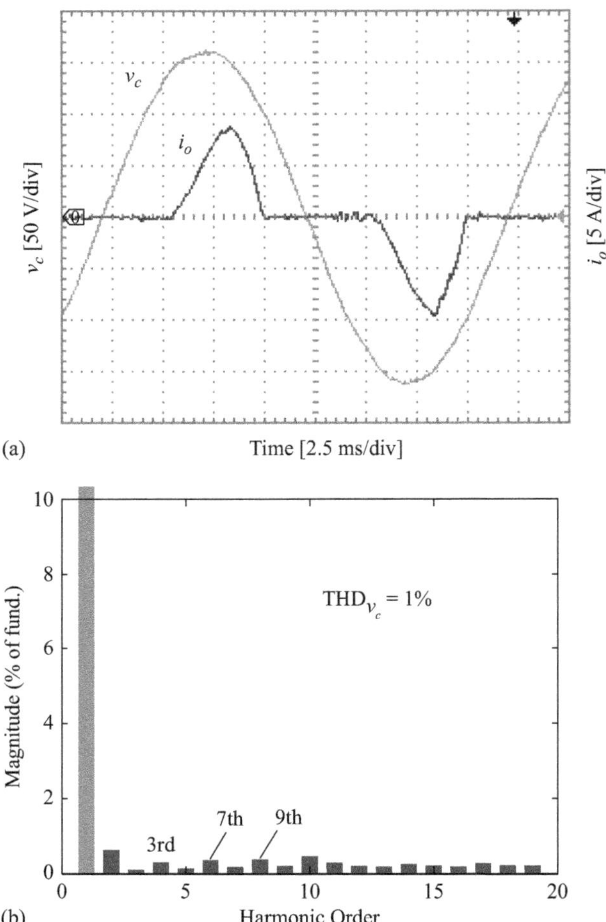

Figure 4.14 Steady-state responses of the inverter system under the rectifier load using the SFC plus the MRSC $G_M(z)$: (a) output voltage v_c and load current i_o and (b) harmonic spectrum of the output voltage v_c

To compare the MRSC scheme with the CRC solution, Figure 4.15 presents the steady-state responses of the inverter system using the SFC plus the CRC, where $Q(z) = (z + 2 + z^{-1})/4$, $k_{rc} = 0.2$ and $G_f(z) = z^4$. Clearly it is shown in Figure 4.15 that the THD of the output voltage v_c is 1.3%. Also compared to those distortions in Figure 4.13., all h-th-order harmonics ($h = 2, 3, 4, \ldots 15$) are significantly reduced in the harmonic spectrum of the output voltage v_c shown in Figure 4.15.

The error-tracking convergence rates are also compared among the three different harmonic compensation schemes – the SFC plus a fundamental-frequency RSC, an MRSC compensator, and a CRC scheme, respectively. Figure 4.16 gives the dynamic responses of the output voltage tracking error $v_c^* - v_c$ of the inverter system under the rectifier load. It can be seen in Figure 4.16 that, the settling-times

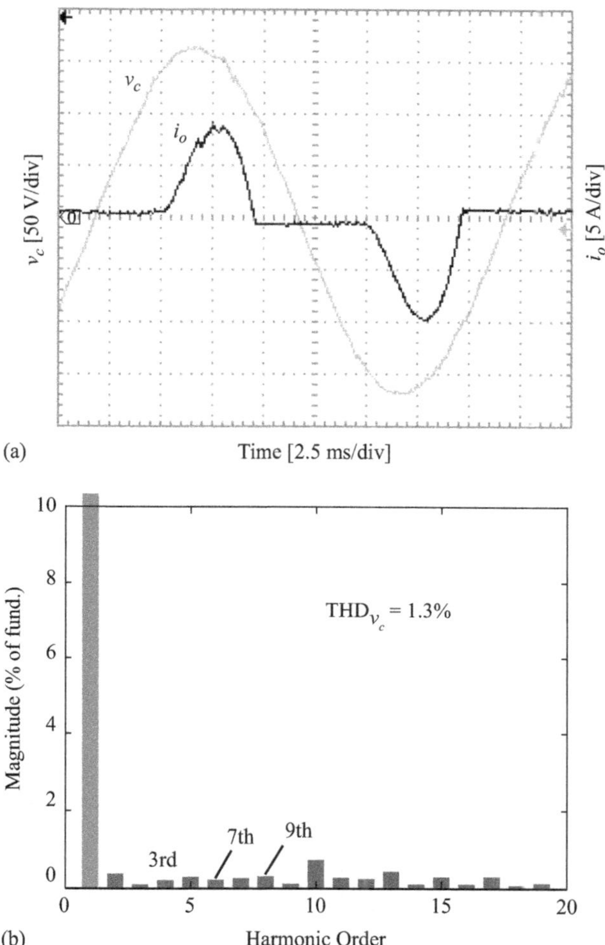

Figure 4.15 Steady-state responses of the inverter system under the rectifier load using the SFC plus the CRC: (a) output voltage v_c and load current i_o and (b) harmonic spectrum of the output voltage v_c

are 0.3 s, 0.2 s, and 0.5 s for the SFC plus a fundamental-frequency RSC $G_1(s)$, an MRSC, and a CRC, respectively. Obviously, the SFC plus the MRSC produces fastest total convergence rate among the three schemes. Furthermore, as shown in Figure 4.16, the SFC plus a fundamental-frequency RSC $G_1(s)$ yields the largest tracking errors among the three schemes, which well matches with its THD value shown in Figure 4.13.

At last the dynamic responses of the inverter system using the SFC plus the MRSC are investigated. Figure 4.17 shows the dynamic responses of the inverter system under step load changes. It is clear that the inverter system is robust to load variations.

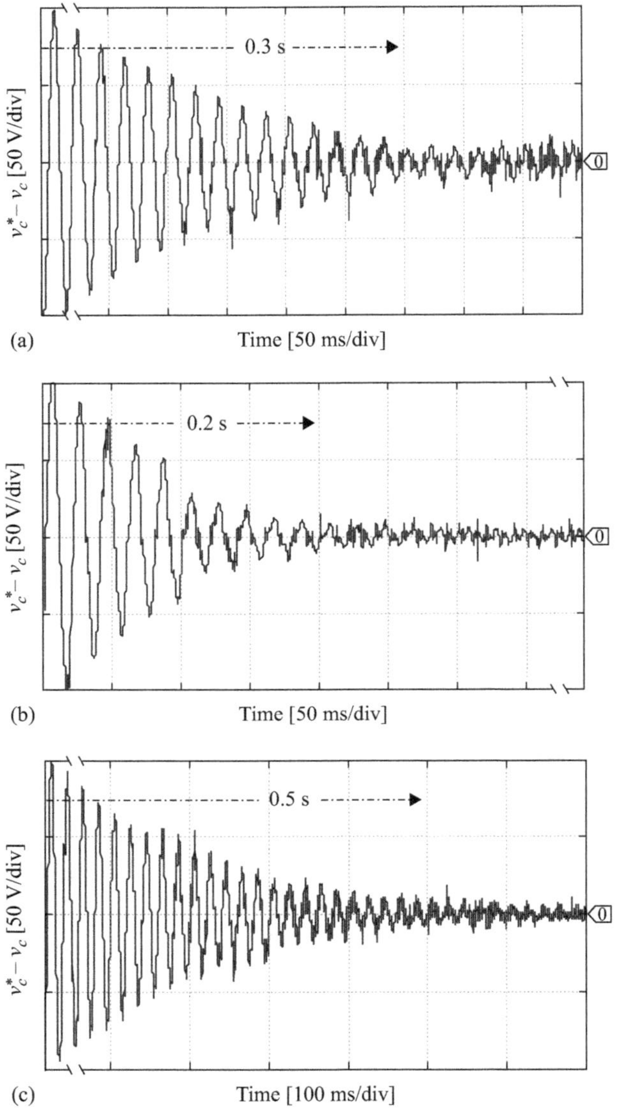

Figure 4.16 Transients of the voltage tracking error $v_c^* - v_c$ using different control schemes: (a) the SFC plus the fundamental frequency RSC $G_1(z)$, (b) the SFC plus the MRSC, and (c) the SFC plus the CRC

4.1.4 Conclusion

The above application cases illustrate that the periodic control provides a simple and very efficient approach for engineers to design high-performance control systems for power converters to produce high-quality sinusoidal signals with

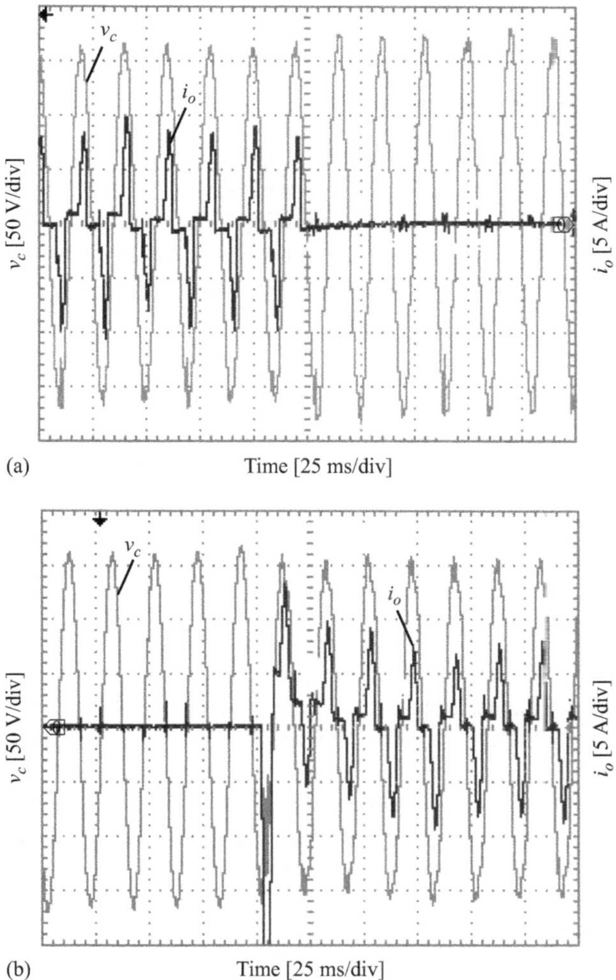

Figure 4.17 Transient response of the inverter system under rectifier load step changes using the SFC plus the MRSC: (a) from full rectifier load to no load and (b) from no load to full rectifier load

minimized distortions, fast dynamic response, and good robustness. The above also demonstrate that:

1. Phase-lead compensation, e.g., the linear phase-lead compensation for the RC schemes and zero-phase compensation for the RSC, enables the periodic control system to reinforce system stability and robustness with a wider range of control gains. A larger control gain at the interested harmonic frequencies will lead to a faster convergence rate.
2. SHC methods, such as the odd-order harmonic RC, the DMRC, and the MRSC, can be used to optimize the convergence rate of power harmonics mitigation by individually tuning the control gains for the interested harmonic frequencies.

4.2 PC of CVCF single-phase high-frequency link (HFL) inverters

4.2.1 Background

The growing requirements for high power density, compact size, and light weight, without compromising efficiency, cost, and reliability, stimulate the development of high-frequency-link (HFL) power converters [28–36]. As shown in Figure 4.18(a), a conventional HFL converter has three power conversion stages: a DC to high-frequency (HF) AC converter with an output galvanic HF transformer, a HF diode rectifier, and a PWM inverter. However, several drawbacks for this configuration have been identified:

- Three-stage conversion may reduce the overall efficiency of the converter;
- The HF diode rectifier only allows unidirectional power flow, and thus it imposes restrictions on the applications of HFL inverters;
- Bulky DC-link capacitors may reduce the reliability, as the capacitors are one of the major life-limiting components in power electronics converters; and
- Two independent control schemes should be applied to the DC/HFAC converter and the DC/AC converter, respectively, thus increasing the control cost and complexity.

Hence, many bidirectional HFL inverters have been introduced [31–33], in order to address the above issues of the conventional HFL inverters. A two-stage HFL single-phase inverter shown in Figure 4.18(b) is a representation, which comprises a DC/HFAC converter at the primary side of the galvanic HF transformer, and a HFAC/DC converter at the secondary side of the galvanic HF transformer. It allows bidirectional power flow between the DC side and AC side, removes the bulky DC-link capacitors, operates in zero voltage switching (ZVS) mode, and requires only one control scheme for the entire inverter. Furthermore, the HFL inverter shown in Figure 4.18(b) is equivalent to a conventional PWM

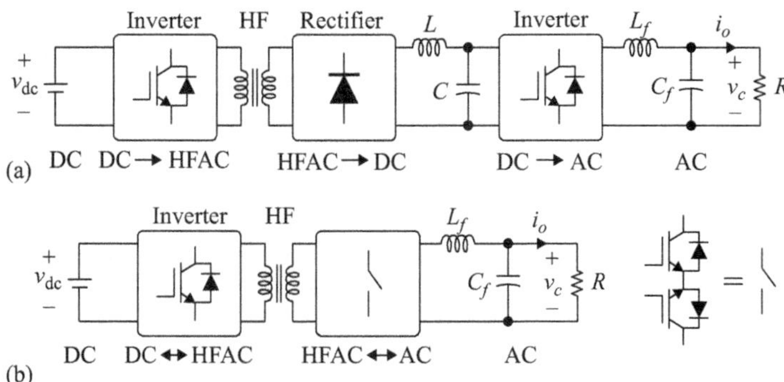

Figure 4.18 Configurations of the high-frequency link inverter systems: (a) unidirectional power flow and (b) bidirectional power flow

inverter [28,29] with a line frequency transformer. Therefore, it offers a high power density, compact size, light weight, high efficiency, high reliability, and low cost converter solution to extensive applications, such as renewable energy systems, grid-connection applications, naval and aerospace power systems, vehicular and transportation applications, high-power telecommunication applications, and so on.

In this section, a two-stage HFL inverter topology is introduced. Based on the equivalent PWM inverter circuit model, a plug-in RC scheme is employed to the two-stage HFL inverter in such a way to generate high-quality CVCF output voltages. A plug-in repetitive-controlled HFL single-phase inverter is built up to validate the proposed circuit model and the RC scheme. Experimental results are provided to demonstrate the effectiveness of the RC scheme in terms of harmonic suppression.

4.2.2 Modeling and control of single-phase HFL inverters

Figure 4.19 shows the circuit diagram of the two-stage HFL inverter, which mainly consists of a HF galvanic transformer with the turn ratio being n, a primary-side H-bridge converter (S_1–S_4), and a secondary-side AC/AC converter (S_5–S_8). Owing to the leakage inductance of the HF transformer, the commutation of the secondary-side AC/AC converter will lead to severe voltage overshoots and ringing across the secondary-side converter, which may induce failures to the power switches and also cause output voltage distortions. As shown in Figure 4.19, an active voltage clamper circuit as highlighted is added to suppress voltage ringing and overshoots [32–36].

The two-stage HFL inverter shown in Figure 4.19 seems to be more complicated than a conventional PWM inverter [28,29]. The modulation method for this HFL inverter has been discussed in [29] and it is also shown in Figure 4.20. Note that the modulation for the voltage clamper is not depicted in Figure 4.20, as the voltage clamper is only used for suppressing voltage ringing and overshoots. It is

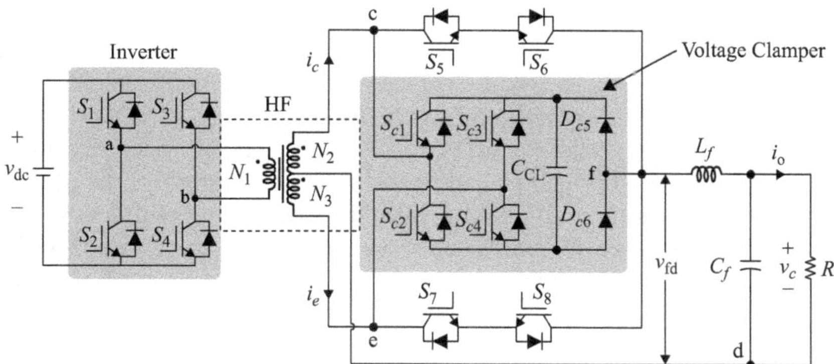

Figure 4.19 Bidirectional HFL single-phase inverter with a voltage clamper circuit to prevent the voltage from ringing and overshoots due to the presence of leakage inductance in the HF transformer

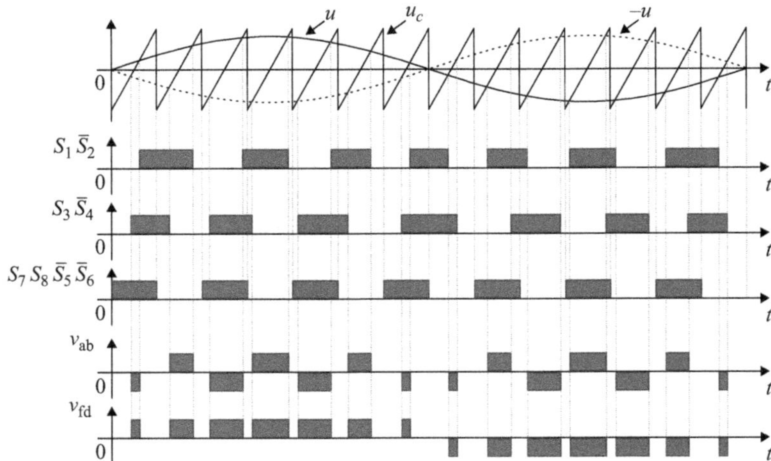

Figure 4.20 Modulation strategy for the single-phase HFL inverter system without the voltage clamper circuit

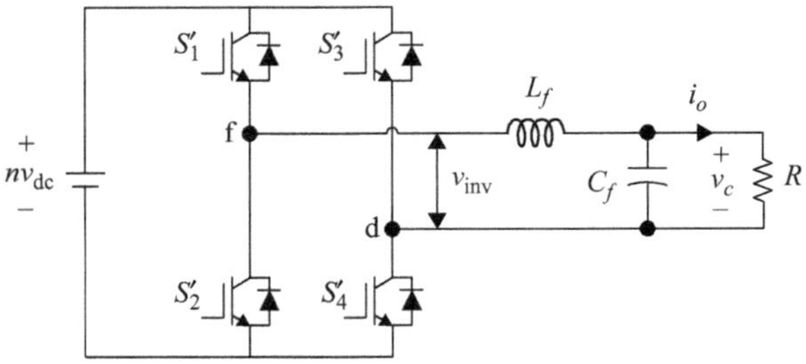

Figure 4.21 Equivalent circuit of the HFL inverter shown in Figure 4.19 without the voltage clamper, where $v_{inv} = v_{fd}$

indicated by [29] and Figure 4.20 that the modulation process from the signal u to the output pulse train v_{fd} is actually equivalent to a conventional unipolar PWM. That is to say, the HFL inverter in Figure 4.19 can be treated as a PWM inverter. Figure 4.21 thus presents the equivalent PWM inverter circuit for the HFL inverter shown in Figure 4.19. Owing to the turn ratio n for the transformer, the equivalent DC source voltage is obtained as nv_{dc}, where the turn ratio $n = N_2/N_1 = N_3/N_1$.

The equivalent system is similar to the one shown in Figure 4.1 with a resistive load R. Thus, the dynamics of the equivalent PWM inverter shown in Figure 4.21 can be described by the sampled-data model given in (4.2), where the inverter output voltage $v_{\mathrm{inv}}(k)$ can take one of the three values: $+nv_{\mathrm{dc}}$, $-nv_{\mathrm{dc}}$, or zero. Accordingly, an SFC scheme is formulated. For convenience, the SFC is given again as:

$$u(k) = -k_1 \frac{v_c(k)}{nv_{\mathrm{dc}}} - k_2 \frac{\dot{v}_c(k)}{nv_{\mathrm{dc}}} + k_{\mathrm{ref}} \frac{v_c^*(k)}{nv_{\mathrm{dc}}} \tag{4.18}$$

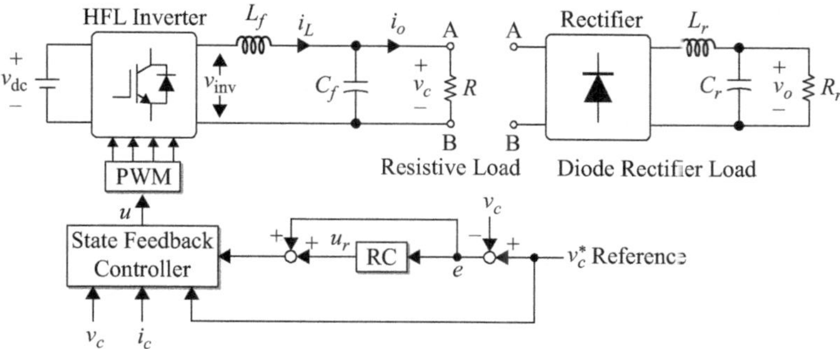

Figure 4.22 Hardware schematic and the overall control scheme for the HFL inverter system

where $u(k) = v_{\text{inv}}(k)/(nv_{\text{dc}}) \in [-1, 1]$. In the same way, the poles of the SFC-controlled HFL inverter system can be arbitrarily assigned by tuning the feedback gains k_1 and k_2.

However, the tracking accuracy of an SFC-based inverter is sensitive to parameter variations and nonlinear loads. Since the tracking errors are periodic signals, as shown in Figure 4.22, a plug-in RC scheme is employed to cancel the periodic errors and thus guarantees an output voltage of high quality in all operating conditions [5,6]. The plug-in CRC can be expressed as:

$$G_{\text{rc}} = \frac{k_{\text{rc}} z^{-N} Q(z)}{1 - z^{-N} Q(z)} G_f(z) \tag{4.19}$$

where k_{rc} is the control gain, $N = f_s/f$ with f being the reference input frequency and f_s being the sampling frequency, a low-pass filter (LPF) $Q(z)$ is included, and a linear phase-lead compensation filter $G_f(z) = z^p$ is also adopted with p being the phase-lead steps [17,18]. Both filters are employed to enhance the controller stability margin. The LPF is designed as $Q(z) = 0.25z + 0.5 + 0.25z^{-1}$.

4.2.3 Experimental validation

Referring to the HFL inverter shown in Figure 4.22, whose parameters are listed in Table 4.6, the SFC in (4.18) is designed as:

$$u(k) = -0.025 \frac{v_c(k)}{nv_{\text{dc}}} - 4.28 \times 10^{-6} \frac{\dot{v}_c(k)}{nv_{\text{dc}}} + 7.2 \frac{v_c^*(k)}{nv_{\text{dc}}} \tag{4.20}$$

According to (4.5), the closed-loop transfer function for the SFC inverter with the nominal parameters can be written as:

$$H(z) = \frac{0.1722z + 0.166}{z^2 - 1.843z + 0.8912} \tag{4.21}$$

In the experiments, the RC gain $k_{\text{rc}} = 0.2$, and $G_f(z) = z^7$ are employed.

Table 4.6 Parameters of the high-frequency link inverter system

Parameter	Nominal value	Unit
Output filter inductor L_f	2.2	mH
Output filter capacitor C_f	23.75	μF
Leakage inductance of transformer	3.16	μH
Transformer turns ratio n	3	–
Input DC bus voltage v_{dc}	60	V
Reference output voltage v_c (peak)	100	V
Clamp capacitor C_{CL}	10.5	μF
Rectifier load inductor L_r	3	mH
Rectifier load capacitor C_r	1100	μF
Rectifier load resistor R_r	50	Ω
Resistive load R	25	Ω
Sampling/Switching frequency f_s	20	kHz

First, the HFL inverter system is tested under different loading conditions, where the RC compensator is not enabled. The experimental results are shown in Figure 4.23, where it can be observed that the output voltage v_c has a low THD of 1.2% with a linear resistive load. However, when the HFL inverter system is connected to the diode rectifier load, the output voltage will be distorted. As shown in Figure 4.23(b), a high THD of 5% for the output voltage has been obtained by the HFL inverter system. Additionally, it is also found in Figure 4.23 that the major harmonics of the output voltage are the low odd-order harmonics (i.e., the 3rd, 5th, 7th, 9th, and 11th). It should be noted that the MRSC scheme can also be applied to cancel out these harmonics, which in return leads to an improved voltage quality.

In order to reduce the harmonics of a wide range of frequencies, the RC compensator is plugged into the SFC system, as shown in Figure 4.22. Experiments have been carried under the same load conditions, and the results are presented in Figure 4.24. It can be seen that the HFL inverter system using the SFC plus a plug-in RC compensator can produce a very low THD of 0.5% for the output voltage v_c under the linear resistive load. Even when the system is feeding the diode rectifier load, a relatively low THD of 1.9% has been achieved in contrast to the high THD in Figure 4.23(b). The above experimental results have verified the effectiveness of the RC scheme for the HFL inverter in terms of harmonic mitigation.

As the dynamic response performance is very important for power converter systems, the HFL inverter system shown in Figure 4.22 is then further tested under load changes (from no load to full load). Figure 4.25 shows the startup experimental results. It can be observed in Figure 4.25 that the HFL inverter with the resistive load takes around 50 ms (i.e., 2~3 cycles of the reference 50 Hz signal) to come to the steady-state, when the SFC with the plug-in RC is activated. In addition, in this case, the HFL inverter can reduce the peak of the tracking error $v_c^* - v_c$ from around 12 V to 0.7 V, as it is shown in Figure 4.25(a). Moreover, it takes the

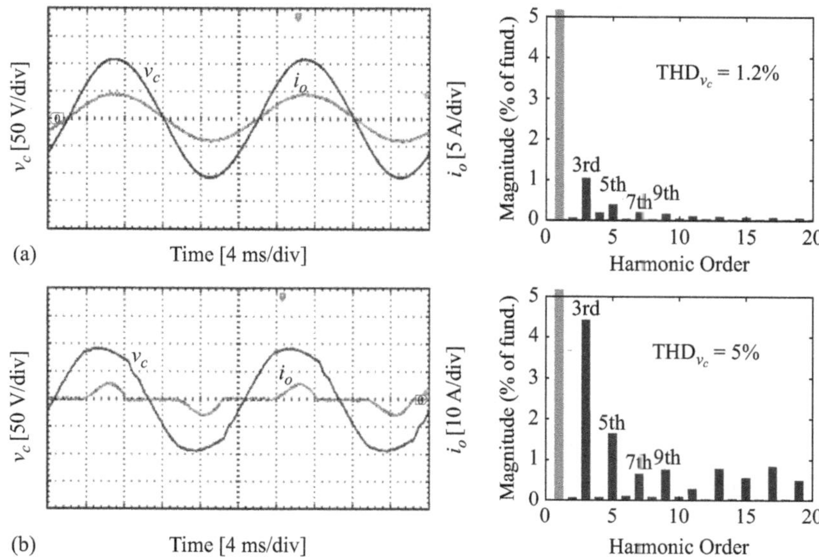

Figure 4.23 Steady-state responses of the high-frequency link inverter system only using the SFC under different loads (left: output voltage v_c and current i_o; right: harmonic distribution of the output voltage v_c): (a) a resistive load and (b) a diode rectifier load

inverter system with the diode rectifier load longer than 80~100 ms (i.e., 4~5 cycles of the reference 50-Hz signal) to settle the dynamic response due to the high non-linearity of the load. Additionally, the steady-state tracking error (peak) under the diode rectifier load shown in Figure 4.25(b) is reduced from 45 V to 4 V, while the steady-state tracking error (peak) under the resistor load shown in Figure 4.25(a) is reduced from 12 V to 1 V. Apparently, offering very high tracking accuracy and very fast dynamic response, the SFC plus a plug-in RC scheme enables the HFL inverter to significantly enhance the performance. It is worthy pointing out that the control gain can be further tuned to reduce the distortions even with nonlinear loads. Alternatively, advanced PC schemes discussed in Chapter 3 can be employed to replace the CRC in the HFL inverter system shown in Figure 4.22.

4.2.4 Conclusion

In this section, the CRC scheme is applied to a HFL single-phase inverter, which is equivalent to a conventional PWM inverter. Based on the equivalent PWM inverter circuit of the HFL inverter, the SFC plus a plug-in RC scheme has been developed so that the HFL inverter system can produce CVCF sinusoidal voltages of high quality under various load conditions. The results of the experiments performed on a 20 kHz HFL inverter test rig have demonstrated the effectiveness of the plug-in RC scheme, which enables the high-performance control of the HFL inverter.

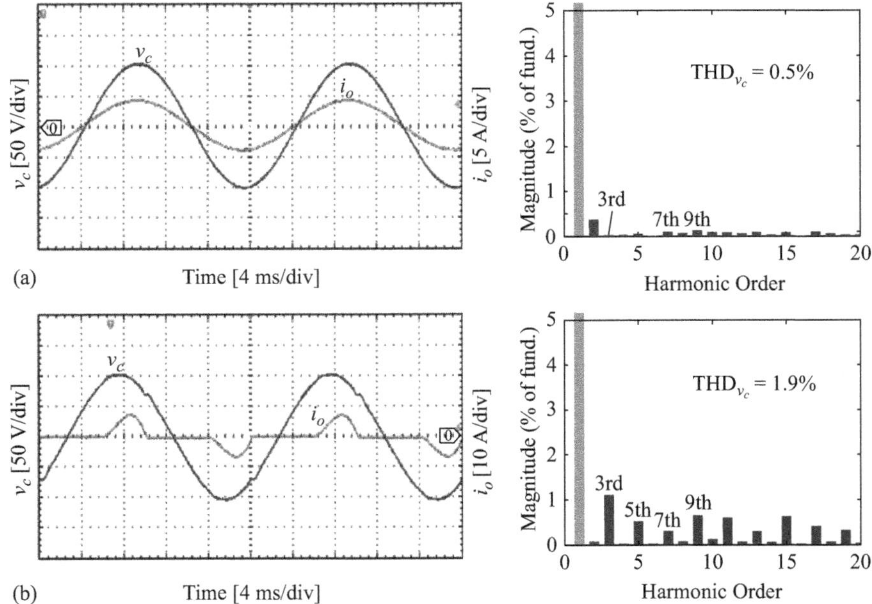

Figure 4.24 Steady-state responses of the high-frequency link inverter system using the SFC plus an RC harmonic compensator under different loads (left: output voltage v_c and current i_o; right: harmonic distribution of the output voltage v_c): (a) a resistive load and (b) a diode rectifier load

In addition, the case-study candidate – the introduced HFL inverter offers bidirectional power flows, high efficiency, high power density, high reliability, and high-performance power conversion solution to extensive applications, such as renewable energy systems, power-grid applications, naval and aerospace power systems, vehicular and transportation applications, high-power telecommunication applications, and so on. The plug-in RC scheme can underpin the applications of the high-performance HFL inverter systems.

4.3 PC of CVCF three-phase PWM inverters

4.3.1 Background

It is well known that, for multiphase PWM DC–AC converters, the harmonic components of their voltages or currents usually concentrate on particular frequencies [23–27,37]. For example, for single-phase PWM inverters, $(4k \pm 1)$ $(k = 1, 2, 3 \ldots)$-order harmonics (i.e., odd-harmonics) are the dominant harmonic distortions [10–15]. For three-phase inverters, $(6k \pm 1)$ $(k = 1, 2, 3, \ldots)$-order harmonics are mainly seen in the harmonic spectrum [26]. For n-phase $(n = 12, 18, 24, \ldots)$ thyristor converters-based power systems, $(nk \pm 1)$-order harmonics are the

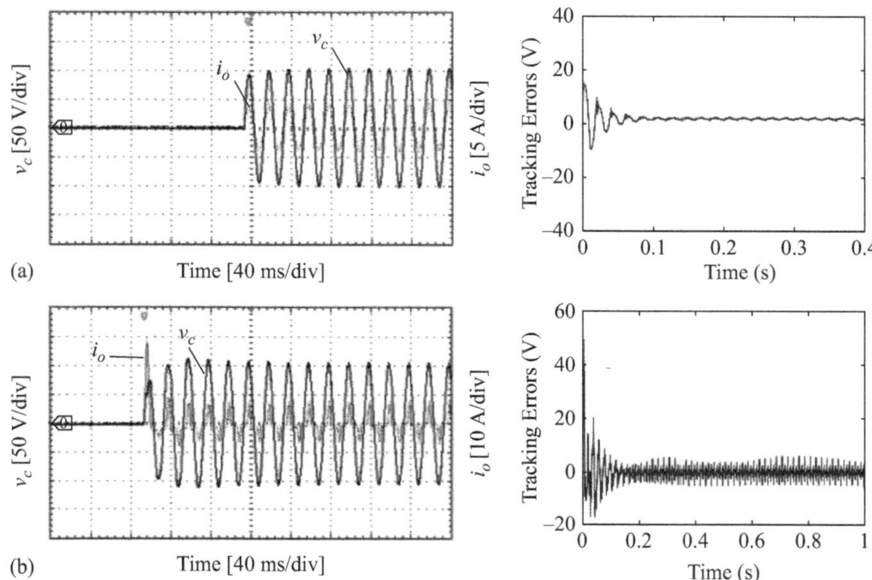

Figure 4.25 Start-up responses of the high-frequency link inverter system using the SFC plus an RC harmonic compensator under different loads (left: output voltage v_c and current i_o; right: tracking errors of the output voltage $v_c^* - v_c$): (a) a resistive load and (b) a diode rectifier load

major. CVCF three-phase PWM inverters are widely used in industry applications, such as uninterruptable power supply, emergence power, and so on. Thus, the CVCF three-phase PWM inverters will introduce ($6k \pm 1$)-order harmonics, which have to be lowered to certain levels by employing advanced control schemes.

For instance, based on the IMP [1], the CRC can be adopted for three-phase PWM inverters [2–7]. Notably, the CRC is equivalent to a parallel combination of a proportional gain, an integrator, and infinite RSC controllers [25,26]. Since the control gains for all RSC controllers in the CRC are identical, as discussed in Chapter 2, it is impossible for the CRC to compensate power harmonics with optimal dynamic responses by tuning these control gains independently. Addressing this issue, the PSRC [24,25] scheme is introduced in Chapter 3, which can be employed in three-phase converters to achieve both high tracking accuracy and fast dynamics responses. Moreover the digital PSRC occupies the same or less data memory as what the digital CRC does. For three-phase PWM inverters, the PSRC scheme that employs a parallel combination of $6k + i$ ($i = 0$, 1, 2, 3, 4, 5)-order selective harmonic controllers with independent control gains, can offer an optimal control solution to the featured power harmonics mitigation.

In this section, a plug-in PSRC scheme is thus developed for three-phase PWM inverters. Experimental tests are also carried out to verify the effectiveness of the PSRC scheme for three-phase PWM inverters.

4.3.2 Modeling and control of CVCF three-phase PWM inverters

A three-phase PWM inverter supplying a resistive or diode rectifier load is shown in Figure 4.26. The dynamics of the three-phase PWM inverter with a resistive load can be described as [5,6]:

$$\begin{bmatrix} \mathbf{W}_1 & \mathbf{Z} \\ \mathbf{Z} & \mathbf{W}_2 \end{bmatrix}\begin{bmatrix} \dot{v}_{ab} \\ \dot{v}_{bc} \\ \dot{v}_{ca} \\ \dot{i}_A \\ \dot{i}_B \\ \dot{i}_C \end{bmatrix} = \begin{bmatrix} \mathbf{A}_1 & \mathbf{A}_2 \\ \mathbf{A}_3 & \mathbf{Z} \end{bmatrix}\begin{bmatrix} v_{ab} \\ v_{bc} \\ v_{ca} \\ i_A \\ i_B \\ i_C \end{bmatrix} + \begin{bmatrix} \mathbf{Z} \\ \mathbf{B}_1 \end{bmatrix}\begin{bmatrix} v_{AB} \\ v_{BC} \\ v_{CA} \end{bmatrix}$$

with

$$\mathbf{A}_1 = \begin{bmatrix} -\frac{1}{RC_f} & 0 & \frac{1}{RC_f} \\ \frac{1}{RC_f} & -\frac{1}{RC_f} & 0 \\ 0 & \frac{1}{RC_f} & -\frac{1}{RC_f} \end{bmatrix}, \quad \mathbf{A}_2 = \begin{bmatrix} \frac{1}{C_f} & 0 & 0 \\ 0 & \frac{1}{C_f} & 0 \\ 0 & 0 & \frac{1}{C_f} \end{bmatrix},$$

$$\mathbf{A}_3 = \begin{bmatrix} -\frac{1}{L_f} & 0 & 0 \\ 0 & -\frac{1}{L_f} & 0 \\ 0 & 0 & -\frac{1}{L_f} \end{bmatrix}$$

$$\mathbf{B}_1 = \begin{bmatrix} \frac{1}{L_f} & 0 & 0 \\ 0 & \frac{1}{L_f} & 0 \\ 0 & 0 & \frac{1}{L_f} \end{bmatrix}, \quad \mathbf{W}_1 = \begin{bmatrix} 1 & 0 & -1 \\ -1 & 1 & 0 \\ 0 & -1 & 1 \end{bmatrix},$$

$$\mathbf{W}_2 = \begin{bmatrix} 1 & -1 & 0 \\ 0 & 1 & -1 \\ -1 & 0 & 1 \end{bmatrix}, \text{ and } \mathbf{Z} = \begin{bmatrix} 0 & 0 & 0 \\ 0 & 0 & 0 \\ 0 & 0 & 0 \end{bmatrix}. \tag{4.22}$$

where the output line-to-line voltages v_{ab}, v_{bc}, and v_{ca} and phase currents i_a, i_b, and i_c are state variables, v_{dc} is the nominal DC-bus voltage, $v_{AB} = v_{dc}u_{AB}$, $v_{BC} = v_{dc}u_{BC}$, and $v_{CA} = v_{dc}u_{CA}$ are PWM output voltages of the inverter with u_{AB}, u_{BC}, and u_{CA} being the normalized control output of the controller, and L_f, C_f, and R are the inductor, capacitor, and resistor of the load.

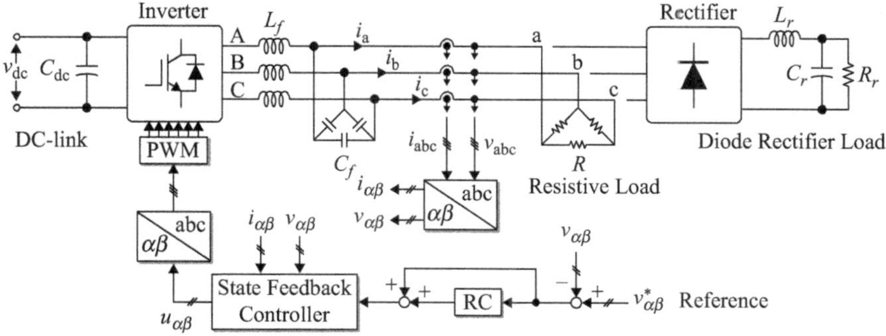

Figure 4.26 A three-phase PWM inverter using the state feedback control plus a plug-in RC scheme

Equation (4.22) can be transformed into a two-phase system in the stationary reference system (i.e., the $\alpha\beta$ reference frame). In that case, the three-phase system can be decoupled into two identical independent single-phase systems in the $\alpha\beta$ reference frame as:

$$\begin{bmatrix} \dot{v}_i \\ \ddot{v}_i \end{bmatrix} = \begin{bmatrix} -\frac{1}{RC_f} & \frac{1}{3C_f} \\ -\frac{1}{L_f} & 0 \end{bmatrix} \begin{bmatrix} v_i \\ \dot{v}_i \end{bmatrix} + \begin{bmatrix} 0 \\ \frac{1}{L_f} \end{bmatrix} u_i \tag{4.23}$$

where $i = \alpha$ or β indicating the phase of the two-phase system. Then, the sampled-data form of (4.23) can be derived as [6]:

$$\begin{bmatrix} \dot{v}_i(k+1) \\ \ddot{v}_i(k+1) \end{bmatrix} = \begin{bmatrix} \varphi_{11} & \varphi_{12} \\ \varphi_{21} & \varphi_{22} \end{bmatrix} \begin{bmatrix} v_i(k) \\ \dot{v}_i(k) \end{bmatrix} + \begin{bmatrix} v_{dc}g_1 \\ v_{dc}g_2 \end{bmatrix} u_i(k)$$

where

$$\varphi_{11} = 1 - \frac{T_s}{RC_f} + \frac{T_s^2}{2R^2C_f^2} - \frac{T_s^2}{6L_fC_f}, \quad \varphi_{12} = \frac{T_s}{3C_f} - \frac{T_s^2}{6RC_f^2}, \quad \varphi_{21} = -\frac{T_s}{L_f} + \frac{T_s^2}{2RL_fC_f},$$
$$\varphi_{22} = 1 - \frac{T_s^2}{6L_fC_f}, \quad g_1 = \frac{T_s^2}{6L_fC_f}, \quad g_2 = \frac{T_s}{L_f}. \tag{4.24}$$

and T_s is the sampling period.

The control objective of the three-phase PWM inverter is to force v_{ab}, v_{bc}, and v_{ca} (i.e., v_α and v_β) to exactly track their reference line voltages v_{ab}^*, v_{bc}^*, v_{ca}^* (i.e., v_α^* and v_β^*) under various loads. As shown in Figure 4.26, the control for the inverter consists of two parts: an SFC and a plug-in RC compensator. Similarly, the SFC can be written as:

$$u_i(k) = -\left(k_1 \frac{v_i(k)}{v_{dc}} + k_2 \frac{\dot{v}_i(k)}{v_{dc}}\right) + k_{ref} \frac{v_i^*(k)}{v_{dc}} \tag{4.25}$$

where $i = \alpha$ or β, and v_i^* is the reference output voltage at the α or β axis.

Clearly, the poles of the SFC system can be arbitrarily assigned by adjusting the feedback gains k_1 and k_2. The transfer function from v_i^* to v_i for the SFC system with nominal parameters can be derived as:

$$H(z) = \frac{m_1 z + m_2}{z^2 + p_1 z + p_2} \tag{4.26}$$

where $p_1 = -(\varphi_{11} - g_1 k_1) - (\varphi_{22} - g_2 k_2)$, $p_2 = (\varphi_{11} - g_1 k_1)(\varphi_{22} - g_2 k_2) - (\varphi_{12} - g_1 k_2)(\varphi_{21} - g_2 k_1)$, $m_1 = g_1 k_{\text{ref}}$, $m_2 = (\varphi_{12} - g_1 k_2) g_2 k_{\text{ref}} - (\varphi_{22} - g_2 k_2) g_1 k_{\text{ref}}$.

In order to accurately, rapidly, and robustly suppress the periodic disturbances, a PSRC $G_{\text{psrc}}(z)$ can be plugged into the SFC system to replace the CRC shown in Figure 4.22 [24,25]. The PSRC can be given as:

$$G_{\text{psrc}}(z) = G_f(z) \sum_{i=0}^{n-1} \left[k_i \frac{e^{j(2\pi i/n)} \cdot z^{-N/n} \cdot Q(z)}{1 - e^{j(2\pi i/n)} \cdot z^{-N/n} \cdot Q(z)} \right] \tag{4.27}$$

where $N = f_s / f_0 \in \mathbb{N}$, $f_0 = 1/T_0$ is the fundamental frequency of references v_{ab}^*, v_{bc}^*, v_{ca}^* (i.e., v_α^* and v_β^*), f_s is the sampling frequency, k_i is the control gain, an LPF $Q(z) = a_1 z + a_0 + a_1 z^{-1}$ with $2a_1 + a_0 = 1$ and $a_0 \geq 0$, $a_1 \geq 0$ is adopted to enhance the system robustness and stability margin, a linear phase-lead filter $G_f(z) = z^p$ is used to compensate the delay of the SFC system with the actual value of p being determined by experiments [17,18]. In order to evaluate the PSRC scheme, the CRC in (2.22) is also developed for comparison.

4.3.3 Experimental validation

In order to verify the control performance, a dSPACE DS1104-based three-phase inverter experiment system is built up referring to Figure 4.26. Figure 4.27 shows the experimental system setup, whose parameters are listed in Table 4.7.

The SFC is chosen as:

$$u_i(k) = -\left(312.12 \frac{v_i(k)}{v_{\text{dc}}} + 355 \frac{\dot{v}_i(k)}{v_{\text{dc}}}\right) + 1080 \frac{v_i^*(k)}{v_{\text{dc}}} \tag{4.28}$$

The closed-loop transfer function can be written as:

$$H(z) = \frac{10z + 9.815}{z^2 + 12.77z - 7.699} \tag{4.29}$$

Determined by experiments, a phase-lead compensation filter $G_f(z) = z^8$ is employed in the CRC and the PSRC to compensate the phase lag and un-modeled delay of the actual $H(z)$.

First, only the SFC has been adopted to control the three-phase CVCF inverter. Figure 4.28 shows the steady-state performance of the three-phase inverter under the three-phase diode rectifier load. It is indicated in Figure 4.28 that the SFC cannot produce good quality output voltage under nonlinear loads. Specifically, the THD of the output line–line voltages is around 6.5%. Furthermore, it can be seen in

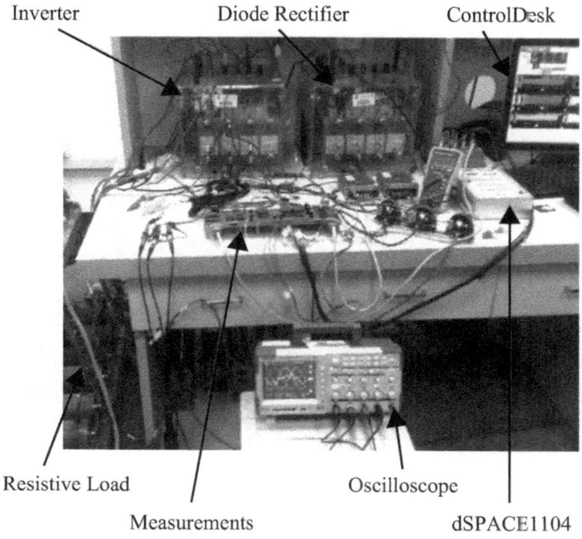

Figure 4.27 Experimental setup for the three-phase CVCF inverter system

Table 4.7 Parameters of the three-phase CVCF inverter system shown in Figure 4.26

Parameters	Nominal value	Unit
DC-link voltage v_{dc}	400	V
Inductor filter L_f	5	mH
Capacitor filter C_f	100	μF
Resistive load R	60	Ω
Rectifier inductor L_r	5	mH
Rectifier capacitor C_r	1100	μF
Rectifier resistor R_r	60	Ω
Switching frequency	6	kHz
Sampling frequency	6	kHz
Output voltage frequency	50	Hz
Output line voltage amplitude (peak)	270	V

Figure 4.28 that the low-order harmonics with the order being up to 15th-order dominate the THD of the output voltage. In order to evaluate the harmonic distribution of the output voltage in details, a harmonic ratio $h_r(j)$ is defined as:

$$h_r(j) = \frac{\left(\sum_{i=1}^{j} M_i\right)}{\left(\sum_{i=1}^{199} M_i\right)} \times 100\% \tag{4.30}$$

in which M_i is the magnitude of the ith-order harmonic. The harmonic ratio $h_r(j)$ for the current spectrum shown in Figure 4.28(b) can be represented in Figure 4.29(a),

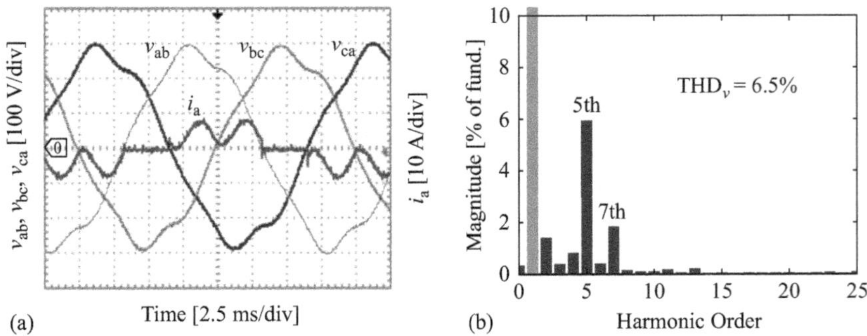

Figure 4.28 Steady-state performance of the three-phase CVCF inverter with only the SFC under a diode rectifier load: (a) output line voltage and line current i_a and (b) harmonic spectrum of the line voltage v_{ab}

which indicates that about 90% of the harmonics is located within the range of [0, 1100] Hz. In addition, Figure 4.29(b) shows the bandwidths of $Q(z) = 0.25z + 0.5 + 0.25z^{-1}$ for the CRC and the actual equivalent LPF $Q^6(z)$ with $Q(z) = 0.05z + 0.9 + 0.05z^{-1}$ for the PSRC [24,25] are about 1100 Hz. Thus, the LPFs allow the RC schemes to eliminate most of the harmonics in the output line voltages. More important, a good trade-off between tracking accuracy and robustness can be achieved. Furthermore, the harmonic ratio of all $(6k \pm 1)$-order harmonics to the total harmonics of the output line voltage shown in Figure 4.29(a) is about 63%. That is to say, the $(6k \pm 1)$-order harmonics dominate the THD of the output line voltages.

Therefore, the PC schemes are plugged in the control system respectively. Namely, the CRC and the PSRC given in (4.27) with $n = 6$ are adopted to enhance the tracking performance of the three-phase CVCF inverter system. According to the harmonics distribution in Figure 4.28(b), if the control gains for the $(6k \pm 1)$-order harmonics are larger than those for other harmonics, i.e. $(k_1 + k_5) > (k_0 + k_2 + k_3 + k_4)$, the error convergence rate can be improved. For comparison, in the following experimental tests, $k_{rc} = 0.6$ is designed for the CRC; $k_0 + k_1 + k_2 + k_3 + k_4 + k_5 = k_{rc}$ with $k_0 = k_2 = k_3 = k_4 = 0.05$, $k_1 = k_5 = 0.40$ are selected for the PSRC.

Figure 4.30 gives the steady-state responses of the three-phase CVCF inverter system using the CRC and the PSRC schemes, where the same rectifier load has been connected. It can be seen in Figure 4.30 that both PC schemes (i.e., the CRC and the PSRC) can achieve perfect tracking of the reference voltages and yield very low THDs of 1.4% and 1.3%, respectively, in contrast to the THD of 6.5% shown in Figure 4.28. Furthermore, for comparison, the tracking errors, the THD levels, and the convergence rates of the inverter using the two PCs are listed in Table 4.8. The data in Table 4.8 indicate that the PSRC achieves almost the same tracking accuracy as what the CRC does, but its error convergence rate is more than two times faster (up to three times). In summary, a PSRC with appropriate control gains and LPFs can achieve almost the same highly accurate tracking of periodic voltages as what the CRC does. Moreover, compared with the CRC, the PSRC enables to

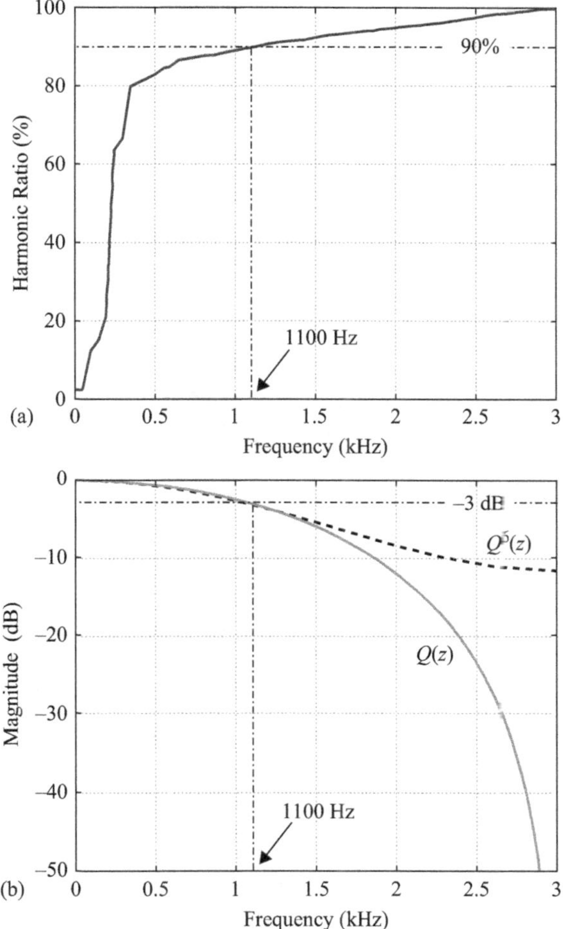

Figure 4.29 Harmonic analysis of the SFC system: (a) harmonic ratio $h_r(j)$ and (b) magnitude response of the LPFs

flexibly improve the convergence rate without any compromise in accuracy, while requiring the same or less memory space.

Next, the dynamic performance of the inverter using the two PC schemes has been studied. Figure 4.31 shows the dynamic responses of the CRC- and the PSRC-controlled inverter under step-resistive load changes between full load and no load. It can be seen in Figure 4.31 that it takes about five cycles (i.e., 100 ms) for the output voltages of the PSRC-controlled inverter to recover its steady-state settling status after slight variations under step load changes, while the recovery time of the CRC system is about eight cycles. It means that, both the PSRC- and the CRC-controlled PWM inverters are robust to load changes, and the PSRC leads to faster dynamic responses than what the CRC does. Moreover, it is noted that the plug-in PCs and the SFC are complementary: the SFC offers fast dynamic response and

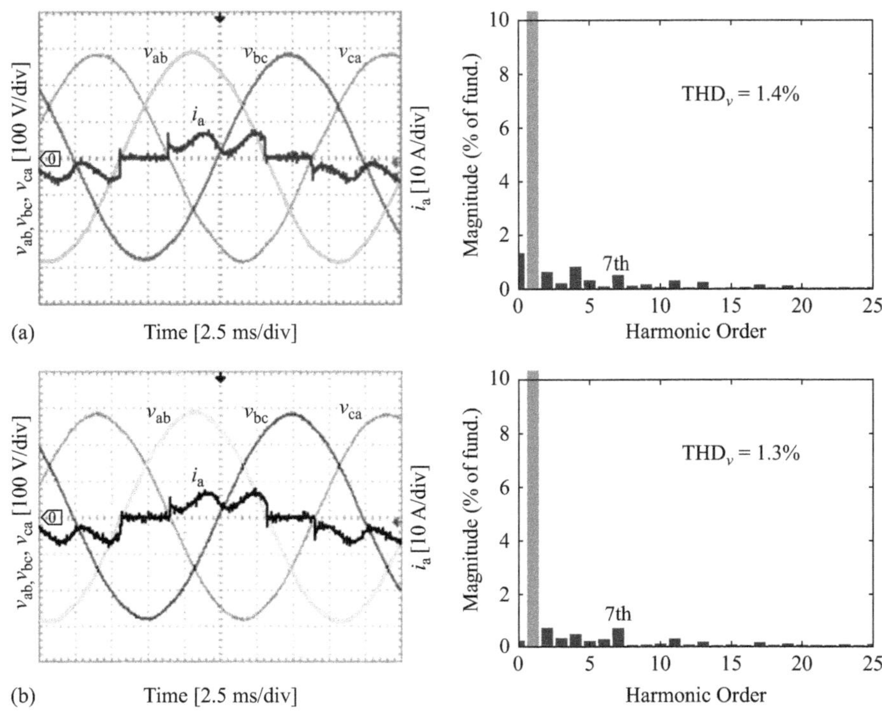

Figure 4.30 Steady-state responses of the three-phase CVCF inverter under the rectifier load using the SFC plus different PC controllers (left: output line voltages v_{ab}, v_{bc}, v_{ca} and phase current i_a; right: harmonic distribution of the output line voltage v_{ab}): (a) the CRC and (b) the PSRC

Table 4.8 Performance comparisons between the CRC and the PSRC

Control schemes	Tracking error (RMS)	THD	Convergence time
SFC	43 V	6.48%	–
SFC + CRC	2.5 V	1.41%	about 0.3 s
SFC + PSRC	3.2 V	1.33%	about 0.14 s

robustness to load changes with "coarse" tracking accuracy; the plug-in PCs ensure "fine" compensation of residual periodic errors of the SFC system but at a relatively slow convergence rate.

4.3.4 Conclusion

In this section, a digital PSRC scheme for multiphase CVCF PWM inverters has been demonstrated on a three-phase PWM inverter with resistive and diode rectifier loads. The PSRC enables the inverter systems to eliminate harmonic distortions in

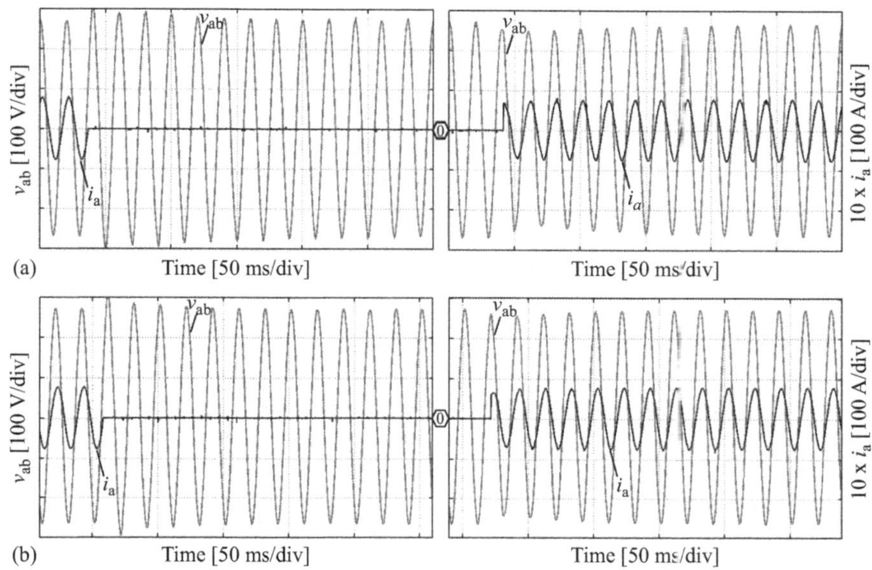

Figure 4.31 Transient responses of the three-phase CVCF inverter using the SFC plus the CRC and PSRC under resistive load step changes (left: from full load to no load; right: from no load to full load): (a) the CRC and (b) the PSRC

the output voltages efficiently and robustly. The PSRC scheme enables the CVCF PWM inverters to significantly improve the total error convergence rate without the loss of tracking accuracy. In addition, the PSRC occupies the same or less memory space as what the CRC does. Experimental results have confirmed the effectiveness and advantages of the PSRC in the control of three-phase CVCF inverter systems.

4.4 PC of grid-connected single-phase photovoltaic (PV) inverters

4.4.1 Background

An intense photovoltaic (PV) integration into the grid challenges the availability, the power quality, and the emerging reliability of the entire electrical power system [40–45]. As an important power quality index, the current distortion level (i.e., THD) should be lower than 5% in normal operation modes with rated output power for PV systems [44]. However, due to the solar irradiance variations, the output currents are usually varying and lower than the rated value. If large-scale PV systems feed such time-varying currents into the grid, the power quality will be challenged again. Furthermore, the PV systems are expected to be active players in future with reactive power compensation and low-voltage ride-through (LVRT) capability [46–49]. In these cases, although the requirements are met by an individual PV inverter under the rated condition, the overall feed-in currents may still violate the regulations. Advanced control solutions thus should be able to maintain the power quality in various operation modes [50–55] to avoid controller redesign.

The proportional–integral (PI) control in the *dq* synchronous reference frame (SRF) is the most popular controller for three-phase inverters. For single-phase PV systems, a proportional-resonant (PR) controller (a PC scheme) in the stationary frame is actually equivalent to a SRF PI controller [20,41]. Such an IMP-based PC scheme takes several advantages over the SRF PI, such as much less computational burden and complexity due to its lack of Park transformations, less sensitiveness to noise, and error in synchronization. Thus, it becomes a popular current regulator for grid-connected single-phase systems. In addition, odd-order harmonics (e.g., the third, fifth, and seventh) are dominant in the output current of single-phase PV inverters [57,58]. As a straightforward way, the MRSC can be plugged into PR controller [56–58] to effectively eliminate the low-order harmonic. However, the MRSC may incur heavy computational burden in compensating a large number of harmonics and design complexity in determining the gains and compensation phases for overmuch RSC components. Alternatively, the CRC can be employed to track or reject all harmonics below the Nyquist frequency in theory [59,60]. Nevertheless, it is a challenging issue to always ensure single-phase grid-connected PV inverters to produce high-quality current in various operational modes.

In this section, in accordance with the mechanism of the harmonic current injection from the single-phase grid-connected inverters in Chapter 1, a high-performance PC solution is developed for single-phase grid-connected PV inverter systems to always produce high-quality current. Experiments on a 1 kVA single-phase grid-connected system are performed to verify the effectiveness of the PC method.

4.4.2 Modeling and control of grid-connected single-phase PV inverters

A single-phase grid-connected PV system based on a single-phase *PQ* theory control method is shown in Figure 4.32, where an inductor-capacitor-inductor (*LCL*) filter is employed to improve the feed-in current quality. By adjusting the active power and reactive power references (P^*, Q^*), the power factor can be controlled flexibly.

Since the filter capacitor C_f in the *LCL* filter is mainly used to filter out the HF switching harmonics, the *LCL* filter shown in Figure 4.32 can be simply modeled as an *L* filter with $L = L_1 + L_2$. According to the discussion in Chapter 1 and Figure 1.8, the entire system can thus be modeled as:

$$\begin{cases} \dfrac{di_g}{dt} = \dfrac{1}{L}\left(v_{\text{inv}} - v_g\right) \\ \dfrac{dv_{\text{dc}}}{dt} = \dfrac{1}{C}\left(i_{\text{pv}} - i_{\text{dc}}\right) \end{cases} \tag{4.31}$$

Following the same analysis in Chapter 1, the DC-link voltage variations can then be approximated as:

$$\tilde{v}_{\text{dc}} \approx -\int \left(\frac{V_{\text{inv}}^1 I_g^1}{CV_{\text{dc}}} \cos(2\omega_0 t + \varphi - \varphi_1) \right) dt \tag{4.32}$$

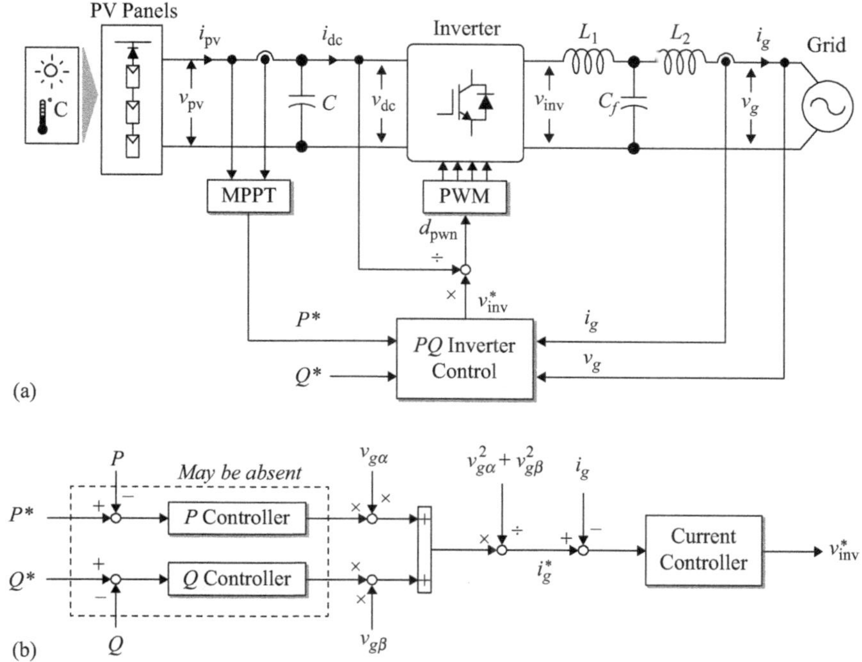

Figure 4.32 A single-phase single-stage grid-connected PV system based on the single-phase PQ theory (MPPT – maximum power point tracking): (a) hardware schematic and (b) inverter control diagram (P, Q are the injected active and reactive power, $v_{g\alpha} = v_g$, and $v_{g\beta}$ is a virtual voltage in quadrature with the grid voltage v_g)

which indicates that the DC-link voltage variations ($\tilde{v}_{dc}$ or $\tilde{v}_{PV}$) are proportional to the fundamental feed-in current amplitude I^1_g. The voltage variations of twice the fundamental grid frequency also vary with the power angle φ, which cannot be zero in the case of reactive power compensation and LVRT with reactive current injection (i.e., nonunity power factor operation). Thus, both the current levels (related to the intermittency of the injected PV power) and the power angle (related to the operation modes) will affect the current quality.

The impacts are further demonstrated on a double-stage system, where in the current controller a PR controller is employed. As shown in Figure 4.33, both the power factor and the current level have a significant impact on the THDs of the injected current. In practice, the current level will vary with the sites and/or the environmental conditions. Since the design and operation modes of single-phase PV systems will be influenced by the grid requirements, it is necessary to examine the clauses in existing grid requirements and prompt the development of upcoming grid codes. More important, it calls for advanced high-performance current control schemes to tolerate the impacts as much as possible. That is to say, the preferred control method is not sensitive to the changes of power factor and output current level.

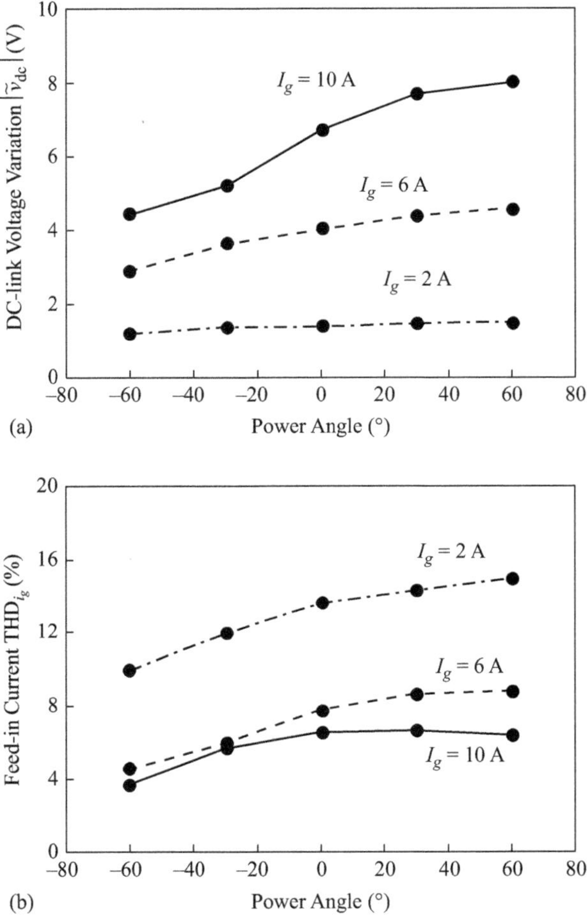

Figure 4.33 Impacts of power angles and current levels (injected power levels) on the DC-link voltage variations and the THD for a single-phase double-stage grid-connected inverter in normal operation, where only the proportional resonant (PR) current control method is adopted: (a) the relationship between the DC-link voltage ripples and the power angle and (b) the relationship between the feed-in current THD and the power angle

In light of the above considerations, a tailor-made hybrid IMP-based PC strategy (PR plus the CRC) shown in Figure 4.34 is developed for the PV inverter shown in Figure 4.32. In the hybrid control system, the PR current controller is used to achieve zero tracking error at the grid fundamental frequency and the CRC is adopted as the harmonic compensator to remove the harmonic components. The combination of the PR and the CRC can make an optimal trade-off between optimal control performance and practical realization complexity.

In the discrete-time domain, the CRC scheme can be expressed as:

$$G_{rc}(z) = \frac{k_{rc}z^{-N}Q(z)}{1 - z^{-N}Q(z)} G_f(z) \tag{4.33}$$

where $N = f_s/f_0$ with f_s being the sampling frequency and $f_0 = 1/T_0$ being the grid fundamental frequency, k_{rc} is the control gain, an LPF $Q(z) = a_1 z + a_0 + a_1 z^{-1}$ with $2a_1 + a_0 = 1$ and $a_0 \geq 0,\ a_1 \geq 0$ is employed to enhance the control performance, a linear phase-lead compensator $G_f(z) = z^p$ is used to compensate the delay of the PR-controlled PV inverter system, where the actual value of p is determined by experiments [17,18]. For the PR controller, using the pre-warping bilinear (Tustin) transformation, its digital form can be obtained as:

$$G_{PR}(z) = k_P + k_R \cdot \frac{\sin(\omega_0 T_s)}{2\omega_0} \cdot \frac{1 - z^{-2}}{1 - 2\cos(\omega_0 T_s)z^{-1} + z^{-2}} \tag{4.34}$$

in which $T_s = 1/f_s$ is the sampling period, k_P and k_R are the control gains, and ω_0 is the fundamental grid angular frequency. Thus, the tailor-made PC scheme can be implemented in digital controllers without much difficulty. In addition, the MRSC is also developed for the grid-connected system for comparison.

4.4.3 Experimental validation

In order to verify the above analysis and to evaluate the hybrid PC scheme, referring to Figure 4.32 and Figure 4.34, a single-phase grid-connected system is tested experimentally. System parameters are listed in Table 4.9. In this test, a commercial inverter is adopted, and it is connected to the grid through an isolation transformer. The control system is implemented in a dSPACE DS1103 system. An *LC* is employed in order to filter out the harmonics at high-switching frequencies, and the voltage across the filter capacitor is measured as the grid voltage. Together with the leakage inductance of the isolation transformer, an *LCL* filter is formed. For the control system shown in Figure 4.32, two PI controllers are employed as the *P* controller and the *Q* controller. The proportional and integral gains for the *P* controller are $k_{pp} = 1.2$, $k_{pi} = 52$, while for the *Q* controller are $k_{qp} = 1$, $k_{qi} = 50$.

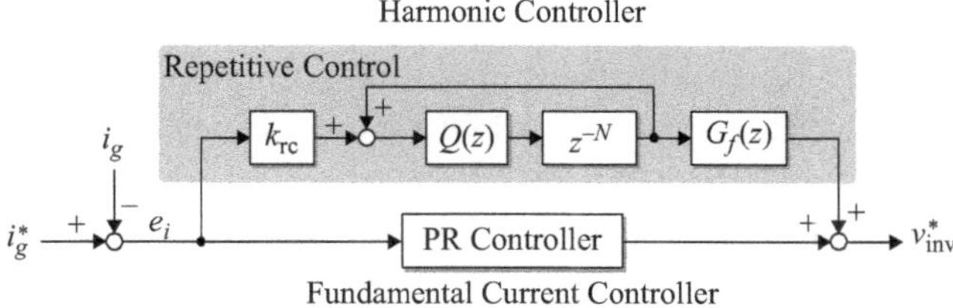

Figure 4.34 PR current controller plus a repetitive control-based harmonic controller for the single-phase grid-connected inverter system

Table 4.9 Parameters of the single-phase grid-connected inverter system

Parameter	Value	Unit
Grid voltage amplitude v_g	325	V
Grid frequency f_0	50	Hz
Inductor of the *LC* filter L_1	3.6	mH
Capacitor of the *LC* filter C_f	2.35	μF
Transformer leakage inductance l_2	4	mH
Sampling and switching frequency f_s, f_{sw}	10	kHz
Rated power	1	kVA

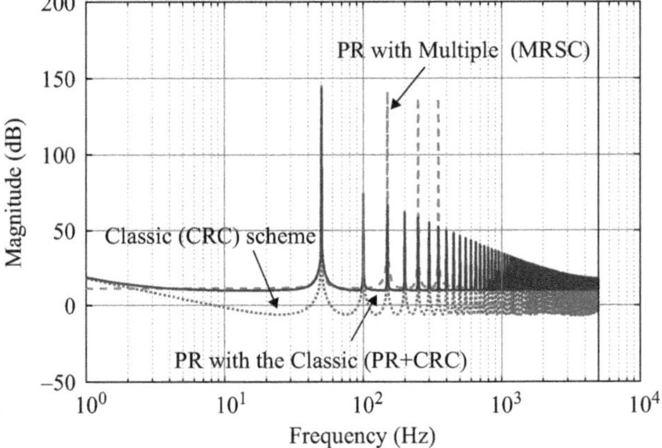

Figure 4.35 Frequency response of the current controllers (i.e., the CRC, the PR + CRC, and the PR + MRSC schemes)

With parameters in Table 4.9, the PR controller in (4.34) with $k_P = 22$, $k_R = 2{,}000$ is designed, which ensures that the closed-loop system is stable (only using PR controller). The plug-in CRC with $k_{rc} = 1.8$ and $Q(z) = 0.05z + 0.9 + 0.05z^{-1}$ is designed to compensate the harmonics. For comparison, a plug-in MRSC controller which includes the third-, fifth-, and seventh-order RSC components with gains being $k_{r3} = k_{r5} = 5000$, $k_{r7} = 7000$ is also developed. Figure 4.35 gives the frequency responses of the PR with the CRC and the PR with the MRSC, where it can be observed that the PR plus the CRC scheme can remove almost all the harmonics below the Nyquist frequency. In contrast, the PR plus the MRSC scheme offers better harmonic suppression capability at the selected harmonic frequencies (i.e., the third-, fifth-, and seventh-order harmonics).

The single-phase grid-connected PV inverter system is then tested, and the results are shown in Figure 4.36, where only the PR current controller is adopted. As shown in Figure 4.36, the experimental results indicate that the THD level of the feed-in current i_g at unity power factor is 8.5%, even the system is running at its rated power level. The resultant THD level already exceeds the limit of 5% in the

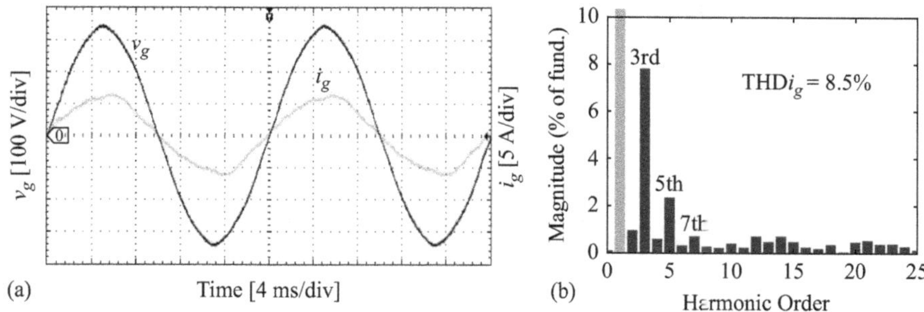

Figure 4.36 Experimental results for the single-phase grid-connected system using the PR scheme: (a) grid voltage v_g and feed-in current i_g and (b) harmonic spectrum of the feed-in current i_g

IEEE 929 standard. Furthermore, it is also observed in Figure 4.36(b) that the major harmonics of the feed-in current are odd-order harmonics within the low-frequency band. Notably, the PR controller fails to eliminate the major harmonic distortions of the feed-in current. As a consequence, harmonic compensators should be developed to improve the feed-in current quality.

It is known that the PC schemes offer a promising harmonics cancelation solution. As shown in Figure 4.34, the CRC can be plugged into the PR current controller to remove most of the harmonics of the feed-in current whose frequencies are below the Nyquist frequency. Alternatively, the MRSC can be used to replace the CRC in order to only suppress the major odd-order harmonics, such as the third-, fifth-, and seventh-order harmonics, in accordance with the harmonic spectrum shown in Figure 4.36(b). In contrast to the harmonic spectrum of the feed-in current shown in Figure 4.36(b), the fast Fourier transform (FFT) analysis of the feed-in currents shown in Figure 4.37 indicates that, the plug-in MRSC scheme has effectively suppressed the major feed-in current harmonics in the low-frequency band, but fail to deal with high-order feed-in current harmonics; the plug-in CRC scheme has effectively reduces most of the feed-in current harmonics. Moreover, the experimental results shown in Figure 4.37 indicates that the PR plus MRSC and PR plus CRC schemes have suppressed the THDs of feed-in current to 3% and 1.7% at unity power factor respectively, which are below the limit of 5%. Notably, compared with the PR plus MRSC scheme, the tailor-made PR plus CRC scheme can produce a better quality feed-in current because it can compensate much more harmonics (e.g., the 13th- and 15th-order harmonics) without increased complexity.

Additionally, the single-phase grid-connected inverter has been tested under various power angles, where the feed-in current amplitudes are fixed. The experimental results shown in Figure 4.38 indicate that the THDs of the feed-in current i_g produced by the PR controller varies with the change of power angle φ, as depicted in the previous discussion in Section 1.1.3 in Chapter 1. By contrast, when the PC compensators are plugged in, the THDs of the feed-in current become almost independent of the power angle: the current THD produced by the PR plus the CRC

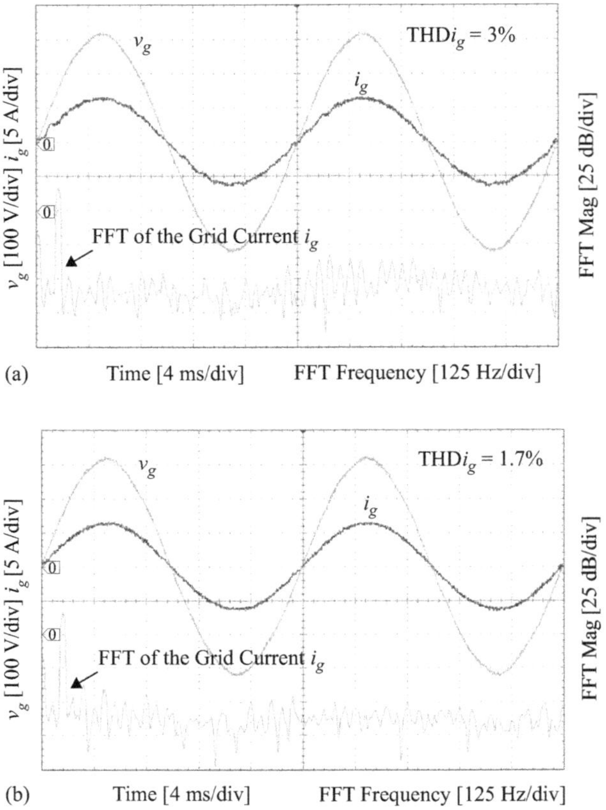

Figure 4.37 Experimental results (grid voltage v_g, feed-in current i_g, and fast Fourier transform of the feed-in current i_g) of the single-phase grid-connected system using various current control methods: (a) the PR controller with MRSC compensator and (b) the PR controller with CRC compensator

scheme is maintained around 2%, and the current THD produced by the PR plus the MRSC scheme is maintained around 3%.

All the above experimental tests have verified the effectiveness of the tailor-made PC current control scheme for grid-connected single-phase inverter systems.

4.4.4 Conclusion

Based on the harmonic injection mechanism, a tailor-made hybrid periodic current control scheme that includes a PR controller and a CRC, is developed for the single-phase grid-connected inverters to effectively suppress the harmonics of the feed-in current with different power factors. The experimental results verify the effectiveness of the tailor-made hybrid PC scheme. That is to say, the PC method is not sensitive to the clause changes of grid requirements on current distortion levels of grid-connected single-phase PV inverters. It enables grid-connected converters to always produce

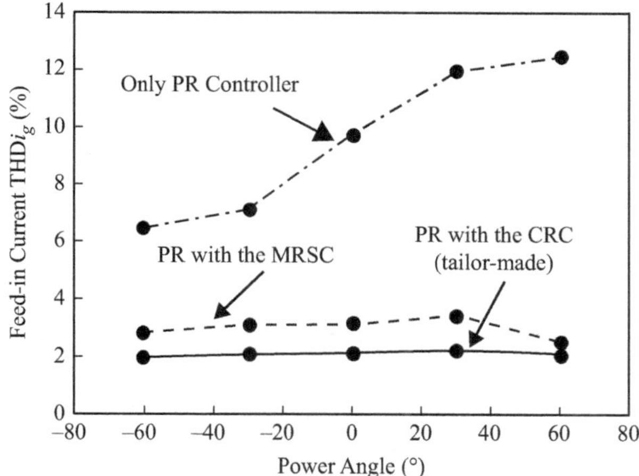

Figure 4.38 Feed-in current THD of the single-phase grid-connected system under different power angles, where the feed-in current amplitude has been fixed as $I_g = 6$ A

high-quality feed-in current while providing "smart" ancillary services, such as reactive power compensation, and so on.

4.5 PC of grid-connected single-phase HFL rectifiers

4.5.1 Background

Light, compact, efficient, and reliable battery chargers are highly demanded in electrical vehicles. HFL power converters provide a promising solution to the on-board battery chargers [1–16], which can operate at HF soft-switching mode. Galvanic isolation HF transformers not only can guarantee safe operation of the HFL converters, but also can increase voltage and current scalabilities. As shown in Figure 4.39, HFL rectifiers with HF isolation transformers can be classified into two types: "DC/DC" and "Cycloconverter" [28–36].

As shown in Figure 4.39(a), the "DC/DC" type HFL rectifiers include an isolated DC/HFAC/DC conversion stage, which comprises a DC/HFAC converter followed by a galvanic isolation HF transformer, and a HFAC/DC rectifier. Such a "DC/DC" HFL rectifier has three cascaded power conversion stages, and can achieve either unidirectional or bidirectional power flow. However, it can also be seen that a bulky DC-link capacitor C should be installed between the first AC/DC stage and the DC/HFAC/DC stage. Clearly, three power conversion stages will reduce the overall system conversion efficiency. At the same time, the bulky DC-link capacitor has become one of the most life-limiting components in the entire system. Nevertheless, for the "DC/DC"-type HFL rectifiers, the AC/DC conversion and the isolated DC/HFAC/DC conversion are decoupled. Two separate feedback control strategies are needed for the above two decoupled conversion stages.

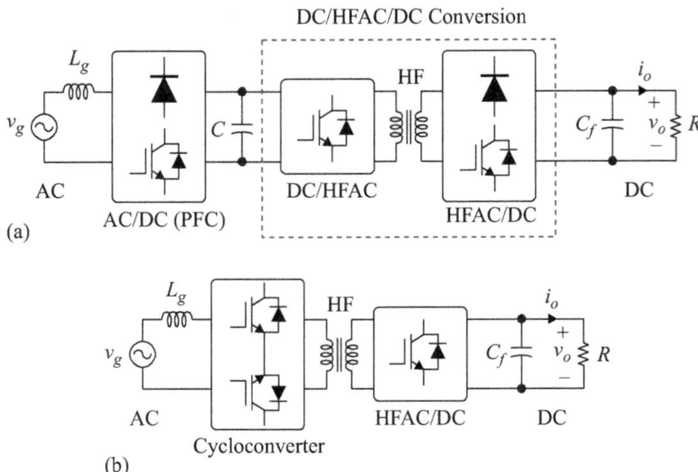

Figure 4.39 High frequency link rectifiers: (a) "DC/DC" type and (b) "Cycloconverter" type

For better system efficiency and reliability, "Cycloconverter-type" HFL rectifiers have been introduced, as shown in Figure 4.39(b). Such HFL rectifiers comprise of two power conversion stages: an AC/AC Cycloconverter followed by a galvanic isolation HF transformer, and an HFAC/DC rectifier stage. Compared with the "DC/DC" type HFL rectifier, the "Cycloconverter" type HFL rectifier is more attractive: higher efficiency due to less cascaded conversion stages, and better reliability due to the removal of the bulky DC-link capacitor. Furthermore, the bidirectional power flow enables "Cycloconverter" type HFL rectifier to serve extensive applications, such as vehicle-to-grid systems. However, the "Cycloconverter" type HFL rectifiers are more complicated due to the coupling between the AC/AC cycloconverter and the HFAC/DC rectifier.

In this section, a "Cycloconverter" type HFL single-phase rectifier is introduced for grid-connected applications. With a proper switching modulation strategy, the "Cycloconverter" type HFL rectifier can be equivalent to a conventional PWM rectifier with a line frequency transformer. Based on the equivalent circuit model, a periodic control scheme – the plug-in MRSC is developed for the "Cycloconverter" type HFL rectifier to not only draw high-quality grid-side currents but also offer flexible reactive power compensation capability. In addition a conventional PI controller is employed to generate a constant output DC voltage. Experiments are carried out to verify the effectiveness of the MSRC scheme for the "Cycloconverter" type HFL rectifier.

4.5.2 Modeling and control of grid-connected single-phase "Cycloconverter" HFL rectifiers

The circuit topology of the "Cycloconverter" type HFL single-phase rectifier is presented in Figure 4.40, where a HF galvanic isolation transformer with the turn

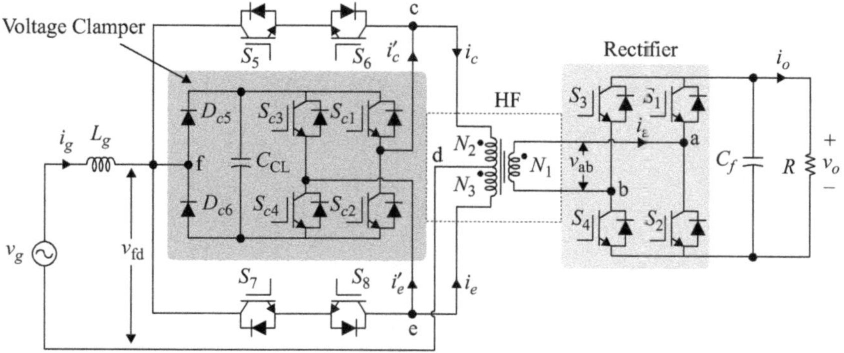

Figure 4.40 Bidirectional "Cycloconverter" type HFL single-phase rectifier, where a voltage camper is adopted to suppress the voltage ringing and overshoots due to the leakage inductance of the high-frequency transformer

ratio being $n = N_2/N_1 = N_3/N_1$ is adopted. The secondary-side of the HF transformer is connected to a single-phase full-bridge rectifier. An active voltage clamper is also employed, as highlighted in Figure 4.40. Note that, since the voltage clamper is only used to suppress the voltage ringing and overshoots due to the transformer leakage inductance, it has little relationship with the power conditioning functionality of the rectifier and thus is omitted in the following sections.

Referring to the modulation strategy shown in Figure 4.20 for the CVCF single-phase HFL inverter shown in Figure 4.19, the same modulation strategy can be applied to the "Cycloconverter" type HFL rectifier shown in Figure 4.40. According to the modulation strategy, it can be identified that the modulation signal u actually can be converted into a typical unipolar PWM signal v_{fd} of a full-bridge rectifier. Consequently, adopting the modulation strategy shown in Figure 4.20, the "Cyclo-converter" type HFL rectifier is equivalent to the conventional PWM rectifier circuit shown in Figure 4.41. Thus, the analysis and control of such a HFL rectifier can be significantly simplified.

The dynamics of the equivalent single-phase PWM rectifier can be described as [30]:

$$v_g = v_{fd} + L_g \frac{di_g}{dt} + R_g i_g \tag{4.35}$$

$$i_o' = \frac{v_o'}{R'} + C_f' \frac{dv_o'}{dt} \tag{4.36}$$

where $v_o' = nv_o$, i_o',$C_f' = C_f/n^2$, and $R' = n^2 R$ denote the equivalent voltage, current, capacitor, and resistor, $n = N_2/N_1 = N_3/N_1$ is the turn ratio of HF transformer, v_o is the output voltage across the capacitor C_f shown in Figure 4.40, and L_g and R_g are the grid-side filter inductance and resistance, respectively.

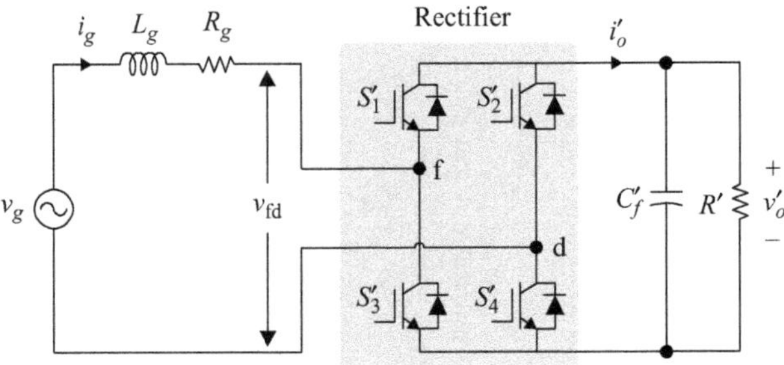

Figure 4.41 Equivalent circuit for the bidirectional "Cycloconverter" type HFL single-phase rectifier, when the modulation scheme in Figure 4.20 is used

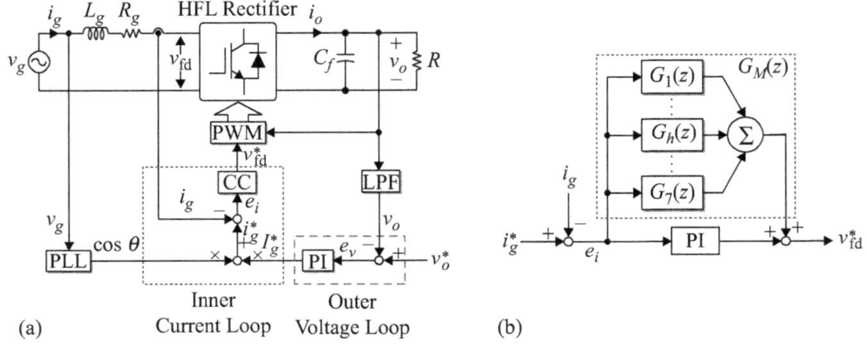

Figure 4.42 Control structure of the "Cycloconverter" HFL rectifier system (PLL – phase-locked loop, CC – current controller, LPF – low-pass filter, PI – proportional–integral controller): (a) overall control diagram and (b) inner current control system

Like a conventional PWM rectifier, the control objective for such a HFL rectifier shown in Figure 4.40 is also to maintain high-quality input current i_g with an adjustable power factor, and a constant output DC voltage v_o. Given that the equivalent circuit shown in Figure 4.41, PC schemes for conventional PWM rectifiers can be directly applied to the "Cycloconverter" type HFL rectifier. Hence, as shown in Figure 4.42, a typical dual-loop structure control scheme [30] is applied to the HFL rectifier, namely an inner AC current loop and an outer DC voltage loop.

Based on the IMP [1], a RSC can achieve zero-error tracking of a sinusoidal signal of the corresponding resonant frequency in the steady-state [15,16]. Considering that the major harmonics of the input current of PWM rectifiers concentrate within low frequency band, an MRSC controller (i.e., a parallel combination of the fundamental-frequency, third-, fifth-, and seventh-order RSC components) can be used to achieve high-quality sinusoidal feed-in currents at the grid side, as shown in Figure 4.42(b). The parallel structure enables that the gain of each RSC component can be tuned independently so that the MRSC can improve its dynamic performance.

According to Figure 4.42(b), a PI plus the MRSC scheme is developed as the current-loop controller. It can be given as:

$$G_{cc}(s) = k_p + \frac{k_i}{s} + \sum_{n=0}^{3} k_{rn} \frac{s}{s^2 + (2n\omega_0 + \omega_0)^2} \tag{4.37}$$

where k_p, k_i, and k_{rn} are the proportional, the integral, and the resonant control gains with n being the harmonic order, and ω_0 is the angular fundamental frequency of the grid voltage. A typical PI controller is used to regulate the output DC voltage v_o to the constant reference v_o^*, which can be given as:

$$G_{vc}(s) = k_{pv} + \frac{k_{iv}}{s} \tag{4.38}$$

where k_{pv} and k_{iv} are the proportional and integral control gains, respectively.

4.5.3 Experimental validation

In order to evaluate the PI plus MRSC current control scheme for the HFL rectifier system shown in Figure 4.40, a digital signal processor (DSP)-based 1 kVA HFL rectifier system has been built up, which is shown in Figure 4.43. The dual-loop control scheme, as illustrated in Figure 4.42, is implemented in a TMS320F28335 DSP. System parameters are listed in Table 4.10.

First, the steady-state performance of the "Cycloconverter" type HFL rectifier is tested, where the dual-loop control strategy shown in Figure 4.42 is adopted. The experimental results shown in Figure 4.44 indicate that the HFL rectifier achieves an almost constant output voltage v_o of 60 V with ripples of $\pm$ 2 V and twice the grid frequency, and draws a sinusoidal grid current with the THD of 2.3% and the power factor angle of 0°.

As stated previously, the "Cycloconverter" HFL rectifier system is able to flexibly adjust the power angle of its grid current for providing reactive power

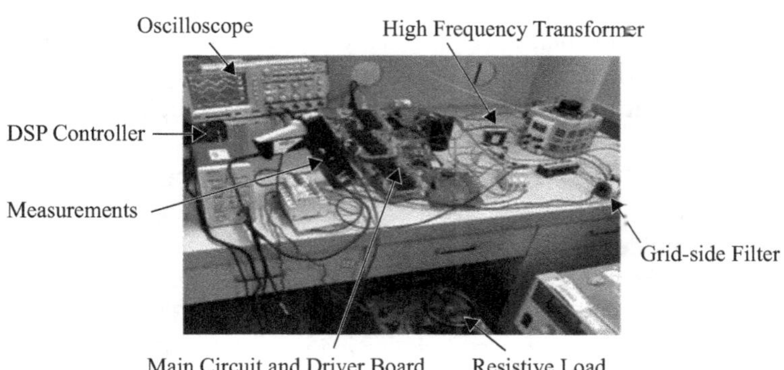

Figure 4.43 Experimental setup of the "Cycloconverter" type HFL rectifier system

Table 4.10 Parameters of the high-frequency link rectifier system

Parameter	Nominal value	Unit
Grid side line filter inductor L_g	2.2	mH
Leakage inductance of transformer	3.16	μH
Transformer turns ratio n	3	–
Reference output DC voltage v_o^*	60	V
Grid peak voltage v_g	100	V
Grid fundamental frequency f_0	50	Hz
Clamp capacitor C_{CL}	10.5	μF
Resistive load R	13	Ω
Sampling frequency f_s	20	kHz
Switching frequency	20	kHz
DC output capacitor C_f	4700	μF

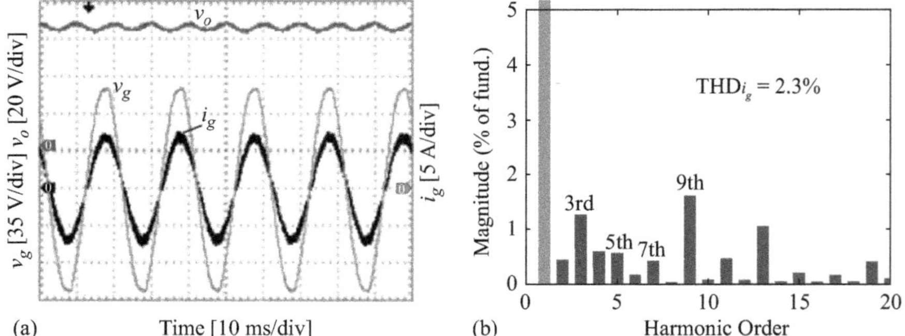

Figure 4.44 Steady-state performance of the HFL rectifier system at unity power factor with a dual-loop control scheme: (a) output DC voltage v_o, grid voltage v_g, and grid current i_g and (b) harmonic spectrum of the grid current i_g

compensation. Hence, experiments have been carried out to test its corresponding capability. The experimental results shown in Figure 4.45 indicate that the grid-connected HFL rectifier system can provide either leading or lagging reactive power compensation to the grid at the point of common coupling (PCC). In addition, the THDs of the resultant grid currents are 2.4% and 2.2% respectively, as shown in Figure 4.45(a) and (b). This confirms that the "Cycloconverter" type HFL rectifier will not generate severe harmonics at the PCC even when the rectifier is required to provide reactive power compensation.

At last, tests have been done to investigate the dynamic performance of the proposed "Cycloconverter" type HFL rectifier system. The start-up dynamic responses of the HFL rectifier system shown in Figure 4.46 indicate that it takes the rectifier system about 0.1 s to force its output DC-link voltage v_o and input current i_g to reach their steady-states. Additionally, dynamic responses of the HFL rectifier

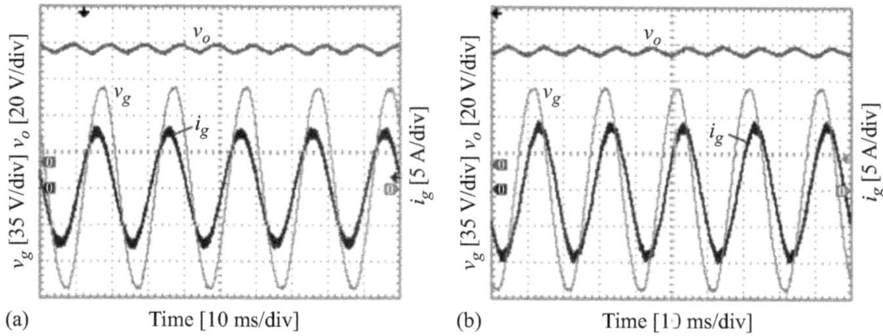

Figure 4.45 Steady-state performance of the HFL rectifier system at difference power factors (output DC voltage v_o, grid voltage v_g, and grid current i_g): (a) leading power factor angle of 30° and (b) lagging power factor angle of 30°

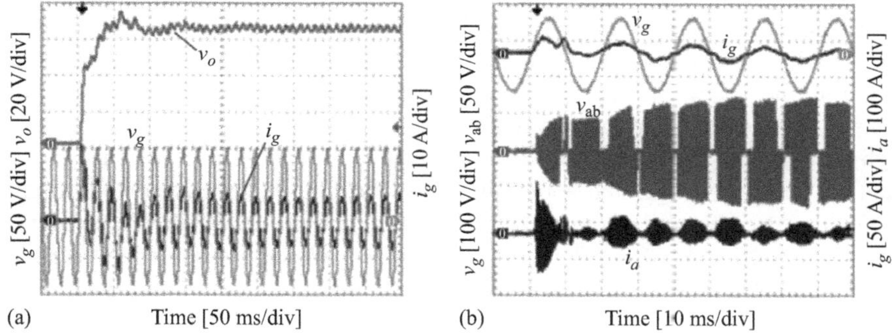

Figure 4.46 Start-up transient performance of the HFL rectifier system: (a) output DC voltage v_o, grid voltage v_g, and grid current i_g and (b) grid voltage v_g, grid current i_g, voltage at the HF transformer secondary side v_{ab}, and its current i_a (see Figure 4.40)

system under step load changes shown in Figure 4.47 indicates that it takes the rectifier around 0.2 s to recover the steady-state when the load is switched from 26 Ω to 13 Ω, and 0.15 s when the load is switched from 26 Ω to 13 Ω. All these experimental results confirm the rapidness and robustness of the "Cycloconverter" type HFL rectifier with the dual-loop control scheme.

In addition, the conversion efficiency of the "Cycloconverter" HFL rectifier system has also been measured. The conversion efficiency is about 84% ~ 89% in the experiments. Generally speaking, the PC-based control scheme enables the HFL rectifier system to draw low-THD input currents with fast dynamic response. Thus the HFL rectifier system with the PC scheme is able to offer a high-performance grid interface for grid-connected devices.

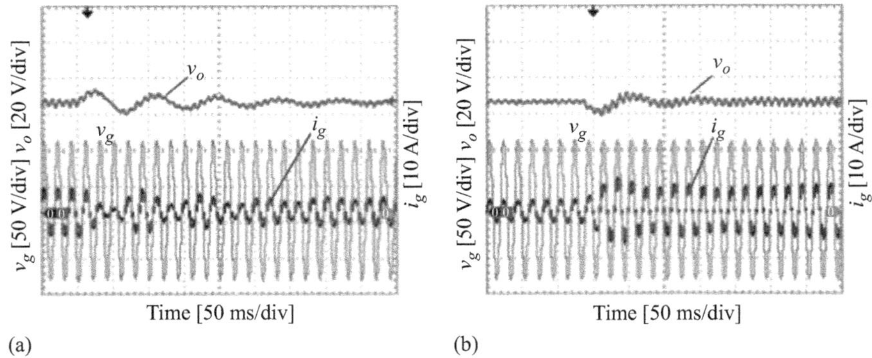

Figure 4.47 Transient responses of the HFL rectifier under step load changes (output DC voltage v_o, grid voltage v_g, and grid current i_g): (a) load changes from 13 Ω to 26 Ω and (b) load changes from 26 Ω to 13 Ω

4.5.4 Conclusion

A "Cycloconverter" type HFL single-phase rectifier has been studied in this section. The HFL rectifier can be equivalent to a conventional PWM rectifier, and thus simplifies the design and analysis of its control scheme. An MRSC-based dual-loop control scheme has been developed to enable the HFL rectifier to function as a high-performance PWM rectifier that can achieve sinusoidal grid currents, a constant DC output voltage, and adjustable power factors. Experiments have demonstrated the effectiveness of the HFL rectifier system with the MRSC-based control scheme. The HFL rectifier offers a light, compact, efficient, reliable, and bidirectional power-flow power conversion to extensive applications.

4.6 PC of grid-connected three-phase PWM inverters

4.6.1 Background

As discussed earlier, three-phase six-pulse PWM converters mainly produce $(6k \pm 1)$-order power harmonics ($k = 1, 2, 3, \ldots$). In order to improve the power quality in the case of grid-connected applications, any associated control strategy should be able to compensate such "featured" harmonics with high control accuracy while maintaining fast dynamic response, guaranteeing robustness, and being feasible for implementation [27].

Among the prior-art control schemes, the IMP-based CRC [2–9] and the RSC [19,20,38,39] provide very simple but effective control solutions in grid-connected applications. However, without considering the harmonic distribution, although the CRC can compensate the harmonics within a wide frequency range, it presents relatively slow dynamic response. In contrast, when the harmonic distribution is taken into account, the MRSC at selected harmonic frequencies can render quite fast dynamic response, but it will increase the computation and design complexity when dealing with a large number of harmonics. Alternatively, the

DFT-based RC [61], which is virtually equivalent to the MRSC, can flexibly and selectively compensate the harmonics of interest. Notably, unlike the MRSC, the outstanding feature of the DFT-based RC is that the computational complexity is independent of the number of selected harmonics. Nonetheless, the DFT-based RC, will also involve a large amount of parallel computation that is proportional to the number of samples per fundamental period. Thus, the DFT-based RC is especially suitable for high-performance fixed-point DSP implementation.

In light of the above issues, the OHC scheme has been adopted to control three-phase grid-connected inverters in this section. Such a PC scheme offers an accurate, fast, robust, and easy-for-implementation solution to three-phase grid-connected power converters. Experiments performed on a three-phase grid-connected PV inverter have validated the effectiveness of the OHC scheme. Notably, in this case study, the designed OHC scheme specifically incorporates the $(6k \pm m)$-order SHC modules in order to mitigate the $(6k \pm 1)$-order harmonics in six-pulse converters [23–27].

4.6.2 *Modeling and control of grid-connected three-phase PWM inverters*

Figure 4.48 shows a grid-connected three-phase six-pulse inverter for PV applications, which is used to feed currents into the grid. As shown in Figure 4.48, the capacitor C_f of the *LCL*-filter is used to eliminate high-order harmonic currents of the switching frequencies. Together with the grid-side inductor L_2, they can be referred to as an "ideal" load, or they can be taken as "model mismatch" [54]. Therefore, the dynamics of the three-phase grid-connected PV inverter shown in Figure 4.48 can be simply described only considering an "L"-filter (inverter-side inductor L_1):

$$\begin{bmatrix} \dot{i}_{\mathrm{a}} \\ \dot{i}_{\mathrm{b}} \\ \dot{i}_{\mathrm{c}} \end{bmatrix} = \begin{bmatrix} -\dfrac{R_1}{L_1} & 0 & 0 \\ 0 & -\dfrac{R_1}{L_1} & 0 \\ 0 & 0 & -\dfrac{R_1}{L_1} \end{bmatrix} \begin{bmatrix} i_a \\ i_b \\ i_c \end{bmatrix} + \begin{bmatrix} \dfrac{v_{sa} - v_{an}}{L_1} \\ \dfrac{v_{sb} - v_{bn}}{L_1} \\ \dfrac{v_{sc} - v_{cn}}{L_1} \end{bmatrix} \tag{4.39}$$

where v_{sa}, v_{sb}, and v_{sc} are the inverter output voltages, i_{a}, i_{b}, and i_{c} are the grid currents, v_{an}, v_{bn}, and v_{cn} are the grid phase voltages, L_1 and R_1 are the inductance and resistance of the inverter-side inductor, respectively.

The corresponding sampled-date model of (4.39) can be expressed as:

$$i_j(k+1) = \frac{b_1 - b_2}{b_1} i_j(k) + \frac{v_{sj}(k)}{b_1} - \frac{v_{jn}(k)}{b_1} \tag{4.40}$$

$$v_{sj}(k) = u_j(k) \frac{v_{\mathrm{dc}}(k)}{2} \tag{4.41}$$

where $j = a, b, c$ indicates the phase, $b_1 = L_1/T_s$ with T_s being the sampling time, $b_2 = R_1$, v_{dc} is the DC bus voltage, and u_j (u_a, u_b, u_c) are the normalized outputs of the system controller.

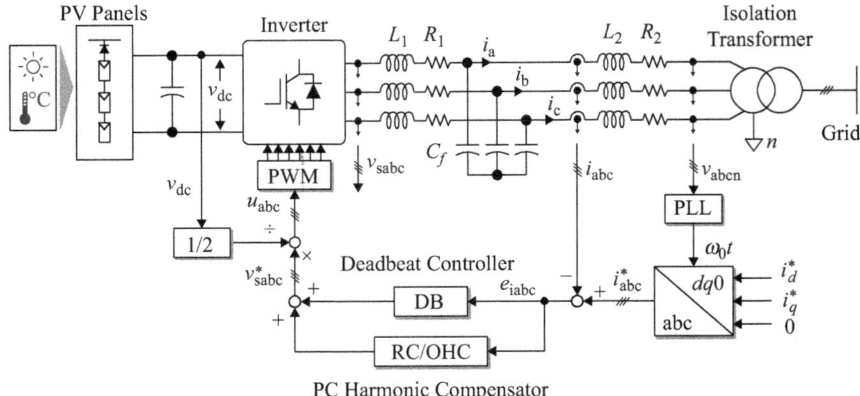

Figure 4.48 A three-phase grid-connected inverter system with an LCL-filter for PV applications

The control objective of the three-phase grid-connected inverter is to achieve a unity power factor in normal operation modes, and feed-in currents of low THD levels. As indicated in Figure 4.48, the dual-loop control scheme includes a dead-beat (DB) plus a plug-in PC harmonic compensator as the inner current control loop, and an outer control loop for generating accurate current references for the inner control loop. Notably, in Figure 4.48, the current references are directly given, while in practice it can be generated by a maximum power point tracking (MPPT)-based output power controller.

If the current controller for the plant in (4.40) is chosen as:

$$u_j(k) = \frac{2}{v_{dc}(k)}\left[v_{jn}(k) + b_1 i_j^*(k) - (b_1 - b_2) i_j(k)\right], \tag{4.42}$$

$i_j(k+1) = i_j^*(k)$ is obtained. It means that a DB current controller is employed. However, as shown in (4.42), the controller parameters are normally derived on the basis of the nominal system parameters. Thus, the DB controller is sensitive to the accuracy of the inverter model as well as system parameter variations. In practice, it is hard to obtain an accurate inverter model due to parameter uncertainties and load disturbances. Hence, a PC (i.e., an OHC) is added to the current controller to improve the control performance. The OHC is given as [27]:

$$G_{OHC}(z) = \sum_{m \in N_m} G_{sm}(z) = \sum_{m \in N_m} \frac{k_m G_f(z)\left(\cos\left(\frac{2\pi m}{n}\right) z^{N/6} Q(z) - Q^2(z)\right)}{z^{N/3} - 2\cos\left(\frac{2\pi m}{n}\right) z^{N/6} Q(z) + Q^2(z)} \tag{4.43}$$

where $N = T_0/T_s \in \mathbb{N}$ with T_s being the sampling period, $N/6 \in \mathbb{N}$, and k_m is the control gain, and N_m is the considered number of harmonic sets, an LPF

$Q(z) = 0.25z + 0.5 + 0.25z^{-1}$ is often chosen for improving the controller robustness, a linear phase-lead compensation filter $G_f(z) = z^p$ is also adopted [17,18]. The control module $G_{\mathrm{sm}}(z)$ for $(6k \pm m)$-order harmonics in (4.43) can be expressed as:

$$G_{\mathrm{sm}}(z) = \frac{k_m G_f(z)\left(\cos\left(\frac{2\pi m}{n}\right) z^{N/6} Q(z) - Q^2(z)\right)}{z^{N/3} - 2\cos\left(\frac{2\pi m}{n}\right) z^{N/6} Q(z) + Q^2(z)} \tag{4.44}$$

which is also called $(6k \pm m)$-order SHC module [26,27]. For comparison, a plug-in CRC current control scheme is also developed for the three-phase grid-connected inverter system.

4.6.3 Experimental validation

In order to validate the PC scheme, a test rig has been built up as shown in Figure 4.49, where a three-phase commercial power converter is connected to the grid through an *LCL*-filter and an isolation transformer. The control system was implemented by using a dSPACE DS1103 rapid prototyping kit. Parameters of the test setup are listed in Table 4.11. As discussed previously, a filter $G_f(\mathrm{z}) = z^p$ is normally used to compensate sampling delays, model mismatches, and un-modeled delay for both the CRC and the OHC systems, where the lead step $p = 3$ is determined by experiments.

First, tests are carried out to determine the harmonic distributions of the only DB-controlled three-phase inverter. The experimental steady-state responses of the grid-connected three-phase inverter shown in Figure 4.50 clearly indicate that, the feed-in currents have a THD of 8.4% (beyond the limit of 5% in the IEEE 929 standard) and contains dominant $(6k \pm 1)$-order harmonics with $k = 1, 2, 3 \ldots$. That is to say, the DB controller cannot force the feed-in currents to perfectly track the sinusoidal references, and thus leads to poor feed-in current quality.

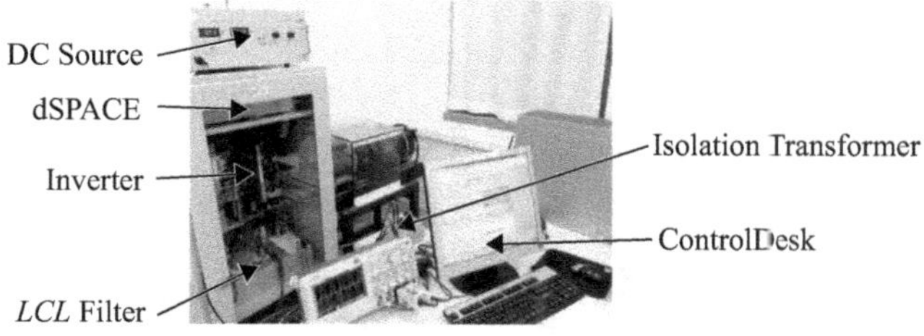

Figure 4.49 Experimental setup of a three-phase grid-connected system with an LCL-filter

Table 4.11 Parameters of the three-phase grid-connected inverter system with an LCL-filter

Parameters	Value	Unit
Inductors of the *LCL*-filter L_1, L_2	1.8	mH
Capacitor of the *LCL*-filter C_f	4.7	μF
Resistance of inductors R_1, R_2	0.02	Ω
Transformer leakage inductance (per phase) L_g	2	mH
Switching and sampling frequency f_s, f_{sw}	9.9	kHz
DC voltage V_{dc}	650	V
Nominal grid voltage (peak) v_{jn}	325	V
Nominal grid frequency ω_0	$2\pi \times 50$	rad/s
Nominal grid current amplitude i_j	4.3	A

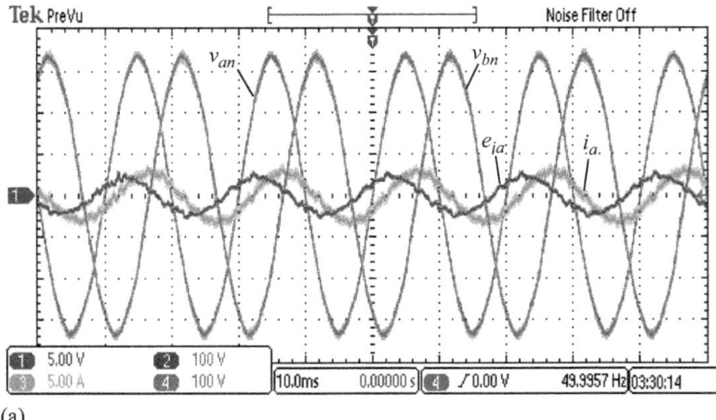

(a)

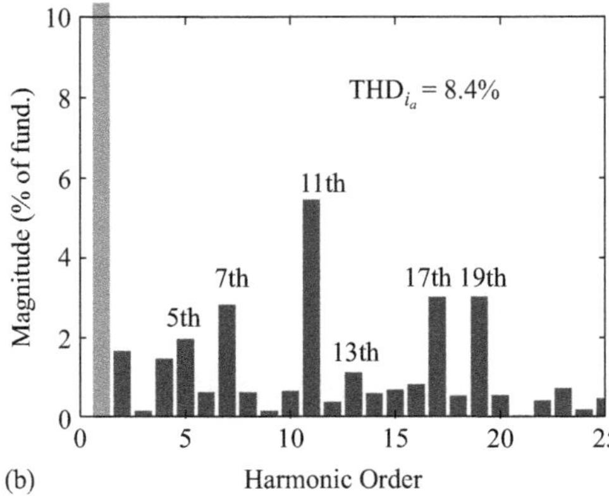

(b)

Figure 4.50 Steady-state performance of the DB-controlled three-phase grid-connected inverter: (a) phase voltages v_{an}, v_{bn} [100 V/div], phase A current i_a [5 A/div] and tracking error $e_{ia} = i_a^* - i_a$ [5 A/div], time [10 ms/div] and (b) harmonic distribution of the grid current i_a

To further evaluate the distribution of harmonics, a harmonic ratio is defined as:

$$H_f(j) = \frac{\sum_{i=0}^{j} M_i}{\sum_{i=0}^{99} M_i} \tag{4.45}$$

where M_i is the magnitude of the i-th-order harmonic. The corresponding harmonic ratio $f(j)$ for the harmonic spectrum of i_a shown in Figure 4.50(b) is given in Figure 4.51(a), which indicates that over 85% of the harmonics are

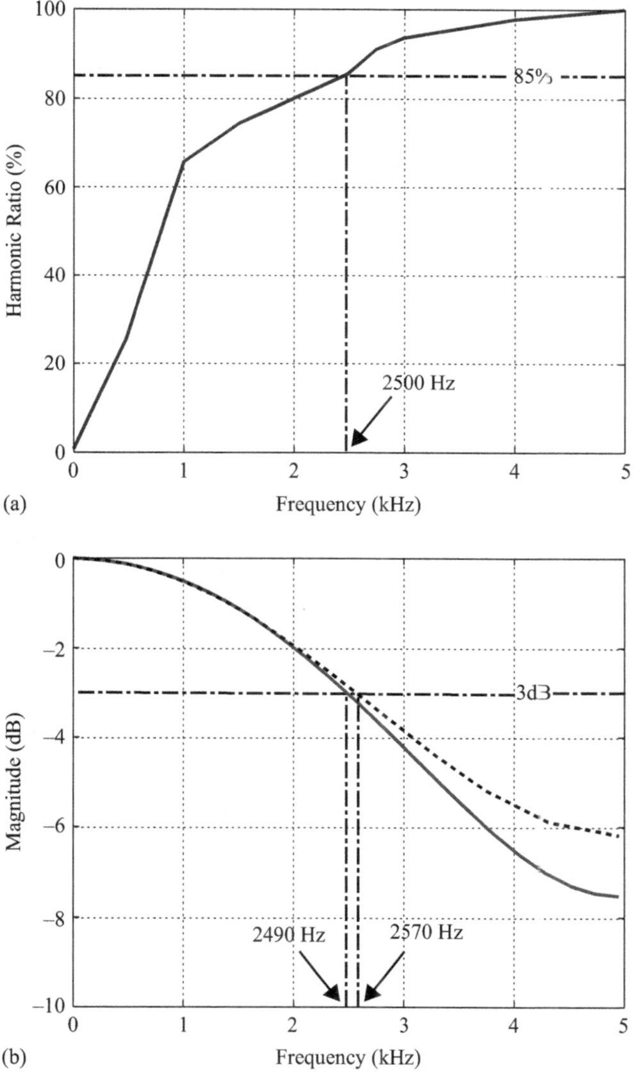

Figure 4.51 Harmonic analysis of the three-phase grid-connected inverter system with a deadbeat current controller: (a) harmonic ratio $h_r(j)$ of the grid current i_a and (b) magnitude response of the LPFs, with $f_s = 9.9$ kHz

located within a frequency range of 0~2.5 kHz. Therefore, the cut-off frequency f_{cutoff} of the LPF $Q(z)$ for the CRC and $Q^2(z)$ for the OSHC should satisfy $f_{\text{cutoff}} \geq 2.5$ kHz for the removal of most of the harmonics. Considering this, the LPFs are designed as $Q(z) = 0.145z + 0.71 + 0.145z^{-1}$ and $Q^2(z) = (0.075z + 0.85 + 0.075z^{-1})^2$, whose cut-off frequencies are 2.5 kHz and 2.6 kHz, respectively.

In addition, according to Figure 4.50(b) and Figure 4.51(a), the ratios of all $6k$-, $(6k \pm 1)$-, $(6k \pm 2)$-, and $(6k \pm 3)$-order harmonics to the total harmonics are nearly 8.8%, 60.6%, 23.7%, and 6.9%. Subsequently, the OHC controller is employed to suppress the harmonics optimally, and it can be given as [27]:

$$G_{\text{OHC}}(z) = \sum_{m \in N_m} G_{nm}(z) = G_{60}(z) + G_{61}(z) + G_{62}(z) + G_{63}(z) \tag{4.46}$$

with $n = 6$. The corresponding control gains for the SHC modules $G_{60}(z)$, $G_{61}(z)$, $G_{62}(z)$, and $G_{63}(z)$ are denoted as k_0, k_1, k_2, and k_3. Since the error convergence rate at any harmonic frequency is proportional to its corresponding control gain, k_0, k_1, k_2 and k_3 are weighted by their ratios in the total harmonics for a better convergence rate. Thus, they satisfy $k_3 < k_0 < k_2 < k_1$. Furthermore, according to the compatible stability criteria discussed in Chapter 3 for the OHC system, the stability range of the control gain k_{rc} for the CRC system is $0 < k_{\text{rc}} < 2$ and that for the OHC system is $0 < k_0 + k_1 + k_2 + k_3 < 2$. For comparison, $k_{\text{rc}} = k_0 + k_1 + k_2 + k_3$, and the designed control parameters are: $k_{\text{rc}} = 2$, $k_0 = 0.2$, $k_1 = 1.2$, $k_2 = 0.4$, and $k_3 = 0.2$.

Next, the CRC compensator with above-designed parameters is plugged into the DB controller to improve the feed-in current quality of the three-phase grid-connected inverter system. In contrast to the experimental results shown in Figure 4.50, the experimental results of the steady-state responses of the DB plus the CRC controlled inverter system shown in Figure 4.52 indicate that, the plug-in CRC successfully reduced the THD of the feed-in current i_a from 8.4% to 2.7% and significantly suppressed the low-order major harmonics. Furthermore, for comparison, the OHC compensator with above design parameters is plugged into the DB controller. The experimental results of the steady-state responses shown in Figure 4.53 indicate that the plug-in OHC compensator also significantly mitigate the low-order major harmonics and produce a good quality feed-in current but with a little higher THD of 3.3%.

At last, in addition to the tests of the steady-state performance, more experiments have been carried to test the dynamic performance of the two PC schemes in the case of a step amplitude change of the feed-in current reference. It can be observed from the experimental results shown in Figure 4.54 that the settling time for the tracking current error $e_{ia}(t) = i_a^*(t) - i_a(t)$ is about 130 ms when the CRC scheme is adopted, while the settling time is about 70 ms when the OHC scheme is

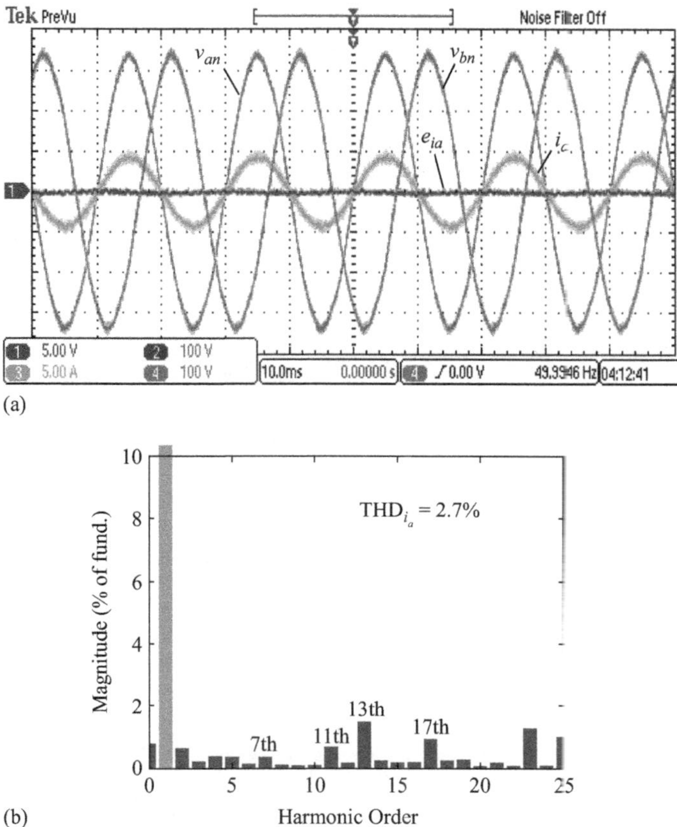

Figure 4.52 *Steady-state response of the three-phase grid-connected inverter system with a DB current controller plus a CRC compensator: (a) phase voltages v_{an}, v_{bn} [100 V/div], phase A current i_a [5 A/div] and tracking error $e_{ia} = i_a^* - i_a$ [5 A/div], time [10 ms/div] and (b) harmonic distribution of the grid current i_a*

employed. The tests have verified that the dynamic response of the OHC can be much faster (up to $n/2$ times) than that of the CRC.

All the experimental results demonstrate that both the CRC and the OHC can enhance the DB control performance, and thus enable the grid-connected three-phase inverter system to produce high-quality currents with low THDs. Moreover, compared with the CRC, the OHC is able to effectively mitigate the power harmonics but with a faster convergence rate. It confirms the effectiveness of the PC harmonic compensators in grid-connected power converter applications.

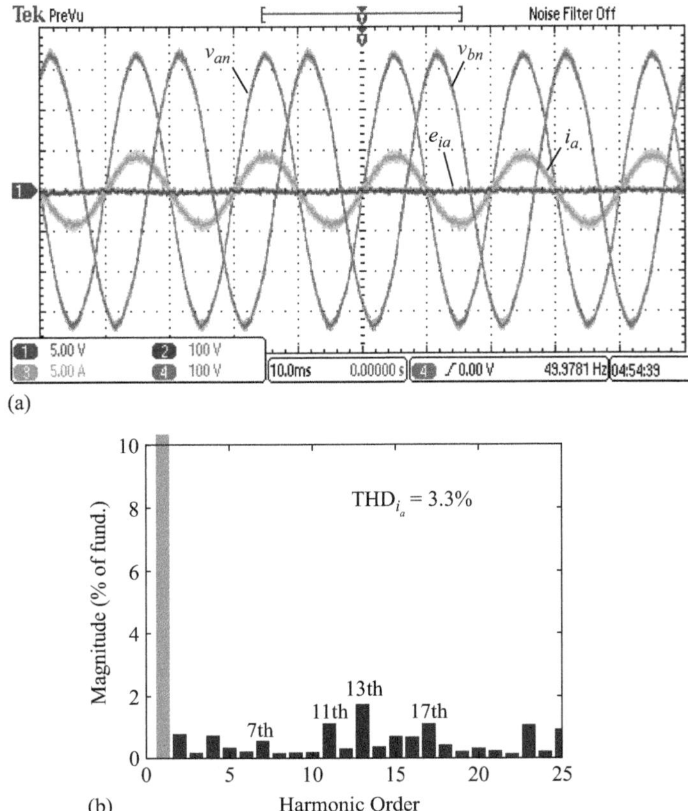

Figure 4.53 Steady-state response of the three-phase grid-connected inverter system with a DB current controller plus a CRC compensator: (a) phase voltages v_{an}, v_{bn} [100 V/div], phase A current i_a [5 A/div] and tracking error $e_{ia} = i_a^* - i_a$ [5 A/div], time [10 ms/div] and (b) harmonic distribution of the grid current i_a

4.6.4 Conclusion

An IMP-based OHC method has been introduced to provide a tailor-made optimal control solution to compensate power harmonics produced by power converters. The OHC can take advantages of both the CRC and the MRSC: high accuracy due to the removal of major harmonics, fast dynamic response due to parallel combination of optimally harmonic-weighted SHC modules, cost-effective and easy real-time implementation due to the universal recursive SHC modules, and compatible design. The application example of a three-phase grid-connected PWM inverter has demonstrated the effectiveness and advantages of the OHC scheme in suppressing the feed-in current harmonics.

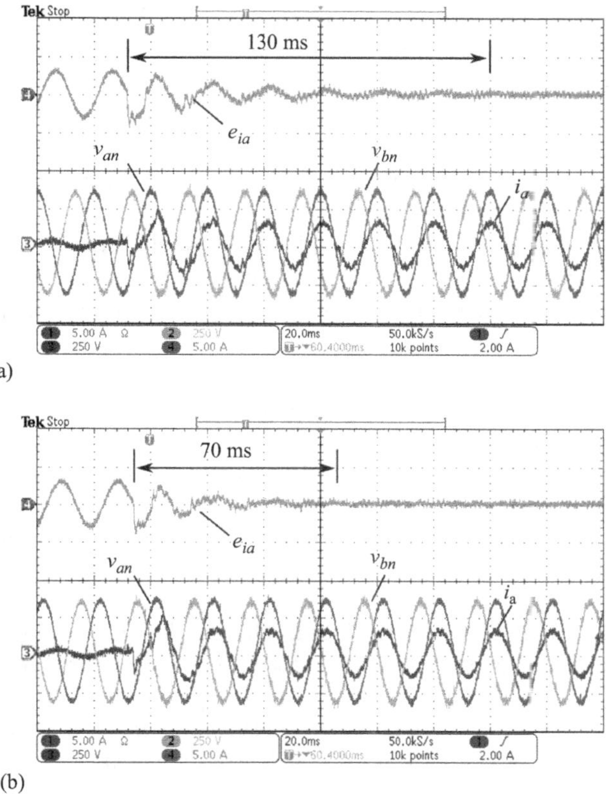

Figure 4.54 Transient performance of the grid-connected three-phase inverter system with different PC schemes (grid voltage v_{an}, v_{bn} [250 A/div], grid current i_a [5 A/div], grid current tracking errors $e_{ia}(t) = i_a^*(t) - i_a(t)$ [5 A/div], grid current reference amplitude: 3A, time [20 ms/div]): (a) the CRC scheme and (b) the OHC scheme

4.7 Summary

Six examples of periodic control (PC) of power converters have been demonstrated in this chapter. Upon specific requirements, the developed PC solutions from the IMP fundamental and advanced PC schemes presented in Chapters 2 and 3 can provide a simple, accurate, fast, and robust control for power converters to compensate periodic voltages or currents.

References

[1] Francis, B.A. and Wonham, W.M., "The internal model principle of control theory," *Automatica*, vol. 12, no. 5, pp. 457–465, 1976.

[2] Haneyoshi, T., Kawamura, A., and Hoft, R.G., "Waveform compensation of PWM inverter with cyclic fluctuating loads," *IEEE Trans. Ind. Appl.*, vol. 24, no. 4, pp. 582–589, 1988.

[3] Tzou, Y.Y., Ou, R.S., Jung, S.L., and Chang, M.Y., "High-performance programmable AC power source with low harmonic distortion using DSP-based repetitive control technique," *IEEE Trans. Power Electron.*, vol. 12, no. 4, pp. 715–725, 1997.

[4] Cardenas, R., Juri, C., Pena, R., Wheeler, P., and Clare, J., "The application of resonant controllers to four-leg matrix converters feeding unbalanced or nonlinear loads," *IEEE Trans. Power Electron.*, vol. 27, no. 3, pp. 1120–1129, 2012.

[5] Zhou, K. and Wang, D., "Zero tracking error controller for three-phase CVCF PWM inverter," *Electron. Lett.*, vol. 36, no. 10, pp. 864–865, 2000.

[6] Zhou, K. and Wang D., "Digital repetitive learning controller for three-phase CVCF PWM inverter," *IEEE Trans. Ind. Electron.*, vol. 48, no. 4, pp. 820–830, 2001.

[7] Zhang, K., Kang, Y., Xiong, J., Chen, J., "Direct repetitive control of SPWM inverter for UPS purpose," *IEEE Trans. Power Electron.*, vol. 18, no. 3, pp. 784–792, 2003.

[8] Ye, Y., Zhang, B., Zhou, K., Wang, D., and Wang, Y. "High-performance cascade-type repetitive controller for CVCF PWM inverter: Analysis and design," *IET Electric Power Appl.*, vol. 1, no. 1, pp. 112–118, 2007.

[9] Yang, Y., Zhou, K. and Lu, W., "Robust repetitive control scheme for three-phase CVCF PWM inverters," *IET Power Electronics*, vol. 5, no. 6, pp. 669–677, 2011.

[10] Zhou, K., Low, K., Tan, S., Wang, D., and Ye, Y., "Odd-harmonic repetitive controlled CVCF PWM inverter with phase-lead compensation," 39th IAS Annual Meeting, Seattle, USA, 2004.

[11] Griñó R. and Costa-Castelló, R., "Digital repetitive plug-in controller for odd-harmonic periodic references and disturbances," *Automatica*, vol. 41, no. 1, pp. 153–157, 2005.

[12] Zhou, K., Low, K.S., Wang, Y., Luo, F.L., Zhang, B., and Wang, Y., "Zero-phase odd-harmonic repetitive controller for a single-phase PWM inverter," *IEEE Trans. Power Electron.*, vol. 21, no. 1, pp. 193–201, 2006.

[13] Zhou, K., Wang, D., Zhang, B., Wang, Y., Ferreira, J. A., and de Haan, S.W.H., "Dual-mode structure digital repetitive control," *Automatica*, vol. 43, no. 3, pp. 546–554, 2007.

[14] Zhou, K., Wang, D., Zhang, B., and Wang, Y., "Plug-in dual-mode-structure repetitive controller for CVCF PWM inverters," *IEEE Trans. Ind. Electron.*, vol. 56, no. 3, pp. 784–791, 2009.

[15] Yang, Y., Zhou, K., and Cheng, M., "Phase compensation resonant controller for single-phase PWM converters," *IEEE Trans. Ind. Informatics*, vol. 9, no. 2, pp. 957–964, 2013.

[16] Yang, Y., Zhou, K., Cheng, M., and Zhang, B., "Phase compensation multi-resonant control of CVCF PWM converters," *IEEE Trans. Power Electron.*, vol. 28, no. 8, pp. 3923–3930, Aug. 2013.

[17] Zhang, B., Zhou, K., Wang, Y., and Wang, D., "Performance improvement of repetitive controlled PWM inverters: A phase-lead compensation solution," *Int. J. Circuit Theory Appl.*, vol. 38, no. 5, pp. 453–469, 2010.

[18] Zhang, B., Wang, D., Zhou, K., and Wang, Y., "Linear phase-lead compensation repetitive control of a CVCF PWM inverter," *IEEE Trans. Ind. Electron.*, vol. 55, no. 4, pp. 1595–1602, Apr. 2008.

[19] Teodorescu, R., Blaabjerg, F., Liserre, M., and Loh, P. C., "Proportional-resonant controllers and filters for grid-connected voltage-source converters," *Proc. IEE*, vol. 153, no. 5, pp. 750–762, Sep. 2006.

[20] Liserre, M. Teodorescu, R., and Blaabjerg, F., "Multiple harmonics control for three-phase grid converter systems with the use of PI-RES current controller in a rotating frame," *IEEE Trans. Power Electron.*, vol. 21, no. 3, pp. 836–841, May 2006.

[21] Kawamura, A., Chuarayapratip, R., and Haneyoshi, T., "Deadbeat control of PWM inverter with modified pulse patterns for uninterruptible power supply," *IEEE Trans. Ind. Electron.*, vol. 35, no. 2, pp. 295–230, May 1998.

[22] Mattavelli, P., "An improved deadbeat control for UPS using disturbance observers," *IEEE Trans. Ind. Electron.*, vol. 52, no. 1, pp. 206–212, 2005.

[23] Zhou, K., Lu, W., Yang, Y., and Blaabjerg, F., "Harmonic control: A natural way to bridge resonant control and repetitive control," Washington, DC, USA: 2013 American Control Conference (ACC), pp. 3189–3193, June 17–19 2013.

[24] Lu, w., Zhou, K., Wang, D., and Cheng, M., "A general parallel structure repetitive control scheme for multiphase DC–AC PWM converters," *IEEE Trans. Power Electron.*, vol. 28, no. 8, pp. 3980–3987, 2013.

[25] Lu, W., Zhou, K., and Wang, D., "General parallel structure digital repetitive control," *Int. J. Control*, vol. 86, no. 1, pp. 70–83, Jan. 2013.

[26] Lu, W., Zhou, K., Wang, D., and Cheng, M., "A generic digital $nk \pm m$ order harmonic repetitive control scheme for PWM converters," *IEEE Trans. Ind. Electron.*, vol. 61, no. 3, pp. 1516–1527, 2014.

[27] Zhou, K., Yang, Y., Blaabjerg, F., and Wang, D., "Optimal selective harmonic control for power harmonics mitigation," *IEEE Trans. Ind. Electron.*, vol. 62, no. 2, pp. 1220–1230, 2015.

[28] Zhu, L., Zhou, K., Zhu, W., and Su, X., "High-frequency link single-phase grid-connected PV inverter," Hefei, China: 2nd IEEE Symposium on Power Electronics for Distributed Generation Systems, pp. 133–137, 2010.

[29] Zhu, W., Zhou, K., Cheng, M., "A bidirectional high-frequency-link single-phase inverter: Modulation, modeling, and control," *IEEE Trans. Power Electron.*, vol. 29, no. 8, pp. 4049–4057, 2014.

[30] Zhu, W., Zhou, K., Cheng, M., Peng, F., "A high-frequency-link single-phase PWM rectifier," *IEEE Trans. Ind. Electron.*, vol. 62, no. 1, pp. 289–298, 2015.

[31] Yamato, I., Tokunaga, N., Matsuda, Y., Amano, H., and Suzuki, Y., "New conversion system for UPS using high frequency link," 19th Annual IEEE Power Electronics Specialists Conference (PESC '88), pp. 658–663, 1988.
[32] Matsui, M., Nagai, M., Mochizuki, N., and Nabae, M., "A high-frequency link DC/AC converter with suppressed voltage clamp circuits – naturally commutated phase angle control with self turn-off devices," *IEEE Trans. Ind. Appl.*, vol. 32, no. 2, pp. 293–300, 1996.
[33] Yamato, I., Tokunaga, N., Matsuda, Y., Amano, H., and Suzuki, Y., "High frequency link DC–AC converter for UPS with a new voltage clamper," 21st Annual IEEE Power Electronics Specialists Conference (PESC '90), pp. 749–756, 1990.
[34] Sood, P.K., Lipo, T.A., and Hansen, I.G., "A versatile power converter for high-frequency link systems," *IEEE Trans. Power Electron.*, vol. 3, no. 4, pp. 383–390, 1988.
[35] Chan, H.L., Cheng, K.W.E., and Sutanto, D., "Bidirectional phase-shifted DC–DC converter," *Electron. Lett.*, vol. 35 no. 7, pp 523–524, 1999.
[36] De, D. and Venkataramanan, R., "Analysis, design, modeling, and implementation of an active clamp HF link converter," *IEEE Trans. Circuits and Systems (I)*, vol. 58, no. 6, pp. 1146–1155, 2011.
[37] Lu, W., Zhou, K., and Yang, Y., "A general internal model principle-based control scheme for CVCF PWM converters," in *Power Electronics for Distributed Generation Systems (PEDG), 2nd IEEE International Symposium on Power Electronics for Distributed Generation Systems*, Hefei, China, pp. 485–489, 2010.
[38] Castilla, M., Miret, J., Camacho, A., Matas, J., de Vicuna, L.G., "Reduction of current harmonic distortion in three-phase grid-connected photovoltaic inverters via resonant current control," *IEEE Trans. Ind. Electron.*, vol. 60, no. 4, pp. 1464–1472, 2013.
[39] Yepes, A.G., Freijedo, F.D., Lopez, O., and Doval-Gandoy, J., "Analysis and design of resonant current controllers for voltage-source converters by means of Nyquist diagrams and sensitivity function," *IEEE Trans. Ind. Electron.*, vol. 58, no. 11, pp. 5231–5250, Nov. 2011.
[40] Winneker, C., "World's solar photovoltaic capacity passes 100-gigawatt landmark after strong year," [Online], Feb. 2013. Available: http://www.epia.org/news/.
[41] Teodorescu, R., Liserre, M., and Rodriguez, P., *Grid Converters for Photovoltaic and Wind Power Systems*, John Wiley & Sons, 2011.
[42] E. ON GmbH, "Grid Code – High and Extra High Voltage." [Online]. Available: http://www.eon-netz.com/.
[43] Comitato Elettrotecnico Italiano, "CEI 0-21: Reference technical rules for connecting users to the active and passive LV distribution companies of electricity." [Online]. Available: http://www.ceiweb.it/.
[44] IEEE-SA Standards Board, "IEEE 929-2000: IEEE recommended practice for utility interface of photovoltaic (PV) systems," Jan. 2000.

[45] IEEE Application Guide for IEEE Std. 1547, "IEEE standard for interconnecting distributed resources with electric power systems," *IEEE Standard 1547. 2–2008*, 2009.

[46] Yang Y. and Blaabjerg, F., "Low voltage ride-through capability of a single-stage single-phase photovoltaic system connected to the low-voltage grid," *Int. J. Photo Energy*, vol. 2013, Article ID 257487, 9 pages, 2013. Available: http://dx.doi.org/10.1155/2013/257487.

[47] Benz, C.H., Franke, W.-T., and Fuchs, F.W., "Low voltage ride-through capability of a 5 kW grid-tied solar inverter," in *Proc. of EPE/PEMC'10*, pp. T12-13-T12-20, 6–8 Sept. 2010.

[48] Kobayashi, H., "Fault ride-through requirements and measures of distributed PV systems in Japan," in *Proc. of IEEE PES General Meeting*, pp. 1–6, July 22–26, 2012.

[49] Yang, Y., Blaabjerg, F., and Zou, Z., "Benchmarking of grid fault modes in single-phase grid-connected photovoltaic systems," *IEEE Trans. Ind. Appl.*, vol. 49, no. 5, pp. 2167–2176, Sept.-Oct. 2013.

[50] Prodanovic, M., De Brabandere, K., Van Den Keybus, J., Green, T., and Driesen, J., "Harmonic and reactive power compensation as ancillary services in inverter-based distributed generation," *IET Gener. Transm. Distrib.*, vol. 1, no. 3, pp. 432–438, May 2007.

[51] Yang, Y., Enjeti, P., Blaabjerg, F., Wang, H., "Wide-scale adoption of photovoltaic energy: Grid code modifications are explored in the distribution grid," *IEEE Ind. Appl. Mag.*, vol. 21, no. 5, pp. 21–31, 2015.

[52] Yang, Y. and Blaabjerg, F., "Overview of single-phase grid-connected photovoltaic systems," *Electric Power Components and Systems*, vol. 43, no. 12, pp. 1352–1363, 2015.

[53] Blaabjerg, F., Teodorescu, R., Liserre, M., and Timbus, A. V., "Overview of control and grid synchronization for distributed power generation systems," *IEEE Trans. Ind. Electron.*, vol. 53, no. 5, pp. 1398–1409, Oct. 2006.

[54] Timbus, A. V., Liserre, M., Teodorescu, R., Rodriguez, P., and Blaabjerg, F., "Evaluation of current controllers for distributed power generation systems," *IEEE Trans. Power Electron.*, vol. 24, no. 3, pp. 654–664, Mar. 2009.

[55] Kjaer, S. B., Pedersen, J. K., and Blaabjerg, F., "A review of single-phase grid-connected inverters for photovoltaic modules," *IEEE Trans. Ind. Appl.*, vol. 41, no. 5, pp. 1292–1306, Sep./Oct. 2005.

[56] Zhou, K., Qiu, Z., Yang, Y., "Current harmonics suppression of single-phase PWM rectifiers," *3rd IEEE International Symposium on Power Electronics for Distributed Generation Systems*, Aalborg, Denmark, 2012.

[57] Zhou, K., Qiu, Z., Watson, N. R., and Liu, Y., "Mechanism and elimination of harmonic current injection from single-phase grid-connected PWM converters," *IET Power Electron.*, vol. 6, no. 1, pp. 88–95, Jan. 2013.

[58] Yang, Y., Zhou, K., and Blaabjerg, F., "Harmonics suppression for single-phase grid-connected PV systems in different operation modes," in *Proc. of IEEE APEC'13*, pp. 889–896, Mar. 2013.

[59] Rashed, M., Klumpner, C., and Asher, G., "Repetitive and resonant control for a single-phase grid-connected hybrid cascaded multilevel converter," *IEEE Trans. Power Electron.*, vol. 28, no. 5, pp. 2224–2234, May 2013.

[60] Rashed, M., Klumpner, C., and Asher, G., "Control scheme for a single-phase hybrid multilevel converter using repetitive and resonant control approaches," in *Proc. EPE. IEEE*, pp. 1–13, 2011.

[61] Mattavelli P. and Marafao, F. P., "Repetitive-based control for selective harmonic compensation in active power filters," *IEEE Trans. Ind. Electron.*, vol. 51, no. 5, pp. 1018–1024, Oct. 2004.

Chapter 5
Frequency-adaptive periodic control

Abstract

Connecting power converters to the electrical networks has become a crucial issue due to the fast-growing penetration of renewable energy sources and distributed generation systems. In these practical applications, the frequency of the periodic signals of interest is not constant always but slowly time-varying due to the imbalance between the power generation and the load demand [21–23].

Both fundamental and advanced PC schemes [1–20] can achieve zero steady-state error tracking of any periodic signal with a known period due to the introduction of high gains at the interested harmonic frequencies by embedding the corresponding internal models. However, frequency variations will lead to mismatch between their embedded nominal internal models and the actual periodic references/ disturbances, and will shift high gains away from the actual frequency of interest. Thus these PC schemes will fail to accurately track the varying-frequency periodic signals. It means that their internal models are sensitive to frequency variations.

Moreover, delay-based PC schemes in their digital forms, such as the digital classic repetitive control (RC), discrete Fourier transform (DFT)-based RC, digital selective harmonic control (SHC), and so on, even require that the period of the references/disturbances can be represented as integer multiple of the sample time of the digital control system. This means that the period of the interested periodic signal should be an integer, but the period will not be an exact integer except by chance. A varying frequency often induces fractional-period harmonics. Such mismatches between the given integer period and actual fractional period will lead these PC strategies to yield low gains at the interested harmonic frequencies and thus produce poor tracking accuracy. For example, the grid frequency is usually varying within a certain range (e.g., 49 Hz ~ 51 Hz) due to the generation-load imbalance and/or continuously connecting and disconnecting of large generation units. The PC schemes may fail to force grid-tied converters to feed good quality power into the grid in the presence of a time-varying grid frequency. Therefore, it calls for frequency-adaptive PC solutions that are able to self-tune the corresponding internal model to match the external signal closely and then accurately compensate the varying-frequency voltages and/or currents for good power quality and also stable operation of the grid-connected systems. The performance of the PC systems depends on how precise the match is between the PC signal generator period and the actual signal period.

Addressing the above issues, this chapter explores the frequency-adaptive internal model principle (IMP)-based PC strategies to compensate frequency-varying harmonics. A direct frequency-adaptive resonant controller is investigated. A fractional-delay filter-based internal model is introduced to provide a general frequency-adaptive PC solution to the compensation of frequency-varying periodic signals. The frequency sensitivity, design, and implementation methodology of frequency-adaptive PC systems are discussed. Compatible synthesis methods for plug-in frequency-adaptive PC schemes are also presented.

5.1 Frequency-adaptive fundamental periodic control

5.1.1 Resonant control (RSC)

A general form of the continuous-time resonant control (RSC) scheme at the harmonic frequency of ω_h can be found in (2.29). In many practical applications, a simplified RSC with $\phi_h = 0$ is employed, which can be expressed as [24–31]:

$$G_h(s) = k_h \frac{s}{s^2 + \omega_h^2} \tag{5.1}$$

where k_h is the control gain for tuning the error convergence rate at the resonance frequency $\pm\omega_h$, $\omega_h = h\omega_0$ is the angular signal frequency with ω_0 being the fundamental angular signal frequency, and $h \in \mathbb{N}$ is the harmonic order. Since the RSC $G_h(s)$ has poles at $\pm j\omega_h$, it exhibits infinite gains at these harmonic frequencies. The infinite gains will lead to zero tracking errors exclusively at $\pm j\omega_h$ if $G_h(s)$ is included into the closed-loop system. From (5.1), it is obvious that, the resonant frequency of the RSC is determined by ω_h. In many cases, if the interested harmonic frequency is almost constant, the resonant frequency ω_h of the RSC in (5.1) can be set simply as equivalent to the interested frequency without any frequency measurement.

In order to compensate multiple selected harmonic frequencies simultaneously, the MRSC scheme has been introduced as shown in (2.30). For convenience, it is given again as [8,9]:

$$G_M(s) = \sum_{h \in N_h} G_h(s) = \sum_{h \in N_h} \left(k_h \frac{s \cos \phi_h - \omega_h \sin \phi_h}{s^2 + \omega_h^2} \right) \tag{5.2}$$

where N_h represents the set of selected harmonic orders and ϕ_h is the phase-lead angle for system delay compensation at the hth-order resonant frequency.

However, in practical applications, the harmonic frequency may be time-varying. For instance, the grid frequency varies around its nominal value due to the generation-load imbalance within a certain range (e.g., 49 Hz ~ 51 Hz for a 50-Hz system). In that case, the actual fundamental frequency can be expressed as [32,56]:

$$\hat{\omega}_0 = \omega_0 + \Delta\omega \tag{5.3}$$

where $\Delta\omega$ is the frequency deviation.

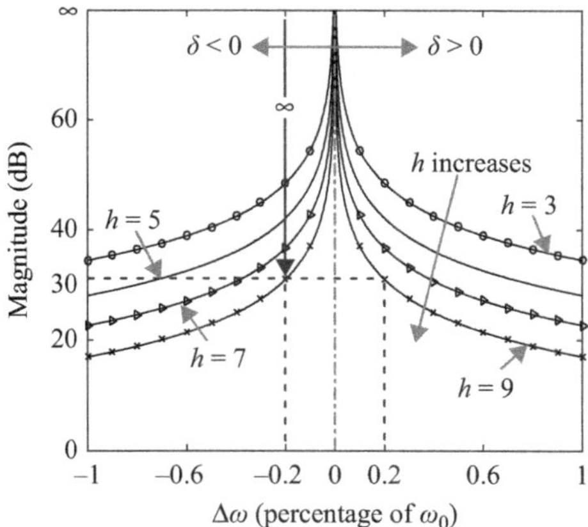

Figure 5.1 Magnitudes of the RSC component $G_h(s)$ as a function of the frequency variation $\Delta\omega$, where $k_3 = 1000$, $k_5 = 800$, $k_7 = 600$, $k_9 = 400$

According to (5.1)–(5.3), the magnitude response of an individual RSC around the harmonic frequency (i.e., $s = jh\hat{\omega}_0 = j\hat{\omega}_h$) can be obtained as:

$$|G_h(jh\hat{\omega}_0)| = \left| \frac{jk_h h\hat{\omega}_0}{-(h\hat{\omega}_0)^2 + (h\omega_0)^2} \right| = \frac{k_h}{h\omega_0} \left| \frac{\delta + 1}{\delta^2 + 2\delta} \right| \tag{5.4}$$

where $\delta = \Delta\omega/\omega_0$. Equation (5.4) indicates that the gain of the RSC at the harmonic frequency ω_h will not be infinite unless $\delta = 0$ (i.e., $\Delta\omega = 0$). The gain reduction of the RSC components due to the frequency variation $\Delta\omega$ is illustrated in Figure 5.1, where it can be observed that even a small frequency variation of $\pm 0.2\%$ can result in a significant performance degradation of the RSC scheme (e.g., the magnitude decreases from ∞ dB to around 30 dB for the ninth-order RSC). It demonstrates that the RSC compensators are sensitive to frequency variations.

5.1.2 Direct frequency-adaptive RSC

If the interested harmonic frequency ω_h can be detected and updated in real time, the RSC in (5.1) can always produce infinite gain at the harmonic frequency $\pm\omega_h$. The infinite gain will lead to zero tracking errors at the interested harmonic frequency. Therefore, the frequency-adaptive RSC can be obtained to accurately compensate the frequency-varying harmonics.

As shown in Figure 5.2, a frequency-adaptive RSC is formed by two integrators, where a frequency detector (such as a phase-locked loop – PLL, which estimates the grid frequency as $\hat{\omega}_0$) should be employed to enable the RSC to directly adapt to the varying frequency. To be differentiated from the conventional RSC, the frequency-adaptive RSC is expressed as [33,34,56]:

$$G_{ah}(s) = \frac{u_{rsc}(s)}{e(s)} = k_h \frac{s}{s^2 + \hat{\omega}_h^2} \tag{5.5}$$

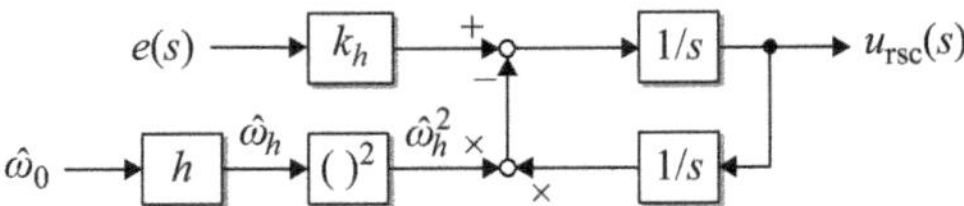

Figure 5.2 Frequency-adaptive resonant controller in the continuous-time domain

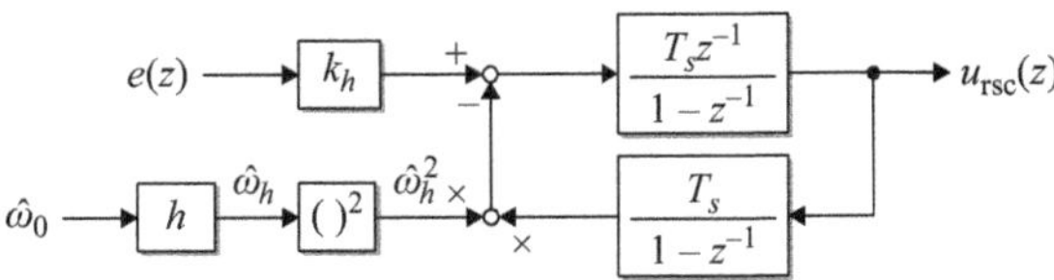

Figure 5.3 Frequency-adaptive resonant controllers in the discrete-time domain

where $\hat{\omega}_h = h\hat{\omega}_0$ denotes the detected interested harmonic frequency and h being the order of harmonics.

One discrete-time form of the frequency-adaptive RSC in (5.5), can be derived as:

$$G_{\mathrm{ah}}(z) = \frac{k_h T_s(z^{-1} - z^{-2})}{1 - (2 - \hat{\omega}_h^2 T_s^2)z^{-1} + z^{-2}} \tag{5.6}$$

which is given in Figure 5.3.

Consequently, according to (5.2) and (5.5), the frequency-adaptive MRSC can be written as [27–31]:

$$G_{aM}(s) = \sum_{h\in N_h} G_{\mathrm{ah}}(s) = \sum_{h\in N_h}\left(k_h \frac{s\cos\phi_h - \hat{\omega}_h \sin\phi_h}{s^2 + \hat{\omega}_h^2}\right) \tag{5.7}$$

where N_h represents the set of selected harmonic orders, and ϕ_h is the phase-lead angle for system delay compensation at the resonance frequency $\pm\hat{\omega}_h$. The analysis and synthesis of the frequency-adaptive MRSC systems are the same as those of the MRSC systems discussed in Section 2.2 in Chapter 2. It is clear that, the frequency-adaptive MRSC can achieve satisfactory tracking performance: high tracking accuracy, fast transient response, and good robustness by independently choosing its gain k_h and compensation phase ϕ_h at each selected harmonic frequency, but also may consume heavy parallel computational burden and tuning difficulty in the implementation of the frequency-adaptive MRSC.

The direct frequency-adaptive RSC [33] provides the simplest frequency-adaptive periodic control solution to varying-frequency harmonics compensation. Compared with the conventional RSC, the frequency-adaptive RSC scheme does not consume any additional computation.

5.1.3 Delay-based classic repetitive control (CRC)

Unlike the RSC, the recursive form PC schemes, such as the classic repetitive control (CRC) [1,2], parallel structure repetitive control (PSRC) [10,11], selective harmonic control (SHC) [12], and optimal harmonic control (OHC) [13,14] are implemented by including time-delays e^{sT_0} or $e^{sT_0/n}$ into the feedback loops, where $T_0 = 2\pi/\omega_0$ is the fundamental period of the periodic signal with ω_0 being the angular frequency. For example, the CRC can be expressed as:

$$G_{\mathrm{rc}}(s) = k_{\mathrm{rc}} \frac{Q(s)e^{-sT_0}}{1 - Q(s)e^{-sT_0}} e^{sT_c} = k_{\mathrm{rc}} \frac{Q(s)e^{-2\pi s/\omega_0}}{1 - Q(s)e^{-2\pi s/\omega_0}} e^{sT_c} \tag{5.8}$$

where k_{rc} is the control gain for tuning error convergence rate, $Q(s)$ with $|Q(j\omega)| \leq 1$ is usually a low-pass filter for selecting the interested harmonics for compensation, and e^{sT_c} with T_c being the compensation time is the phase-lead compensator for the system delay compensation. In practical applications, k_{rc}, $Q(s)$ and e^{sT_c} are employed to shape the RC system performance, such as system robustness, dynamic response, etc.

In the same manner, the actual fundamental frequency can be expressed as $\hat{\omega}_0 = \omega_0 + \Delta\omega = \omega_0(1 + \delta)$, where $\Delta\omega$ is the frequency deviation and $\delta = \Delta\omega/\omega_0$. Using the Euler's identity, the magnitude response of the CRC around the actual harmonic frequency (i.e., $s = jh\hat{\omega}_0$) can be obtained as [32,56]:

$$|G_{\mathrm{rc}}(jh\hat{\omega}_0)| = \left| k_{\mathrm{rc}} \frac{Q(jh\hat{\omega}_0)e^{-2\pi(jh\hat{\omega}_0)/\omega_0}}{1 - Q(jh\hat{\omega}_0)e^{-2\pi(jh\hat{\omega}_0)/\omega_0}} \right| = \frac{k_{\mathrm{rc}}}{\sqrt{2 - 2\cos(2\pi h\delta)}} \Bigg|_{|Q(jh\hat{\omega}_0)|=1} \tag{5.9}$$

where the phase-lead compensator e^{sT_c} has been ignored for simplicity. Equation (5.9) implies that the CRC no longer approaches an infinite control gain at the actual harmonic frequency when there is a frequency variation, i.e. $\delta \neq 0$. Figure 5.4 further illustrates the effect of a frequency deviation on the harmonic rejection ability of the CRC compensator, where a remarkable gain drop for the third-order harmonic (e.g., the magnitude decreases from ∞ dB to 28.5 dB) occurs due to a frequency change of $\delta = \pm 0.2\%$ (i.e., corresponding to a frequency variation of ± 0.1 Hz in 50-Hz systems), and consequently, the rejection ability is significantly degraded. Obviously, the CRC with a fixed time-delay T_0 is sensitive to frequency variations.

Furthermore, the PC schemes are commonly employed in their digital forms in practical applications. In the discrete-time domain, the CRC can be expressed as:

$$G_{\mathrm{rc}}(z) = k_{\mathrm{rc}} \frac{Q(z)z^{-N}}{1 - Q(z)z^{-N}} G_f(z) \tag{5.10}$$

where the integer $N = f_s/f_0 \in \mathbb{N}$ with f_0 being the fundamental harmonic frequency and f_s being the sampling rate, $G_f(z)$ is the phase-lead compensator, and the filter $Q(z)$ with $|Q(j\omega)| \leq 1$ is used to select the interested harmonics for compensation.

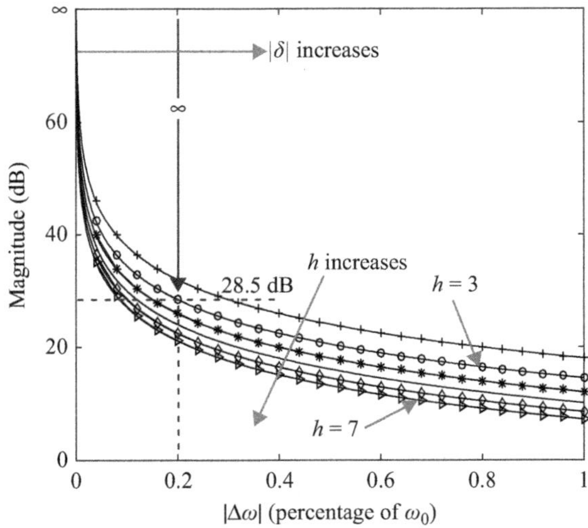

Figure 5.4 Magnitudes of the CRC $G_{rc}(s)$ as a function of the frequency variation $\Delta\omega$, where $k_{rc} = 1$

It is clearly seen that the amplitudes of $G_{rc}(z)$ at frequencies $2\pi m f_0$ with $m = 0, 1, 2, \ldots, M$ and $M = [N/2]$ approach infinity if $Q(z) = 1$. That is to say, the CRC provides zero steady-state error tracking of all harmonic components below the Nyquist frequency if $Q(z) = 1$ [2].

To ensure exact compensation of the interested harmonics, the CRC must have an integer order N. The time delay element z^{-N} with an integer order N is actually corresponding to the time-delay e^{-sT_0} in (2.3), and can be easily implemented in practice. However, in the case of a time-varying fundamental frequency f, the order $N = f_s/f_0$ would often be fractional with a fixed sampling rate f_s. It is impossible to directly implement a time-delay element z^{-n} if n is fractional. Hence, high control gains of the CRC with an integer order N will be shifted away from the interested harmonic frequencies in the presence of a time-varying frequency f. This means that the digital CRC with an integer order N is sensitive to the change of the fundamental harmonic frequency f.

In fact, due to the mismatch between the time delay length and the varying period of harmonics, all the delay-based PC strategies, such as the CRC, PSRC, SHC, and OHC, are sensitive to the harmonic frequency variations. In addition, the digital DFT-based RC is sensitive to the harmonic frequency variations since it assumes an integer order N of its polynomial that is also equal to the period of the fundamental harmonic frequency. The DFT-based RC in (2.45) can actually be treated as one kind of the time-delay-based PC schemes.

5.1.4 Frequency-adaptive CRC with a fixed sampling rate

The key to frequency-adaptive delay-based PC schemes is how to accurately produce required fractional delay for the internal models embedded in the PC.

Moreover, the implementation feasibility and complexity of the fractional delay would also determine the success of these frequency-adaptive PC schemes.

If the time T_0 in the delay element e^{-sT_0} can be adjusted to track the actual period of the fundamental frequency harmonics in real time, the continuous-time CRC in (5.8) can always produce infinite gains at the interested harmonic frequencies. Therefore, the frequency-adaptive CRC in the continuous-time domain can be obtained to accurately compensate the frequency-varying harmonics. However, it is not easy to implement variable delay elements in the continuous-time domain. Moreover, the PC strategies are usually employed in their digital forms in practical applications.

To enable the digital CRC to be frequency-adaptive, a variable sampling rate f_s [29,30] can be used to ensure a constant integer order $N = f_s/f$ in the presence of a time-varying frequency f. By keeping infinite gains at the interested harmonic frequencies, a variable sampling rate based CRC [34–36,56] can always mitigate the interested harmonics completely. Compared with the conventional CRC, the frequency-adaptive CRC with a variable sampling rate will not make any difference in terms of performance, design, and system stability. However, a variable sampling rate will significantly increase the real-time implementation complexity of the control systems [37], such as online timer interrupt updates, online controller redesign, and thus is not widely adopted in practical applications.

Owing to the implementation infeasibility of the delay z^{-N} with a fractional value $N=f_s/f$, a digital CRC with a fixed sampling rate f_s cannot exactly compensate the harmonics of a varying-frequency f. Nevertheless, the fractional digital delay can be expressed as:

$$z^{-N} = z^{-N_i-F} \tag{5.11}$$

with $N_i = [N]$ being the integer part of N and $F = N - N_i$ ($0 \leq F < 1$) being the fractional part of N. To enable the CRC to be frequency-adaptive, it is the most conventional and easiest way [38,39] to neglect the fractional delay z^{-F}, and replace the fractional delay z^{-N} with the nearest integer delay z^{-N_i} or $z^{-(N_i+1)}$ [38,39]. However, the harmonic rejection capability of such an integer-delay-based CRC might decay due to the omittance of the fractional delay z^{-F}, especially when the sampling rate f_s is low.

Fortunately, fractional delay (FD) filters with a fixed sampling rate provide a systematical approach to accurate band-limited approximation [40–42] of the fractional digital delay z^{-F}. With the fractional digital delay z^{-F} approximated by FD filters, the CRC will become immune to frequency variations. Among various FD filter design techniques, a traditional *Lagrange* interpolation method offers a simple but efficient approach to fast coefficient update and continuous control of the delay value. Using the *Lagrange* interpolation polynomial finite-impulse-response (FIR) FD filter, the fractional digital delay z^{-F} can be approximated as [43–51,56]:

$$z^{-F} \approx \sum_{k=0}^{n} A_k z^{-k} \tag{5.12}$$

where $k = 0, 1, \ldots, n$, and the coefficient A_k can be obtained as:

$$A_k = \prod_{i=0,\, i \neq k}^{n} \frac{F - i}{k - i} \quad k, i = 0, 1, \ldots, n \tag{5.13}$$

Specifically, if $n = 1$ in (5.13), a linear interpolation polynomial $z^{-F} \approx (1 - F) + Fz^{-1}$ will be obtained. Equations (5.12) and (5.13) indicate that a *Lagrange* interpolation-based FIR FD filter only needs a small number of multiplications and additions for coefficient updates, and it is well suitable for fast online tuning of the fractional delay value.

Substituting (5.11) and (5.12) into (5.10), a frequency-adaptive digital CRC can be obtained as [33–36]:

$$G_{\mathrm{arc}}(z) = k_{\mathrm{rc}} \frac{\left(z^{-N_i} \sum_{k=0}^{n} A_k z^{-k}\right) Q(z)}{1 - \left(z^{-N_i} \sum_{k=0}^{n} A_k z^{-k}\right) Q(z)} G_f(z) \tag{5.14}$$

which can be structured as shown in Figure 5.5.

The frequency-adaptive CRC in (5.14) will become the CRC in (5.10) when $F = 0$. It provides a general way to track or eliminate any periodic signal with an arbitrary fundamental frequency. According to the properties of the *Lagrange* interpolation method, the approximation remainder term of the fractional delay can be derived as (5.15):

$$R_n = z^{-F} - \sum_{k=0}^{n} A_k z^{-k} = \frac{\xi^{-F-n} \prod_{i=0}^{n-1} (-F - i)}{(n+1)!} \prod_{i=0}^{n} (F - i) \tag{5.15}$$

where $\xi \in [T_k, T_{k+1}]$ with T_k and T_{k+1} being the k-th and $(k+1)$-th sampling instants, respectively. With an increase of the degree n, a more accurate but complicated approximation can be acquired.

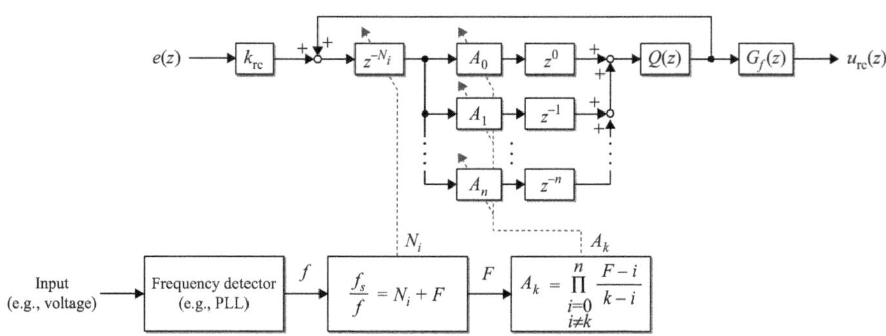

Figure 5.5 Frequency-adaptive digital classic repetitive controller $G_{arc}(s)$

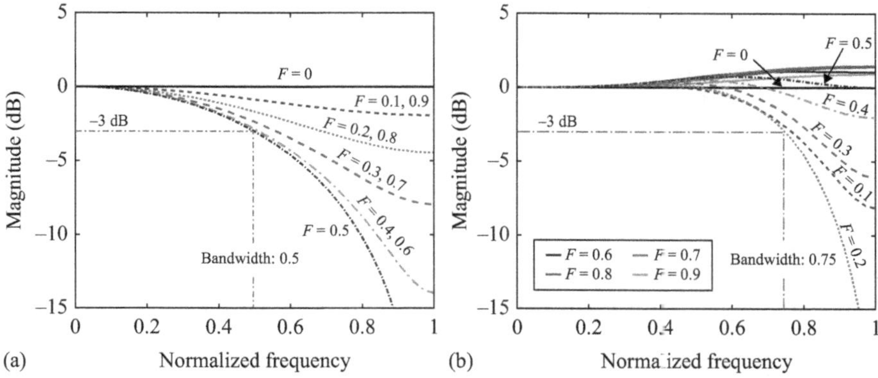

Figure 5.6 Frequency responses of the Lagrange interpolation-based FD filters: (a) $n = 1$ and (b) $n = 3$

Figure 5.6 shows the magnitude responses of the *Lagrange* interpolation-based FIR FD filter of (5.12) with the order $n = 1$ and $n = 3$ for various fractional F from 0 to 0.9. It is seen that the FD filter of (5.12) with the order $n = 3$ gives a good approximation of the fractional delay z^{-F} at low frequencies within the bandwidth of 75% of the Nyquist frequency, while a bandwidth of 50% of the Nyquist frequency for the FD filter of order $n = 1$. Within its pass band, the magnitude of the *Lagrange* interpolation-based FD filter of (5.12) is very close to that of the fractional delay z^{-F}. Then, the FD filter-based frequency-adaptive CRC can exactly compensate the fractional period signals within its bandwidth. Moreover, the coefficient in (5.13) for the FD filter only consumes a small number of additions and multiplications.

Besides the FIR FD filters, compact digital infinite-impulse-response (IIR) filters can also be used to approximate the fractional delay with a smaller number of multiplications than what the FIR FD filters do. Unfortunately, the design of an IIR FD filter with the prescribed magnitude and phase response is much more complicated. The design of a FIR filter is greatly eased by the fact that the filter coefficients are equal to the samples of the filter impulse response so that (in full-band approximation) the frequency-domain specifications can be turned into the "coefficient domain" by the inverse discrete-time Fourier transform. This is not possible for the IIR filters. Another major disadvantage of a digital IIR filter is the possible instability: updated coefficients may cause the poles of the digital IIR filter to move out of the unit circle. Therefore, the IIR FD filters are not recommended to be used in the construction of the fractional delay-based PC schemes.

Generally speaking, the *Lagrange* interpolation-based FIR FD filter offers an attractive way to enable frequency-adaptive delay-based PC schemes (including the CRC): easy explicit formulas for the coefficients, very accurate response at low frequencies, and a smooth magnitude response. However, it should be pointed out that such low-pass FD filters only provide satisfactory approximation of the

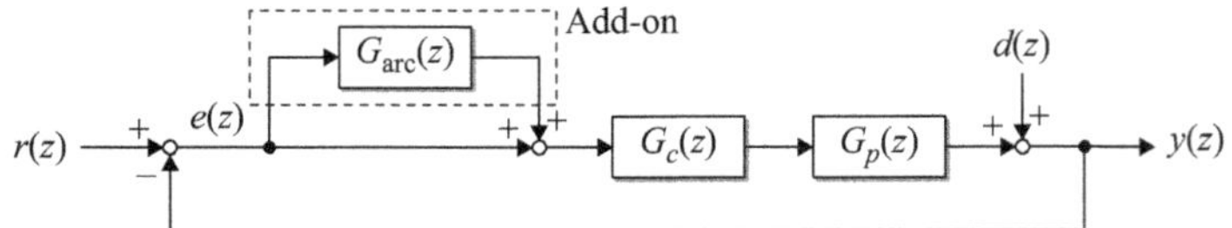

Figure 5.7 Plug-in (add-on) frequency-adaptive CRC system

fractional delay within its bandwidth. It implies that, at the high-frequency band, such FD filter-based frequency-adaptive PC strategies may not perform as well as what their corresponding conventional ones can do.

5.1.5 Frequency-adaptive CRC system and design

Figure 5.7 shows the typical closed-loop control system with a plug-in frequency-adaptive digital CRC, where $r(z)$ is the reference input, $y(z)$ is the output, $e(z) = r(z) - y(z)$ is the tracking error, $d(z)$ is the disturbance, $G_p(z)$ is the plant, $G_c(z)$ is the conventional feedback controller, $G_f(z)$ is a phase lead compensation filter to stabilize the overall closed-loop system, and $Q(z) = a_1 z + a_0 + a_1 z^{-1}$ with $2a_1 + a_0 = 1$ is a low-pass filter to enhance the robustness of the system [34].

Before the plug-in frequency-adaptive CRC is employed, the transfer function of the closed-loop system can be written as:

$$H(z) = \frac{y(z)}{r(z)} = \frac{G_c(z)G_p(z)}{1 + G_c(z)G_p(z)} = \frac{z^{-c}B^{+}(z^{-1})B^{-}(z^{-1})}{A(z^{-1})} \tag{5.16}$$

where $c \in \mathbb{R}$ is the known delay steps of the system, $B^{+}(z^{-1})$ and $B^{-}(z^{-1})$ are the cancelable and un-cancelable parts of the numerator, and $A(z^{-1}) = 0$ is the system characteristic equation.

To achieve zero-phase compensation [1], the compensating filter $G_f(z)$ can be chosen as:

$$G_f(z) = \frac{z^{c}A(z^{-1})B^{-}(z^{-1})}{B^{+}(z^{-1})b} \tag{5.17}$$

where $b \geq \max|B^{-}(z^{-1})|^2$. The delay steps c can be determined by experiments in practical applications. With the plug-in frequency-adaptive CRC in (5.11), the error transfer function of the overall system can be derived as:

$$\frac{e(z)}{r(z) - d(z)} = \frac{\left(1 + G_c(z)G_p(z)\right)^{-1}\left(1 - z^{-N_i}\left(\sum_{k=0}^{n} A_k z^{-k}\right)\right)}{1 - z^{-N_i}\left(\sum_{k=0}^{n} A_k z^{-k}\right)Q(z)(1 - k_r G_f(z)H(z))} \tag{5.18}$$

which indicates that the closed-loop frequency-adaptive CRC system will be asymptotically stable if the following two conditions hold:

1. Roots of $1 + G_c(z)G_p(z) = 0$ are inside the unit circle.
2. Roots of $1 - z^{-N_i}\sum_{k=0}^{n} A_k z^{-k} Q(z)\left(1 - k_r G_f(z)H(z)\right) = 0$ are inside the unit circle, then

$$\left|\left(1 - k_r G_f(z)H(z)\right)\right| < |Q(z)|^{-1}\left|\sum_{k=0}^{n} A_k z^{-k}\right|^{-1} \tag{5.19}$$

The above stability criteria for the frequency-adaptive CRC system is fully compatible to that of the CRC system [58–61]: the stability condition is the same as that for the CRC system; within the pass band of the FD filter in (5.12), when $\left|\sum_{k=0}^{n} A_k z^{-k}\right| \to 1$, the stability criterion of (5.19) for the frequency-adaptive CRC system is almost equivalent to that for the CRC system. Furthermore, assuming that the bandwidth of the FD filter in (5.12) is larger than the bandwidth of the low-pass filter $Q(z)$ in practical applications, the frequency-adaptive CRC in (5.14) would be almost the same as that of the CRC in (5.10) due to $|Q(z)|\left|\sum_{k=0}^{n} A_k z^{-k}\right| \to 1$. When $G_f(z)$ in (5.17) is applied to achieve zero-phase compensation, the stability range for the frequency-adaptive CRC gain will be $0 < k_r < 2$. It is clear that the synthesis of the frequency-adaptive CRC system can be almost the same as that of the well-known CRC system [58–61].

5.2 Frequency-adaptive advanced periodic control

5.2.1 Frequency adaptive parallel structure repetitive control (PRSC)

The PSRC is also a time-delay based PC scheme. The digital PSRC can be expressed as [10,11]:

$$G_{\text{psrc}}(z) = \sum_{m=0}^{n-1}\left[k_{\text{pm}}\frac{e^{j(2\pi m/n)}\cdot z^{-N/n}Q_m(z)}{1 - e^{j(2\pi m/n)}\cdot z^{-N/n}Q_m(z)}G_f(z)\right] \tag{5.20}$$

where $N = T_0/T_s \in \mathbb{N}$ with T_s being the sampling period, $G_f(z)$ is a phase-lead compensator, and the filter $Q_m(z)$ often employs an FIR low-pass filter.

In (5.20), the time-delay element can be generally written as:

$$z^{-N/n} = z^{-N_i - F} \tag{5.21}$$

where $N_i \in \mathbb{N}$ is the integer part of N/n, and F is the fractional part of N/n that can also take the *Lagrange* interpolation-based FD filter:

$$z^{-F} \approx \sum_{k=0}^{n} A_k z^{-k} = \left(\prod_{\substack{i=0\\ i\neq k}}^{n}\frac{F-i}{k-i}\right)z^{-k} \tag{5.22}$$

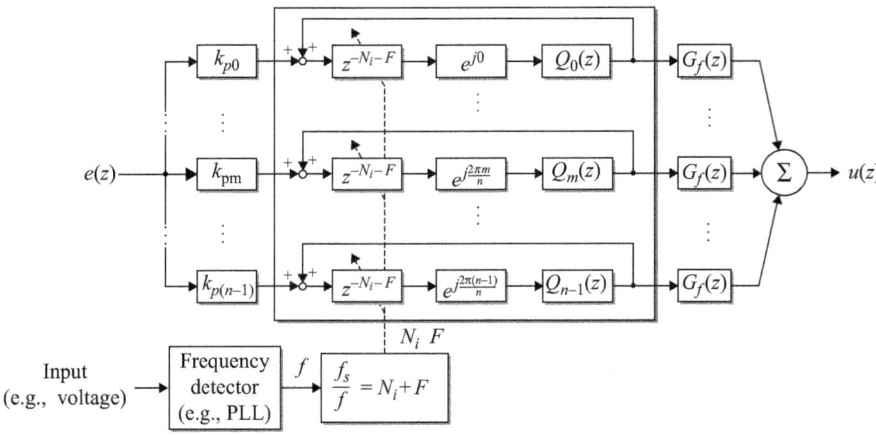

Figure 5.8 Frequency-adaptive digital parallel structure repetitive controller

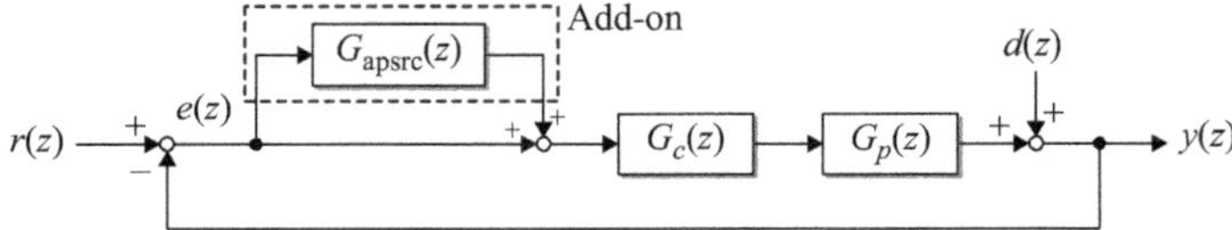

Figure 5.9 Plug-in (add-on) frequency-adaptive parallel structure repetitive control system

Substituting (5.21) and (5.22) into (5.20), a frequency-adaptive PSRC can be given as [37]:

$$G_{\text{apsrc}}(z) = \sum_{m=0}^{n-1}\left[k_{\text{pm}}\frac{e^{j(2\pi m/n)}(z^{-N_i}z^{-F})Q_m(z)}{1 - e^{j(2\pi m/n)}(z^{-N_i}z^{-F})Q_m(z)}G_f(z)\right]$$

$$= \sum_{m=0}^{n-1}\left[k_{\text{pm}}\frac{e^{j(2\pi m/n)}\left(z^{-N_i}\sum\limits_{k=0}^{n}A_k z^{-k}\right)Q_m(z)}{1 - e^{j(2\pi m/n)}\left(z^{-N_i}\sum\limits_{k=0}^{n}A_k z^{-k}\right)Q_m(z)}G_f(z)\right] \tag{5.23}$$

Figure 5.8 shows the implementation of the frequency-adaptive PSRC. The plug-in frequency-adaptive PSRC system is given in Figure 5.9. Assuming that the bandwidth of the FD filter in (5.22) is larger than the bandwidth of the low-pass filter $Q(z)$, the frequency-adaptive PSRC in (5.23) would be equivalent to the PSRC in (5.20) due to $|Q(z)|\left|\sum_{k=0}^{n}A_k z^{-k}\right| \to 1$ within the bandwidth of $Q(z)$. Under such conditions, the design and synthesis of a frequency-adaptive PSRC system are the same as those of the conventional PSRC system that are presented in Section 3.1 in Chapter 3.

5.2.2 Frequency-adaptive selective harmonic control (SHC)

Similarly, as shown in Figure 5.10, using the *Lagrange* interpolation-based FD filter, the frequency-adaptive digital SHC can be written as [52]:

$$G_{\text{asm}}(z) = \frac{k_m G_f(z)\left(\cos(2\pi m/n)Q(z)\left(z^{-N_i}\sum_{k=0}^{n} A_k z^{-k}\right) - Q^2(z)\right)}{\left(z^{-N_i}\sum_{k=0}^{n} A_k z^{-k}\right)^2 - 2\cos(2\pi m/n)Q(z)\left(z^{-N_i}\sum_{k=0}^{n} A_k z^{-k}\right) + Q^2(z)} \tag{5.24}$$

where $N = T_0/T_s \in \mathbb{N}$ with T_s being the sampling period, $N_i = [N/n] \in \mathbb{N}$ is the integer part of N/n, and F is the fractional part of N/n. The FIR FD filter is:

$$\sum_{k=0}^{n} A_k z^{-k} = \sum_{k=0}^{n}\left(\prod_{\substack{i=0\\ i\neq k}}^{n} \frac{F-i}{k-i}\right) z^{-k} \tag{5.25}$$

In practice, to simplify the design of frequency-adaptive digital SHC, a linear phase-lead compensator $G_f(z) = z^c$ is usually employed in the SHC (5.24) instead of the inverse transfer function of $H(z)$ [58–61]; and an FIR low-pass filter $Q(z) = \alpha_1 z + \alpha_0 + \alpha_1 z^{-1}$ with $2\alpha_1 + \alpha_0 = 1$, $\alpha_0 \geq 0$ and $\alpha_1 \geq 0$ is also usually employed. The implementation of the frequency-adaptive digital SHC with linear phase-lead compensation is illustrated in Figure 5.11.

Similarly, when the bandwidth of the FD filter in (5.25) is larger than the bandwidth of the low-pass filter $Q(z)$, the frequency-adaptive SHC in (5.24) would be equivalent to the conventional SHC in (3.29) due to $|Q(z)|\left|\sum_{k=0}^{n} A_k z^{-k}\right| \to 1$. Under such conditions, the design and synthesis of the plug-in frequency-adaptive

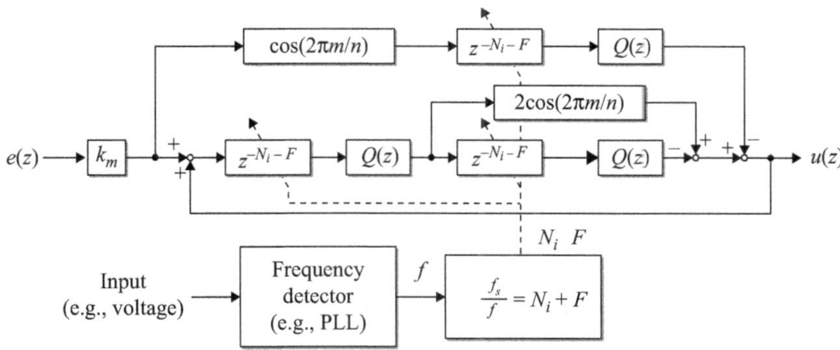

Figure 5.10 Frequency-adaptive digital selective harmonic controller

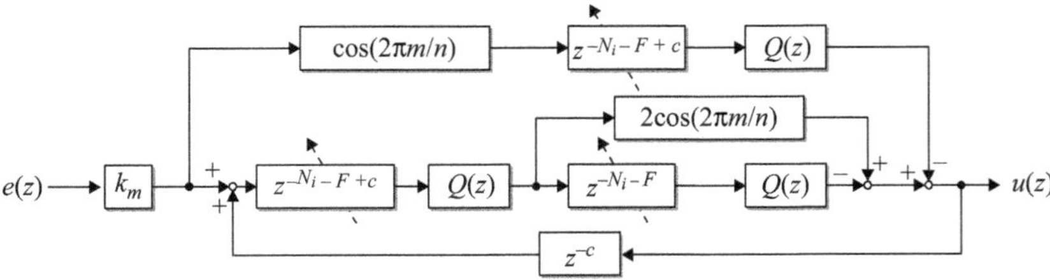

Figure 5.11 Frequency-adaptive digital selective harmonic controller with a linear phase-lead compensator z^c

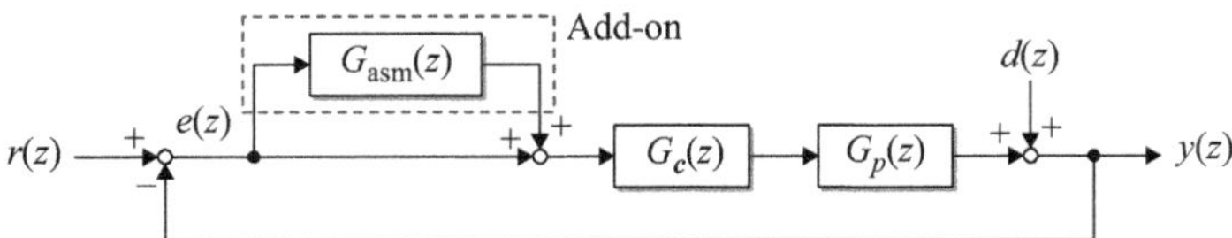

Figure 5.12 Plug-in (add-on) frequency-adaptive digital selective harmonic control system

SHC system shown in Figure 5.12 are the same as those of the conventional SHC system that are presented in Section 3.2 in Chapter 3.

5.2.3 Frequency-adaptive optimal harmonic control (OHC)

From Section 3.3, a digital OHC scheme comprises paralleled digital SHC modules tailored for the selected harmonics can be given as [52]:

$$G_{\mathrm{OHC}}(z) = \sum_{m\in N_m} G_{\mathrm{sm}}(z) = \sum_{m\in N_m} k_m \frac{\left[\cos(2\pi m/n) z^{N/n} Q(z) - Q^2(z)\right] G_f(z)}{z^{2N/n} - 2\cos(2\pi m/n) z^{N/n} Q(z) + Q^2(z)} \tag{5.26}$$

Likewise, substituting the frequency-adaptive SHC scheme of (5.24) into (5.26), the frequency-adaptive digital OHC can be obtained as:

$$G_{\mathrm{aOHC}}(z) = \sum_{m\in N_m} \frac{k_m G_f(z)\left(\cos(2\pi m/n)\left(z^{N_i}\sum_{k=0}^{n} A_k z^k\right)Q(z) - Q^2(z)\right)}{\left(z^{N_i}\sum_{k=0}^{n} A_k z^k\right)^2 - 2\cos(2\pi m/n)\left(z^{N_i}\sum_{k=0}^{n} A_k z^k\right)Q(z) + Q^2(z)} \tag{5.27}$$

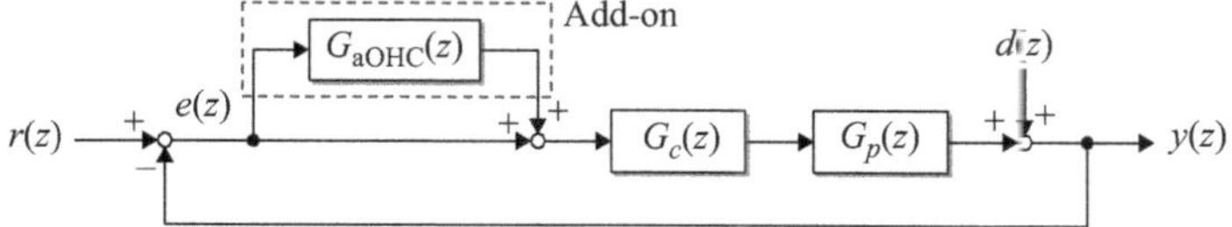

Figure 5.13 Plug-in (add-on) frequency-adaptive digital optimal harmonic control system

where $N = T_0/T_s \in \mathbb{N}$ with T_s being the sampling period, $N_* = [N/n] \in \mathbb{N}$ is the integer part of N/n, and F is the fractional part of N/n. The FIR FD filter is:

$$\sum_{k=0}^{n} A_k z^{-k} = \sum_{k=0}^{n} \left(\prod_{\substack{i=0 \\ i \neq k}}^{n} \frac{F-i}{k-i} \right) z^{-k} \tag{5.28}$$

Likewise, when the bandwidth of the FD filter in (5.28) is larger than the bandwidth of the low-pass filter $Q(z)$, the frequency-adaptive OHC in (5.27) would be almost the same as that of the conventional SHC in (3.38) due to $|Q(z)|\left|\sum_{k=0}^{n} A_k z^{-k}\right| \rightarrow 1$. Under such conditions, the design and synthesis of the plug-in frequency-adaptive OHC system shown in Figure 5.13 are the same as those of the conventional OHC system that are presented in Section 3.3.

5.3 Frequency-adaptive discrete Fourier transform-based RC

As discussed in Chapter 2, any periodic signal can be decomposed into the summation of its harmonics. The DFT [35,36] can be adopted to flexibly construct the digital internal models of all harmonic components of any interested periodic signal. Based on the IMP, the closed-loop digital control system with these DFT-based internal models being included, can achieve zero error tracking of the interested periodic signal. Such a control method is named DFT-based RC [54].

From Section 2.3, as also shown in Figure 5.14, a DFT-based RC shown in Figure 2.12 can be written as [53,54]:

$$G_{\mathrm{DFT}}(z) = \frac{u_{\mathrm{rc}}(z)}{e(z)} = \frac{k_F F_{\mathrm{DFT}}(z)}{1 - F_{\mathrm{DFT}}(z) z^{-N_a}} \tag{5.29}$$

in which the integer $N = f_s/f \in \mathbb{N}$ with f being the fundamental harmonic frequency and f_s being the sampling rate, k_F is the control gain for tuning the dynamic response, N_a represents the number of leading steps for phase-lead compensation, and the DFT-based FIR comb filter $F_{\mathrm{DFT}}(z)$ can be given as [54]:

$$F_{\mathrm{DFT}}(z) = \sum_{h \in N_h} F_{\mathrm{dh}}(z) = \frac{2}{N} \sum_{i=1}^{N} \left(\sum_{h \in N_h} \cos\left[\frac{2\pi}{N} h(i+N_a)\right] \right) z^{-i} = \left(\sum_{i=1}^{N} b_h(i) z^{-i} \right) z^{N_a} \tag{5.30}$$

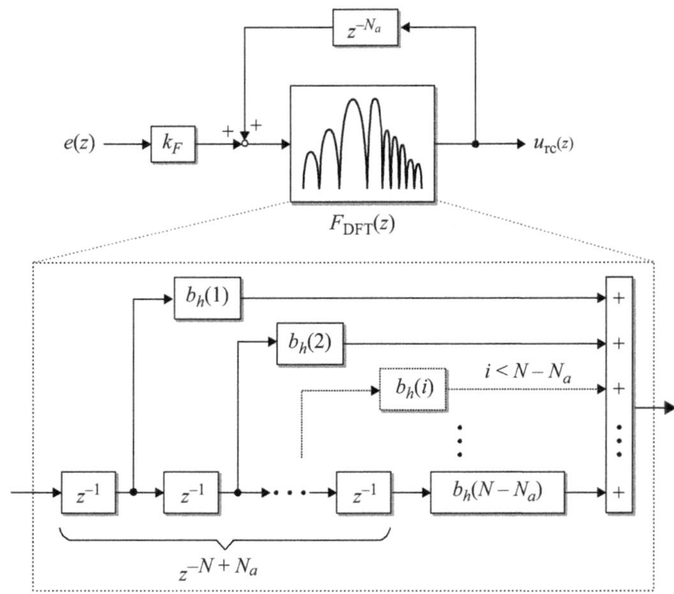

Figure 5.14 Implementation of the DFT-based RC

and

$$F_{dh}(z) = \frac{2}{N}\sum_{i=1}^{N}\cos\left(\frac{2\pi}{N}h(i+N_a)\right)z^{-i} = \left(\sum_{i=1}^{N}a_h(i)z^{-i}\right)z^{N_a} \tag{5.31}$$

Obviously, from (5.30) and (5.31), it can be seen that it will take a large amount of complex computations to calculate the coefficients $b_h(i)$. To facilitate the real-time implementation of the comb filter $F_{DFT}(z)$ (and thus the DFT-based RC), the coefficients $b_h(i)$ should be pre-calculated and stored in lookup tables in advance. Figure 5.14 indicates that the implementation of the comb filter $F_{DFT}(z)$ will further involve lots of multiplications and additions. The DFT-based RC is thus suitable for the fixed-point DSP implementation.

It can be found in Figure 5.14 that, if the number of $N = f_s/f_0$ change with the fundamental frequency f of the interested periodic signal, it is infeasible to consume a large amount of complex computations to online recalculate and update all the coefficients $b_h(i)$ for the implementation of the DFT-based RC. Hence, it is not feasible either to directly employ the simple and efficient *Lagrange* interpolation-based FD method shown in Figure 5.5 to transform the DFT-based RC in (5.29) into a frequency-adaptive DFT-based RC.

A variable sampling rate can enable delay-based digital PC (including the DFT-based RC) to be frequency-adaptive [29,30], as aforementioned. With a variable sampling rate f_s, $N = f_s/f$ can maintain a constant integer in the presence of a varying frequency f. Since the unit delay z^{-1} with a variable sampling rate actually is a variable fractional delay element, it can be used to form any expected fractional delay. Therefore, without online updating coefficients, the delay-based digital PC schemes with a variable sampling rate are frequency-adaptive despite having other implementation complexity.

Hereinafter, a new method named *virtual unit delay* is introduced to enable the DFT-based RC with a fixed sampling rate to be frequency-adaptive. It combines the advantages of both the *Lagrange* interpolation FD-filter-based frequency-adaptive PC and the variable-sampling-rate-based frequency-adaptive RC.

In the presence of a varying frequency f with a fixed sampling rate f_s, the actual value $N_F = f_s / f_0$ would be fractional, and can be written as:

$$N_F = N + F = N\left(1 + \frac{F}{N}\right) = N(1 + F_N) \tag{5.32}$$

where F with $|F| \ll N$ is the difference between actual value N_F and the nominal integer number N ($N = f_s / f_0$ with f_0 being the nominal frequency), and $|F_N| = |F/N| \ll 1$. Thus, with a fixed sampling rate, the comb filter $F_{\mathrm{DFT}}(z)$ in (5.30) can be mathematically written as:

$$F_{\mathrm{DFT}}(z) = \sum_{h \in N_h} F_{\mathrm{dh}}(z) = \frac{2}{N+F} \sum_{i=1}^{N+F} \left(\sum_{h \in N_h} \cos\left[\frac{2\pi}{N+F} h(i + N_a)\right] \right) z^{-i} \tag{5.33}$$

When F is fractional, it is impossible to online calculate the coefficients at the right side of (5.33). This implies that the DFT-based RC in (5.29) with a fixed sampling rate is sensitive to frequency variations.

Imaging that, a virtual variable sampling rate f_v enables $N_F = f_v / f_0$ to always maintain its nominal integer value N in the presence of a varying frequency f, the comb filter $F_{\mathrm{DFT}}(z)$ in (5.33) can then be rewritten as:

$$F_{\mathrm{DFT}}(z_v) = \frac{2}{N} \sum_{i=1}^{N} \left(\sum_{h \in N_h} \cos\left[\frac{2\pi}{N} h(i + N_a)\right] \right) z_v^{-i} = \left(\sum_{i=1}^{N} b_h(i) z_v^{-i} \right) z_v^{N_a} \tag{5.34}$$

where the virtual unit delay z_v^{-1} is a fractional delay [35,57], that can be written as:

$$z_v^{-1} = z^{-(1+F_N)} \tag{5.35}$$

Also employing the *Lagrange* interpolation method, the fractional delay z^{-F_N} in (5.35) can be well approximated by a linear interpolation (i.e., the *Lagrange* interpolation with order $n = 1$) based FIR FD filter as [40]:

$$z_v^{-1} = z^{-(1+F_N)} \approx \begin{cases} |F_N| z^0 + (1 - |F_N|) z^{-1} & -1 < F_N < 0 \\ (1 - |F_N|) z^{-1} + |F_N| z^{-2} & 0 \le F_N < 1 \end{cases} \tag{5.36}$$

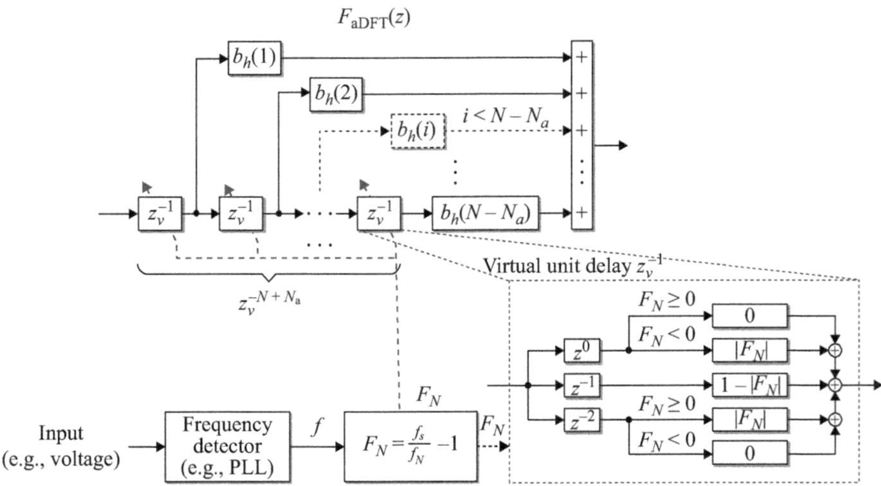

Figure 5.15 Implementation of the frequency-adaptive comb filter $F_{aDFT}(z)$

Therefore, substituting (5.36) into (5.34), a frequency-adaptive DFT-based comb filter $F_{\mathrm{aDFT}}(z)$ with a fixed sampling rate can be obtained as:

$$F_{\mathrm{aDFT}}(z_v) = \frac{2}{N}\sum_{i=1}^{N}\left(\sum_{h\in N_h}\cos\left[\frac{2\pi}{N}h(i+N_a)\right]\right)z_v^{-i} = \left(\sum_{i=1}^{N} b_h(i)z_v^{-i}\right)z_v^{N_a} \tag{5.37}$$

The frequency-adaptive DFT-based comb filter $F_{\mathrm{aDFT}}(z)$ in (5.37) with a fixed sampling rate is approximately equivalent to a frequency-adaptive DFT-based comb filter $F_{\mathrm{DFT}}(z)$ in (5.30) with a variable sampling rate. Figure 5.15 illustrates the implementation of the frequency-adaptive comb filter $F_{\mathrm{aDFT}}(z)$. Compared with the comb filter $F_{\mathrm{DFT}}(z)$ in (5.30), the frequency-adaptive comb filter $F_{\mathrm{aDFT}}(z)$ in (5.37) will add $2N$ multiplications, N additions, and N unit delays in its implementation. By replacing $F_{\mathrm{DFT}}(z)$ with $F_{\mathrm{aDFT}}(z)$ in (5.29), a frequency-adaptive DFT-based RC can be obtained as:

$$G_{\mathrm{aDFT}}(z) = \frac{u_F(z)}{e(z)} = \frac{k_F F_{\mathrm{aDFT}}(z)}{1 - F_{\mathrm{aDFT}}(z)z_v^{-N_a}} \tag{5.38}$$

The major benefits of such a virtual unit delay method include: the coefficients $b_h(i)$ of the frequency-adaptive filter $F_{\mathrm{aDFT}}(z)$ are the same as those of the conventional filter $F_{\mathrm{DFT}}(z)$. This means that all coefficients $b_h(i)$ can be calculated offline and do not need to be updated online.

As shown in Figure 5.16, the smaller the fractional value F_N is, the closer the FD filter in (5.36) approximates the expected fractional delay. That is to say, the frequency-adaptive DFT-based RC in (5.38) is suitable for compensating harmonics with a very small frequency variation range (e.g., $|f-f_0|/f_0 \leq 1\%$) [55]. Higher sampling rates would lead to a better compensation accuracy.

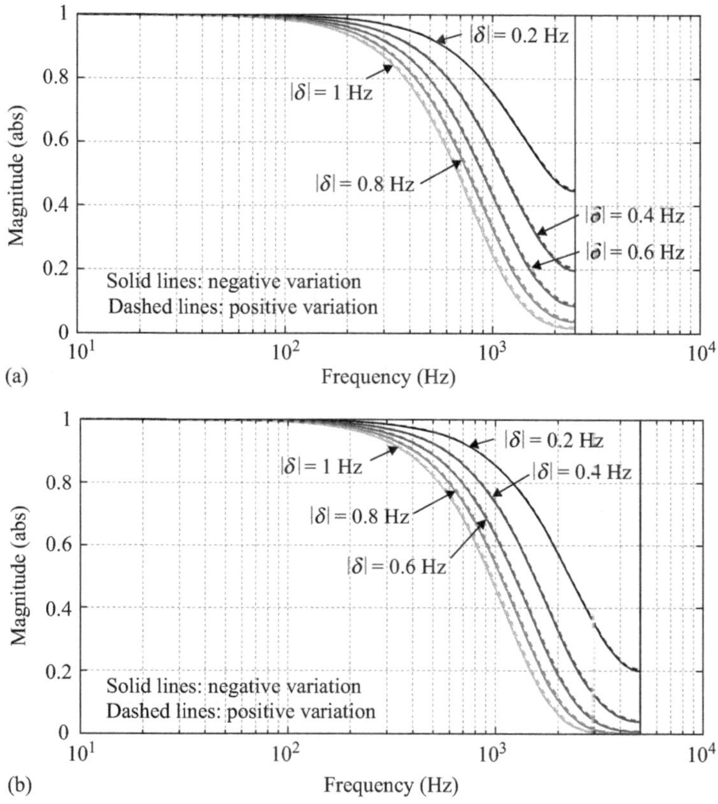

Figure 5.16 Magnitude responses of cascaded virtual unit delays z_v^{-N}, where $f_0 = 50$ and $\delta = f - 50$ representing the frequency variation: (a) $f_s = 5$ kHz, $N = 100$ and (b) $f_s = 10$ kHz, $N = 200$

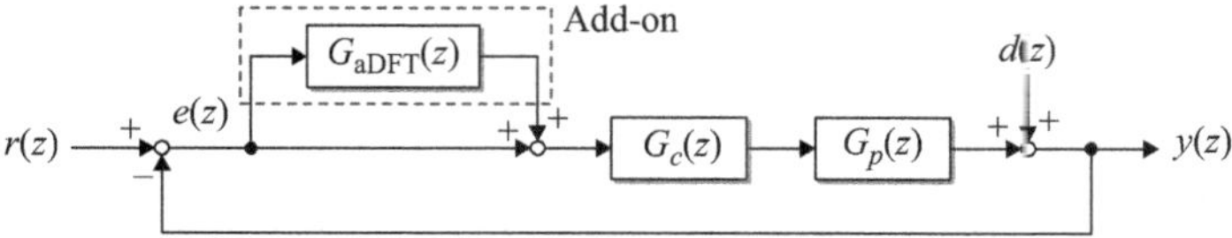

Figure 5.17 Plug-in (add-on) frequency-adaptive digital DFT-based RC system

It should be pointed out that such a virtual unit delay z_v^{-1} can also be employed to develop the frequency-adaptive CRC, PSRC, SHC, and OHC, but at the same time it will significantly increase the computational burden and needs more memory elements in implementation. Nevertheless, Figure 5.17 exemplifies a plug-in virtual unit delay-based frequency-adaptive DFT-RC system.

For the plug-in frequency-adaptive DFT-based RC system shown in Figure 5.17, likewise, when the FIR FD filter in (5.36) accurately approximates the fractional delay z_v^{-1}, the frequency-adaptive DFT-based RC in (5.38) would be equivalent to the conventional DFT-based RC in (5.29). Under such conditions, the design and synthesis of the plug-in frequency-adaptive DFT-based RC system shown in Figure 5.17 are the same as those of the DFT-based system that is presented in Section 3.4. It should be noted that the DFT-based RC is only suitable for fixed-point DSP implementation due to its large amount of computation.

It should also be pointed out that, unlike the unit delay with a variable sampling rate, the virtual unit delay with a fixed sampling rate can only be used to approximate a fractional delay within a very limited frequency range (e.g., $|f-f_0|/f_0 \leq 1\%$), as shown in Figure 5.16). Like the previous FD-filter-based frequency-adaptive PC schemes, at high frequency bands, the virtual unit delay-based frequency-adaptive PC schemes cannot perform as well as what the corresponding conventional one can do. They demand a large amount of additional computation in their implementation.

5.4 Fractional-order phase-lead compensator

Most modern control systems are employed in their digital forms with a fixed sampling rate in practical applications. Now it is well known that, due to the frequency sensitivity of the delay, integer-delay-based PC schemes with a fixed sampling rate may be significantly degraded in terms of control accuracy in the presence of frequency variations for references and/or disturbances. Fractional delays will enable the periodic controllers to become frequency-adaptive.

Likewise, the delay-based linear phase-lead compensator for the delay-based PC systems is also sensitive to the frequency variation and the ratio of the sampling frequency to the periodic signal frequency. Fractional order implementation will enable periodic control systems to achieve better performance against frequency variations and/or the sampling ratio, such as high accuracy, large stability range, and fast dynamic response.

A phase-lead compensator $G_f(z)$ for the plug-in CRC in Figure 5.18 can be used to improve the system stability by providing phase lead to cancel out the phase lag of the closed-loop system [20]. In practice, to simplify the system design and implementation, a linear phase-lead compensator of $G_f(z) = z^c$ is often adopted in the CRC, where c is an integer-phase-lead step. For a reference signal with the frequency of f, the phase-lead compensator z^c produces a phase lead $\phi = c \times (f/f_s) \times 360°$ in the frequency domain and a time lead $t_d = c \times (1/f_s)$ in the time domain, where f_s represents the sampling frequency, to compensate the system phase lag. The phase compensation resolution of z^c is $(f/f_s) \times 360°$. Obviously,

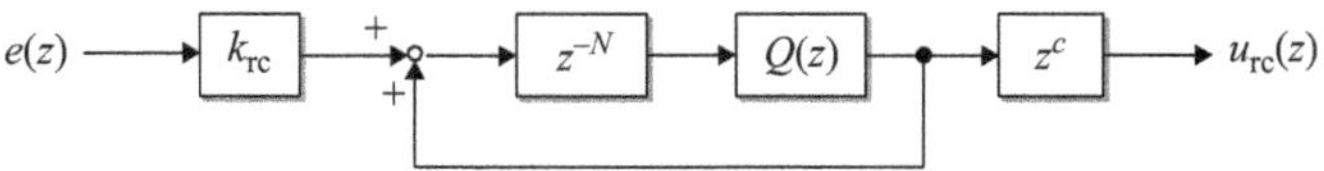

Figure 5.18 Classic repetitive controller with a linear phase-lead compensator

a lower sampling frequency f_s or a higher reference signal frequency f will lead to a lower phase compensation resolution.

Linear phase-lead compensation $G_f(z) = z^c$ [17,18] is an effective way to improve the CRC performance. It is clear that a lower ratio of f/f_s could lead to higher compensation resolution and z^c can generate more accurate phase-lead compensation. However, it cannot provide satisfactory compensation accuracy with a low compensation resolution in the case of high ratio of f/f_s, such as onboard AC power supplies with 400 Hz reference signals on ships and aircrafts [62], high power low switching frequency (sampling frequency) grid converters for wind power and PV generation [63], wide frequency range programmable AC power supplies for grid simulators [47], and multirate repetitive control [64] with a reduced sampling frequency $f_m = f_N/m$ and $m = 2, 3, \ldots$. Imprecise phase lead may degrade control performance or even make the RC systems unstable.

To offer accurate phase compensation, a fractional-order phase-lead compensator is developed as [65]:

$$G_f(z) = z^{\gamma} = z^{n_i+F} \tag{5.39}$$

where $\gamma \in \mathbb{R}^+$ is the fractional lead step, $n_i = \lfloor \gamma \rfloor$ is the integer part, and F is its fractional part.

The fractional-order phase-lead compensator can be approximated by a first-order *Lagrange* interpolation polynomial FIR filter as:

$$G_f(z) = z^{\gamma} = z^{n_i+F} \approx (1-F)z^{n_i} + Fz^{n_i+1} \tag{5.40}$$

Such a fractional-order lead step γ will produce a linear phase lead:

$$\phi = \gamma \times \frac{f}{f_s} \times 360^\circ \tag{5.41}$$

Therefore, the fractional-order phase-lead compensator of (5.40) offers more accurate phase compensation than what the phase-lead compensator $G_f(z) = z^c$ with an integer step c does.

Like (2.23) for the integer-order phase-lead compensation CRC system, the corresponding stability criterion for the fractional-order one can be derived as:

$$|(1 - k_{rc} z^{\gamma} H(z))Q(z)| < 1, \quad \text{for } z = e^{j\omega} \text{ with } \omega < \frac{\pi}{T_s} \tag{5.42}$$

where

$$H(z) = \frac{G_c(z)G_p(z)}{1 + G_c(z)G_p(z)} = \left|H(e^{j\omega})\right| e^{j\theta_H(\omega)}, \quad \text{for } z = e^{j\omega} \text{ with } \omega < \frac{\pi}{T_s}$$

Following the same frequency-domain analysis approach in Section 2.1, (5.42) becomes:

$$k_{rc}^2 < \frac{1 - |Q(e^{j\omega})|^2}{|H(e^{j\omega})|^2 |Q(e^{j\omega})|^2} + \frac{2k_{rc}\cos(\theta_H + \gamma\omega)}{|H(e^{j\omega})|} \tag{5.43}$$

$$\text{If } 2k\pi - \frac{\pi}{2} < \theta_H + \gamma\omega < 2k\pi + \frac{\pi}{2}, \quad k = 0, 1, 2, \ldots, \text{ then}$$

$$0 < k_{rc} < \frac{2\cos(\theta_H + \gamma\omega)}{|H(e^{j\omega})|} \tag{5.44}$$

which will sufficiently enable the CRC system to be asymptotically stable;

$$\text{if } 2k\pi + \frac{\pi}{2} < \theta_H + \gamma\omega < 2k\pi + \frac{3\pi}{2}, \quad k = 0, 1, 2, \ldots, \text{ then}$$

$$\frac{2\cos(\theta_H + \gamma\omega)}{|H(e^{j\omega})|} < k_{rc} < 0 \tag{5.45}$$

which will enable the closed-loop CRC system to be asymptotically stable.

It should be pointed out that both the integer-order CRC and the frequency-adaptive RC can adopt the fractional-order linear phase-lead compensator to further improve their system stability and dynamic responses. Moreover, the fractional-order linear phase-lead compensator provides a universal phase-lead compensation method to PC schemes, e.g., the SHC and the PSRC.

5.5 Summary

To efficiently mitigate varying frequency harmonics, frequency-adaptive PC schemes with a fixed sampling rate have been comprehensively investigated: direct frequency adaption method for the RSC, the *Lagrange* interpolation-based FIR FD filter for delay-based PC strategies (such as the CRC, PSRC, SHC, and OHC), and the virtual variable sampling rate (i.e., virtual unit delay) FIR FD filter for the DFT-based RC. Frequency sensitivity, design, and implementation methodology of the frequency-adaptive PC schemes have been explored. Also, compatible synthesis methods for plug-in frequency-adaptive PC systems have been presented in this chapter. In addition, the FIR FD filter is used to create fractional-order phase-lead compensation for PC schemes.

The main features of the three frequency-adaptive methods for the PC schemes in this chapter are listed as follows:

1. Direct frequency adaption method –
   - Suitable for the RSC.
   - Simple. It does not almost bring any added computations.
   - Stability criteria and system design rules-of-thumb for the frequency-adaptive RSC systems are the same as those for the conventional RSC systems.

2. *Lagrange* interpolation-based FIR FD filter
   - Suitable for time-delay-based PC schemes.
   - Easy for implementation and online tuning.
   - Light computation. It only brings a few added computations.
   - Stability criteria and system design rules-of-thumb for the frequency-adaptive PC systems are compatible to those for the corresponding PC systems.

3. Virtual variable sampling rate FIR FD filter
   - Suitable for both DFT-based RC and time-delay-based PC systems.
   - Stability criteria and system design rules-of-thumb for the frequency-adaptive DFT-based RC systems are compatible to those for the conventional DFT-based RC systems.
   - Increased computational burden is proportional to the ratio of the sampling rate to the periodic signal frequency.
   - Suitable for fixed-point DSP implementation.
4. Fractional-order phase-lead compensation
   - A universal phase-lead compensation method for PC schemes, e.g. the CRC, SHC, PSRC, etc., in the presence of frequency variations and/or a low ratio of sampling frequency to signal frequency.

It should be pointed out that the FIR FD-filter-based frequency-adaptive digital PC schemes are mainly used to compensate harmonics within a small range of frequency variation (e.g., less than 1% of the nominal frequency) due to the bandwidth limitation of the low-pass FD filter.

References

[1] Tomizuka, M., Tsao, T.-C., and Chew, K.-K., "Analysis and synthesis of discrete-time repetitive controllers," *J. Dyn. Syst., Meas. Control*, vol. 111, no. 3, pp. 353–358, 1989.

[2] Tzou, Y.-Y., Jung, S.-L., and Yeh, H.-C., "Adaptive repetitive control of PWM inverters for very low THD AC-voltage regulation with unknown loads," *IEEE Trans. Power Electron.*, vol. 14, no. 5, pp. 973–981, Sep. 1999.

[3] Liserre, M., Teodorescu, R., and Blaabjerg, F., "Multiple harmonics control for three-phase grid converter systems with the use of PI-RES current controller in a rotating frame," *IEEE Trans. Power Electron.*, vol. 21, no. 3, pp. 836–841, May 2006.

[4] Lascu, C., Asiminoaei, L., Boldea, I., and Blaabjerg, F., "High Performance Current Controller for Selective Harmonic Compensation in Active Power Filters," *IEEE Trans. Power Electron.*, vol. 22, no. 5, pp. 1826–1835, 2007.

[5] Lascu, C., Asiminoaei, L., Boldea, I., and Blaabjerg, F., "Frequency Response Analysis of Current Controllers for Selective Harmonic Compensation in Active Power Filters," *IEEE Trans. Ind. Electron.*, vol. 56, no. 2, pp. 337–347, 2009.

[6] Yepes, G., Freijedo, F.D., Lopez, O., and Doval-Gandoy, J., "High performance digital resonant controllers implemented with two integrators," *IEEE Trans. Power Electron.*, vol. 6, no. 2, pp. 563–576, 2011.

[7] Miret, J., Castilla, M., Matas, J., Guerrero, J.M., and Vasquez, J.C., "Selective harmonic-compensation control for single-phase active power filter with high harmonic rejection," *IEEE Trans. Ind. Electron.*, vol. 56, no. 8, pp. 3117–3127, Aug. 2009.

[8] Yang, Y., Zhou, K., Cheng, M., and Zhang, B., "Phase compensation multi-resonant control of CVCF PWM converters," *IEEE Trans. Power Electron.*, vol. 28, no. 8, pp. 3923–3930, Aug. 2013.

[9] Yang, Y., Zhou, K., Cheng, M., "Phase compensation resonant controller for single-phase PWM converters," *IEEE Trans. Ind. Inform.*, vol. 9, vol .2, pp. 957–964, 2013.

[10] Lu, W., Zhou, K., Wang, D., and Cheng, M., "A general parallel structure repetitive control scheme for multiphase DC–AC PWM converters", *IEEE Trans. Power Electron.*, vol. 28, no. 8, pp. 3980–3987, 2013.

[11] Lu, W., Zhou, K., and Wang, D., "General parallel structure digital repetitive control," *Int. J. Control*, vol. 86, no. 1, pp. 70–83, Jan. 2013.

[12] Lu, W., Zhou, K., Wang, D., and Cheng, M., "A generic digital nk±m order harmonic repetitive control scheme for PWM converters," *IEEE Trans. Ind. Electron.*, vol. 61, no. 3, pp. 1516–1527, 2014.

[13] Zhou, K., Yang, Y., Blaabjerg, F., and Wang, D., "Optimal selective harmonic control for power harmonics mitigation," *IEEE Trans. Ind. Electron.*, vol. 62, no. 2, pp. 1220–1230, 2015.

[14] Zhou, K., Lu, W., Yang, Y., and Blaabjerg, F., "Harmonic control: A natural way to bridge resonant control and repetitive control," *in Proc. Amer. Control Conf.*, pp. 3189–3193, Jun. 2013.

[15] Broberg, H.L. and Molyet, R.G., "A new approach to phase cancellation in repetitive control," *IEEE Industry Applications Society Annual Meeting*, 1994.

[16] Zhou, K., Low, K.S., Wang, Y., Luo, F.L., Zhang, B., and Wang, Y., "Zero-phase odd-harmonic repetitive controller for a single-phase PWM inverter". *IEEE Trans. Power Electron.*, vol. 21, no.1, pp. 193–201, 2006.

[17] Zhang, B., Wang, D., Zhou, K., and Wang, Y., "Linear phase-lead compensation repetitive control of a CVCF PWM inverter," *IEEE Trans. Ind. Electron.*, vol. 55, no. 4, pp. 1595–1602, Apr. 2008.

[18] Zhang, B., Zhou, K., Wang, Y., and Wang, D., "Performance improvement of repetitive controlled PWM inverters: A phase-lead compensation solution," *Int. J. Circuit Theory Appl.*, vol. 38, no. 5, pp. 453–469, Jun. 2010.

[19] Zhou, K. and Wang, D., "Digital repetitive controlled 3-phase PWM rectifier," *IEEE Trans. Power Electron.*, vol. 18, no. 1, pp. 309–316, 2003.

[20] Zhou, K., Yang, Y., Blaabjerg, F., and Wang, D., "Optimal selective harmonic control for power harmonics mitigation," *IEEE Trans. Ind. Electron.*, vol. 62, no. 2, pp. 1220–1230, 2015.

[21] Gonzalez-Espin, F., Garcera, G., Patrao, I., and Figueres, E., "An adaptive control system for three-phase photovoltaic inverters working in a polluted and variable frequency electric grid," *IEEE Trans. Power Electron.*, vol. 27, no. 10, pp. 4248–4261, Oct. 2012.

[22] Chen, D., Zhang, J., and Qian, Z., "An improved repetitive control scheme for grid-connected inverter with frequency-adaptive capability," *IEEE Trans. Ind. Electron.*, vol. 60, no. 2, pp. 814–823, Feb. 2013.

[23] Rashed, M., Klumpner, C., and Asher, G., "Repetitive and resonant control for a single-phase grid-connected hybrid cascaded multilevel converter," *IEEE Trans. Power Electron.*, vol. 28, no. 5, pp. 2224–2234, May 2013.

[24] Teodorescu, R., Blaabjerg, F., Liserre, M., and Loh, P.C., "Proportional-resonant controllers and filters for grid-connected voltage-source converters," *Proc. IEE*, vol. 153, no. 5, pp. 750–762, Sep. 2006.

[25] Liserre, M., Teodorescu, R., and Blaabjerg, F., "Multiple harmonics control for three-phase grid converter systems with the use of PI-RES current controller in a rotating frame," *IEEE Trans. Power Electron.*, vol. 21, no. 3, pp. 836–841, May 2006.

[26] Teodorescu, R., Liserre, M., and Rodriguez, P., *Grid Converters for Photovoltaic and Wind Power Systems*. Hoboken, NJ, USA: Wiley, 2011.

[27] Wells, J.R., Nee, B.M., Chapman, P.L., and Krein, P.T., "Selective harmonic control: A general problem formulation and selected solutions," *IEEE Trans. Power Electron.*, vol. 20, no. 6, pp. 1337–1345, Nov. 2005.

[28] Dahidah, S.A. and Agelidis, V.G., "Selective harmonic elimination PWM control for cascaded multilevel voltage source converters: A generalized formula," *IEEE Trans. Power Electron.*, vol. 23, no. 4, pp. 1620–1630, Jul. 2008.

[29] Ni, R., Li, Y.W., Zhang, Y., Zargari, N.R., and Cheng, Z., "Virtual impedance-based selective harmonic compensation (VI-SHC) PWM for current source rectifiers," *IEEE Trans. Power Electron.*, vol. 29, no. 7, pp. 3346–3356, Jul. 2014.

[30] Miret, J., Castilla, M., Matas, J., Guerrero, J.M., and Vasquez, J.C., "Selective harmonic-compensation control for single-phase active power filter with high harmonic rejection," *IEEE Trans. Ind. Electron.*, vol. 56, no. 8, pp. 3117–3127, Aug. 2009.

[31] Ama, N.R.N., Martinz, F.O., Matakas, L., and Kassab, F., "Phase-locked loop based on selective harmonics elimination for utility applications," *IEEE Trans. Power Electron.*, vol. 28, no. 1, pp. 144–153, Jan. 2013.

[32] Yang, Y., Zhou, K., and Blaabjerg, F., "Frequency adaptability of harmonics controllers for grid-interfaced converters," *Int. J. Control*, 2015. Page 1–12 | Received 16 Oct 2014, Accepted 21 Feb 2015, Accepted author version posted online: 27 Feb 2015, Published online: 02 Apr 2015 http://dx.doi.org/10.1080/00207179.2015.1022957

[33] Kortabarria, I., Andreu, J., de Alegria, I.M., Santamaria, V., "Design of a frequency adaptive control for a grid-connected single-phase voltage source inverter", in *Proc. of IEEE IECON 2011*, pp. 1289–1294, 7–10 Nov. 2011.

[34] Bodson, M. and Douglas, S., "Adaptive algorithm for the rejection of sinusoidal disturbances with unknown frequency," *Automatica*, vol. 33, no. 12, pp. 2213–2221, 1997.

[35] Olm, J.M., Ramosc, G.A., Costa-Castelló, R., "Adaptive compensation strategy for the tracking/rejection of signals with time-varying frequency in digital repetitive control systems", *J. Process Control*, vol. 20, no. 4, pp. 551–558, Apr. 2010.

[36] Hornik, T. and Zhong, Q.-C., "H∞ repetitive voltage control of grid-connected inverters with a frequency adaptive mechanism," *IET Power Electron.*, vol. 3, no. 6, pp. 925–935, Mar. 2010.

[37] Atmel Corporation. AVR 133: Long-delay generation using the AVR microcontroller, Jan. 2004. Application Note. [Online]. Available: http://www.atmel.com/Images/doc1268.pdf.

[38] Liu, T. and Wang, D., "Parallel structure fractional repetitive control for PWM inverters," *IEEE Trans. Ind. Electron.*, vol. 62, no. 8, pp. 5045–5054, Feb. 2015.

[39] Liu, T., Wang, D., and Zhou, K. "High-performance grid simulator using parallel structure fractional repetitive control," *IEEE Trans. Power Electron.*, vol. 31, no. 3, pp. 2669–2679, 2016.

[40] Laakso, T.I., Valimaki, V., Karjalainen, M., and Laine, U.K., "Splitting the unit delay [FIR/all pass filters design]," *IEEE Signal Process. Mag.*, vol. 13, no. 1, pp. 30–60, Jan. 1996.

[41] Tavazoei, M.S. and Tavakoli-Kakhki, M., "Compensation by fractional order phase-lead/lag compensators," *IET Control Theory Appl.*, vol. 8, no. 5, pp. 319–329, Mar. 2014.

[42] Kwan, H.K., Jiang, A., "FIR, all-pass, and IIR variable fractional delay digital filter design," *IEEE Trans. Circuits Syst. I*, Regul. Pap., vol. 56, no. 9, pp. 2064–2074, 2009.

[43] Herran, M.A., Fischer, J.R., Gonzalez, S.A., Judewicz, M.G., Carugati, I., and Carrica, D.O., "Repetitive control with adaptive sampling frequency for wind power generation systems," *IEEE J. Emerging Sel. Topics Power Electron.*, vol. 2, no. 1, pp. 58–69, Mar. 2014.

[44] Charef, A. and Bensouici, T., "Digital fractional delay implementation based on fractional order system," *IET Signal Process.*, vol. 5, no. 6, pp. 547–556, Sep. 2011.

[45] Cao, Z. and Ledwich, G.F., "Adaptive repetitive control to track variable periodic signals with fixed sampling rate," *IEEE Trans. Mechatronics*, vol. 7, no. 3, pp. 378–384, 2002.

[46] Kimura, Y., Mukai, R., Kobayashi, F., and Kobayashi, M., "Interpolative variable – speed repetitive control and its application to a deburring robot with cutting load control," *Adv. Robot.*, vol. 7, no. 1, pp. 25–39, 1993.

[47] Zou, Z., Zhou, K., Wang, Z., and Cheng, M. "Fractional-order repetitive control of programmable AC power sources," *IET Power Electron.*, vol. 7, no. 2, pp. 431–438, 2014.

[48] Hu, J. and Tomizuka, M., "Adaptive asymptotic tracking of repetitive signals – a frequency domain approach", *IEEE Trans. Automatic Control*, vol. 38, no. 10, pp. 1572–1579, 1993.

[49] Zou, Z.-X., Zhou, K., Wang, Z., and Cheng, M., "Frequency adaptive fractional order repetitive control of shunt active power filters," *IEEE Trans. Ind. Electron.*, vol. 62, no. 3, pp. 1659–1668, 2015.

[50] Nazir, R., Zhou, K., Watson, N., and Wood, A., "Analysis and synthesis of fractional order repetitive control for power converters," *Electric Power Systems Research*, vol. 124, pp. 110–119, 2015.

[51] Wang, Y., Wang, D., Zhang, B., and Zhou, K., "Fractional delay based repetitive control with application to PWM DC/AC converters," in *Proc. IEEE Int. Conf. Control Appl.*, 2007, pp. 928–933.

[52] Yang, Y., Zhou, K., Wang, H., Blaabjerg, F., Wang, D., and Zhang, B., "Frequency adaptive selective harmonic control for grid-connected inverters," *IEEE Trans. Power Electron.*, vol. 30, no. 7, pp. 3912–3924, 2015.

[53] McGrath, B.P., Holmes, D.G., and Galloway, J.J.H., "Power converter line synchronization using a discrete Fourier transform (DFT) based on a variable sample rate," *IEEE Trans. Power Electron.*, vol. 20, no. 4, pp. 877–884, Jul. 2005.

[54] Mattavelli, P. and Marafao, F.P., "Repetitive-based control for selective harmonic compensation in active power filters," *IEEE Trans. Ind. Electron.*, vol. 51, no. 5, pp. 1018–1024, Oct. 2004.

[55] IEEE Application Guide for IEEE Std. 1547, "IEEE standard for interconnecting distributed resources with electric power systems," *IEEE Standard*, 1547.2–2008, 2009.

[56] Yang, Y., Zhou, K., and Blaabjerg, F., "Enhancing the frequency adaptability of periodic current controllers with a fixed sampling rate for grid-connected power converters," *IEEE Trans. Power Electron.*, vol. 31, no. 10, pp. 7273–7285, Oct. 2016.

[57] Yang, Y., Zhou, K., and Blaabjerg, F., "Virtual unit delay for digital frequency adaptive T/4 delay phase-locked loop system," in *Proc. IPEMC 2016 – ECCE Asia*, pp. 1–7, 22–25, May 2016.

[58] Zhou, K. and Wang, D., "Zero Tracking Error Controller for Three-phase CVCF PWM Inverter," *Electronics Letters*, vol. 36, no. 10, pp. 864–865, 2000.

[59] Zhou, K., Wang, D., and Low, K.S., "Periodic errors elimination in CVCF PWM DC/AC converter systems: A repetitive control approach," *IEE Proc. – Control Theory and Appl.*, vol. 147, no. 6, pp. 694–700, 2000.

[60] Zhou, K. and Wang, D., "Digital repetitive learning controller for three-phase CVCF PWM inverter," *IEEE Trans. Ind. Electron.*, vol. 48, no. 4 pp. 820–830, 2001.

[61] Zhou, K. and Wang, D., "Unified robust zero error tracking control of CVCF PWM converters," *IEEE Trans. Circuits and Systems (I)*, vol. 49, no. 4, pp. 492–501, 2002.

[62] Jensen, U.B., Blaabjerg, F., and Pedersen, J.K., "A new control method for 400-Hz ground power units for airplanes," *IEEE Trans. Ind. Appl.*, vol. 36, no. 1, pp. 180–187, 2000.

[63] Blaabjerg, F., Chen, Z., and Kjaer, S.B., "Power electronics as efficient interface in dispersed power generation systems," *IEEE Trans. Power Electron.*, vol. 19, no. 5, pp. 1184–1194, 2004.

[64] Zhang, B., Zhou, K., and Wang, D., "Multirate repetitive control for PWM DC/AC converters," *IEEE Trans. Ind. Electron.*, vol. 61, no. 6, pp. 2883–2890, 2014.

[65] Liu, Z., Zhang, B., and Zhou, K., "Fractional-order phase lead compensation for multi-rate repetitive control on three-phase PWM DC/AC inverter," in *Proc. IEEE Annual Applied Power Electron. Conf. Exp. (APEC)*, pp. 1155–1162, 20–24, Mar. 2016.

Chapter 6
Frequency-adaptive periodic control of power converters

Abstract

This chapter will give three application examples of frequency-adaptive periodic control (FAPC) of power converters under frequency variations, which include voltage control for programmable AC power sources, current control for grid-connected photovoltaic (PV) inverters, and current harmonics compensation for shunt active filters.

6.1 Frequency-adaptive periodic control (FAPC) of programmable AC power sources

6.1.1 Background

Programmable AC power sources that are capable of simulating a number of voltage waveforms with various amplitudes and frequencies, e.g., user-programmable sine, square, or harmonic waveforms, are widely used in automatic testing equipment and bench-top applications, such as 400-Hz avionics and shipboard system testing, 50-Hz and 60-Hz electrical network simulations. Advanced control technologies are needed to enable high-performance programmable AC power sources to accurately synthesize sine/cosine or harmonic-distorted signals under various loading conditions [1]. Conventional feedback control schemes, such as proportional–integral–derivative (PID) control, proportional-resonant (PR) control, deadbeat (DB) control, and sliding mode control [2–5], often fail to force pulse-width modulation (PWM) inverter to produce high-quality sinusoidal outputs under nonlinear rectifier loads and/or parameter uncertainties. Plenty of applications demonstrate that the internal model principle (IMP)-based classic periodic controllers [6–18] can achieve zero steady-state error tracking of any periodic signal with a known period. Since the frequency of the reference signal for AC programmable power supplies can be set within a wide range, the conventional digital repetitive control (RC) scheme cannot exactly track or reject the references/disturbances with a varying frequency, and may lead to significant performance degradation. To address this issue, a frequency-adaptive classic repetitive control (FACRC) strategy with a fixed sampling rate can be employed to enable programmable AC power sources to precisely produce various variable-frequency variable-magnitude periodic output voltage waveforms under various loading conditions.

6.1.2 Modeling and control of three-phase PWM inverters

Figure 6.1 shows a three-phase voltage source PWM inverter for the programmable AC power source, where v_{ab}, v_{bc}, and v_{ca} are the output line-to-line voltages, i_{La}, i_{Lb}, and i_{Lc} are the inductor currents of each phase, v_{dc} is the DC bus voltage, L, L_{r}, C, C_r, and R, R_r are the nominal values of filter inductors, filter capacitors, and load resistors. The inverter is used to accurately generate clean sinusoidal or harmonic-distorted voltages under various load conditions. To fulfill the requirements, a state feedback plus plug-in FACRC scheme is employed to achieve zero steady-state tracking error with good dynamics.

The three-phase inverter with a linear resistive load R shown in Figure 6.1 can be described by the following equation in the stationary $\alpha\beta$ frame [6,7]:

$$\begin{bmatrix} \dot{v}_\alpha \\ \dot{i}_\alpha \\ \dot{v}_\beta \\ \dot{i}_\beta \end{bmatrix} = \begin{bmatrix} -\dfrac{1}{RC} & \dfrac{1}{3C} & 0 & 0 \\ -\dfrac{1}{L} & 0 & 0 & 0 \\ 0 & 0 & -\dfrac{1}{RC} & \dfrac{1}{3C} \\ 0 & 0 & -\dfrac{1}{L} & 0 \end{bmatrix} \begin{bmatrix} v_\alpha \\ i_\alpha \\ v_\beta \\ i_\beta \end{bmatrix} + \begin{bmatrix} 0 & 0 \\ \dfrac{v_{dc}}{L} & 0 \\ 0 & 0 \\ 0 & \dfrac{v_{dc}}{L} \end{bmatrix} \begin{bmatrix} u_\alpha \\ u_\beta \end{bmatrix} \tag{6.1}$$

where v_α, v_β, i_α, i_β are the output voltages and inductor currents in the $\alpha\beta$ frame, and u_α, u_β are the corresponding control vectors with $\boldsymbol{v}_{\alpha\beta} = [v_\alpha \ v_\beta]$, $\boldsymbol{i}_{\alpha\beta} = [i_\alpha \ i_\beta]$, and $\boldsymbol{u}_{\alpha\beta} = [u_\alpha \ u_\beta]$. Equation (6.1) can be treated as two identical independent subsystems. The sampled-data model for each subsystem in (6.1) with f_s being the sampling frequency can be approximated as [6,7]:

$$\begin{bmatrix} v(k+1) \\ i(k+1) \end{bmatrix} = \begin{bmatrix} \varphi_{11} & \varphi_{12} \\ \varphi_{21} & \varphi_{22} \end{bmatrix} \begin{bmatrix} v(k) \\ i(k) \end{bmatrix} + \begin{bmatrix} g_1 \\ g_2 \end{bmatrix} u(k) \tag{6.2}$$

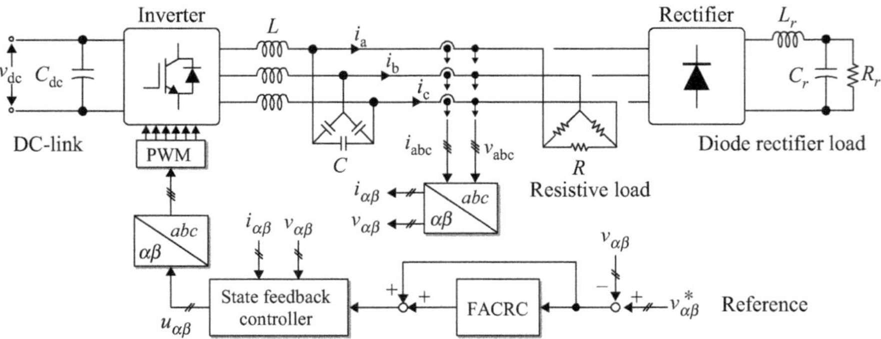

Figure 6.1 Frequency-adaptive classic repetitive control (FACRC) system in a three-phase PWM inverter

where $v = v_\alpha$ or v_β, $i = i_\alpha$ or i_β, $u = u_\alpha$ or u_β, and the coefficients are

$$\varphi_{11} = 1 - 1/(RCf_s) + 1/(2R^2C^2f_s^2) - 1/(6LCf_s^2),$$
$$\varphi_{22} = 1 - 1/(6LCf_s^2),$$
$$\varphi_{12} = 1/(3Cf_s) - 1/(6RC^2f_s^2),$$
$$\varphi_{21} = -1/(Lf_s) + 1/(2RLCf_s^2),$$
$$g_1 = v_{dc}/(6LCf_s^2), \text{ and } g_2 = v_{dc}/(Lf_s).$$

In order to force the output voltage v to track the reference signal v^* under various load conditions, a state feedback controller is chosen as follows:

$$u = -(k_1 v + k_2 i) + k v^* \tag{6.3}$$

where v^* is the reference voltage. With this feedback controller, (6.2) can be rewritten as:

$$\begin{bmatrix} v(k+1) \\ i(k+1) \end{bmatrix} = \begin{bmatrix} \varphi_{11} - g_1k_1 & \varphi_{12} - g_1k_2 \\ \varphi_{21} - g_2k_1 & \varphi_{22} - g_2k_2 \end{bmatrix} \begin{bmatrix} v(k) \\ i(k) \end{bmatrix} + \begin{bmatrix} g_1 k \\ g_2 k \end{bmatrix} v^*(k) \tag{6.4}$$

The poles of the closed-loop system (6.4) can be arbitrarily assigned by adjusting the feedback gains k_1 and k_2. From (6.4) the closed-loop transfer function from v^* to v with the nominal parameters can be written as [6,7]:

$$H(z) = \frac{m_1 z + m_2}{z^2 + p_1 z + p_2} \tag{6.5}$$

with

$$p_1 = -(\varphi_{22} - g_2k_2) - (\varphi_{11} - g_1k_1),$$
$$p_2 = (\varphi_{11} - g_1k_1)(\varphi_{22} - g_2k_2) - (\varphi_{12} - g_1k_2)(\varphi_{21} - g_2k_1),$$
$$m_1 = g_1k, \text{ and } m_2 = g_2k - g_1k(\varphi_{22} - g_2k_2).$$

In order to control the periodic signals with frequency variations, a FACRC $G_{arc}(z)$ is plugged into the state feedback control system as [19]:

$$G_{arc}(z) = k_{rc} \frac{\left(z^{-(N_i+F)}\right)Q(z)}{1 - \left(z^{-(N_i+F)}\right)Q(z)} G_f(z) \tag{6.6}$$

where k_{rc} is the control gain, $Q(z)$ is a low pass filter (LPF), $G_f(z) = z^c$ is a linear phase-lead filter with the phase-lead step $c \in \mathbb{N}$ being determined by experiments [19–21], and z^{-F} is a *Lagrange* interpolation-based fractional delay (FD) filter that can be written as [19,22]:

$$z^{-F} = \sum_{k=0}^{n} A_k z^{-k} = \sum_{k=0}^{n} \prod_{\substack{i=0 \\ i \neq k}}^{n} \frac{F - i}{k - i} z^{-k} \quad k, i = 0, 1, \ldots, n \tag{6.7}$$

with F being the fractional part of the fractional number of $N = f_s/f = N_i + F$, f_s being the sampling frequency and f being the fundamental frequency of the output signals.

6.1.3 Experimental validation

A dSPACE DS1104-based three-phase PWM inverter system has been built up to verify the effectiveness of the FACRC, and the experimental setup is shown in Figure 6.2. System parameters are listed in Table 6.1, where the bandwidth of the output LC filter being 1470 Hz allows the inverter to efficiently produce output voltages of a fundamental frequency less than 750 Hz.

With these parameters shown in Table 6.1, a state feedback controller with $k_1 = 0.0087$, $k_2 = 0.0495$, and $k = 0.0169$ is chosen as [19]:

$$u(k) = -(0.0087v(k) + 0.00495i(k)) + 0.0169v^*(k) \tag{6.8}$$

Then, a DB controller is adopted, and its transfer function for the closed-loop system is obtained as:

$$H(z) = \frac{z + 0.8336}{z^2 + 0.8336z + 0} = \frac{1}{z} \tag{6.9}$$

For the FACRC in (6.6), $n = 3$ is chosen to be the *Lagrange* polynomial degree, leading to:

$$z^{-N} = z^{-N_i - F} = z^{-N_i}\left(A_0 + A_1 z^{-1} + A_2 z^{-2} + A_3 z^{-3}\right) \tag{6.10}$$

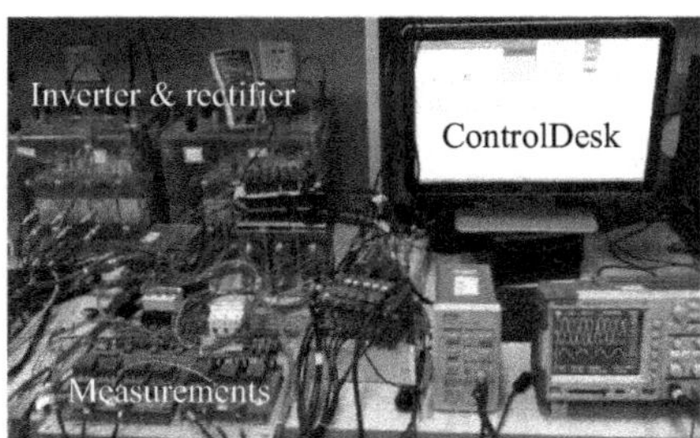

Figure 6.2 Experimental setup of a three-phase inverter system

Table 6.1 System parameters of a three-phase inverter system

Parameters	Value	Unit
DC-link voltage V_{dc}	312	V
Inductor filter L	1.875	mH
Capacitor filter C	18.7	μF
Linear resistive load R	30	Ω
Rectifier inductor L_r	1	mH
Rectifier capacitor C_r	4400	μF
Rectifier resistor R_r	30	Ω
PWM switching frequency	10	kHz
Sampling frequency	10	kHz

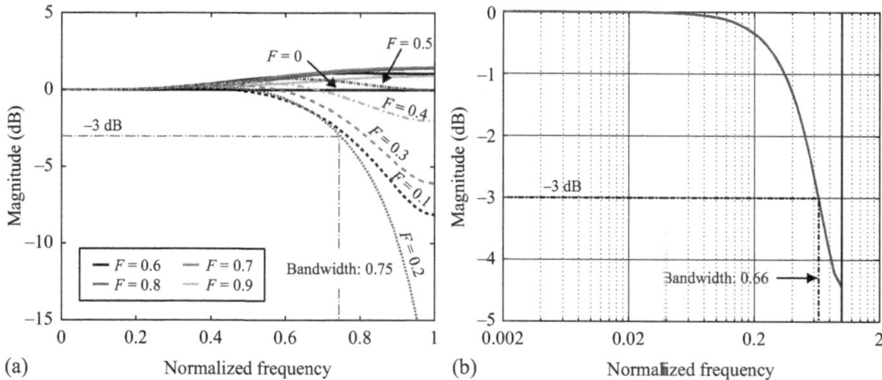

Figure 6.3 Magnitude responses of (a) the fractional delay filter z^{-F} in (6.10) and (b) the low-pass filter Q(z)

As shown in Figure 6.3(a), the bandwidth of the FD filter z^{-F} in (6.10) is about $0.75f_s/2 = 3.75$ kHz. For the FACRC, $Q(z) = 0.1z + 0.8 + 0.1z^{-1}$ results in the cut-off frequency being $0.66f_s/2 = 3.3$ kHz, as shown in Figure 6.3(b), and $G_f(z) = z^6$ is selected in this case study. Since the bandwidth of the FD filter is larger than the bandwidth of the LPF $Q(z)$, the FACRC in (6.6) will be almost the same as the CRC in (2.5) within the bandwidth of the LPF $Q(z)$ due to $|Q(z)|\left|\sum_{k=0}^{n} A_k z^{-k}\right| \to 1$. Under such conditions, the design and synthesis of the FACRC system are the same as those of the CRC system that have been presented in Chapter 2. A CRC scheme with identical parameters is also developed for the inverter system for comparison.

Figure 6.4 shows that the CRC with a fixed sampling rate successfully enables the inverter to produce high-quality 50 Hz 104 V (peak) and 400 Hz 52 V (peak) sinusoidal line voltages under a rectifier load as shown in Figure 6.1 in the case of an integer period $N = 200$ and $N = 25$, respectively. Note that the FACRC of an integer order N is the same as the CRC.

However, in the case of a noninteger period N, the performance of the CRC will be degraded. Figure 6.5(a) shows that the output voltages of the CRC-controlled PWM inverter system (with $N = 217$ being rounded from 217.4) under a rectifier load contain considerable harmonic distortions in the case of a fractional period $N = 217.4$ and reference sinusoidal voltages of 46 Hz 104 V (peak). Figure 6.5(b) shows the output voltages of the FACRC-controlled PWM inverter system under the same rectifier load that produces high-quality sinusoidal voltages in the case of the fractional period $N = 217.4$ and the reference sinusoidal voltage of 46 Hz 104 V (peak).

Furthermore, Figure 6.6(a) shows the steady-state responses of the CRC-controlled inverter (with $N = 26$) with a reference sinusoidal voltage of 390 Hz 52 V (peak) under the rectifier load with its parameters listed in Table 6.1, where the amplitude of the output voltage is much less than that of the reference signal, due to the difference between the given integer period of $N = 26$ for the CRC and the actual fractional period of $N = 25.6$ of the output voltage. For comparison, the

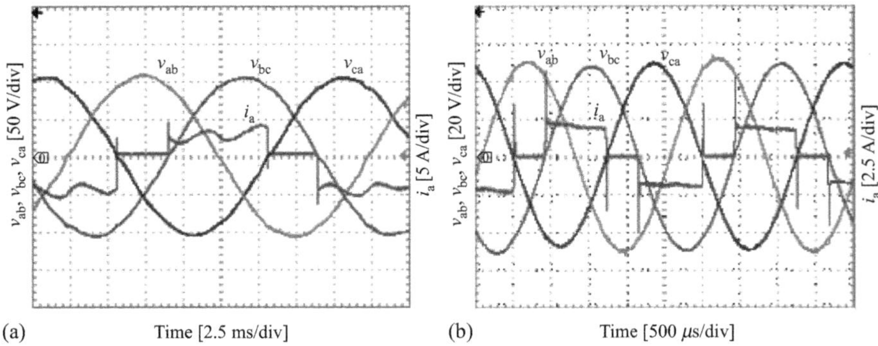

Figure 6.4 Steady-state responses of the three-phase inverter system with the classic repetitive control scheme at a reference frequency of (a) 50 Hz and (b) 400 Hz

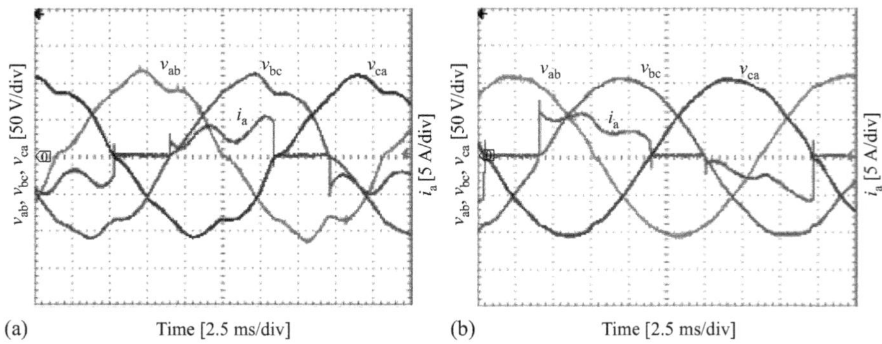

Figure 6.5 Steady-state responses of the three-phase inverter system under a nonlinear load at the reference frequency of 46 Hz using (a) the CRC ($N=217$) and (b) the FACRC ($N=217.4$)

FACRC is applied to the inverter system. Figure 6.6(b) shows the steady-state responses of the FACRC-controlled inverter with the reference sinusoidal voltage of 390 Hz 52 V (peak) under the same rectifier load, where the output voltages are high-quality sinusoidal voltages with the desired amplitude and frequency.

In addition, Table 6.2 summarizes the experimental results with the nonlinear rectifier load shown in Figures 6.4–6.6, and also with a linear resistor load. The results in Table 6.2 further confirm that the inverter system with the FACRC can accurately track the reference sinusoidal voltages under various loading conditions and fractional periods, but the CRC may fail to do so in the case of noninteger periods.

When considering harmonics in the reference $v^* = 104 \cdot \sin(46 \times 2\pi t) + 10.4 \cdot \sin(5 \times 46 \times 2\pi t) + 10.4 \cdot \sin(7 \times 46 \times 2\pi t)$, i.e., a sinusoidal voltage of 46 Hz fundamental frequency with 10% fifth- and seventh-order harmonics, Figure 6.7(a) shows the steady-state responses of the CRC-controlled inverter (N is rounded to 217)

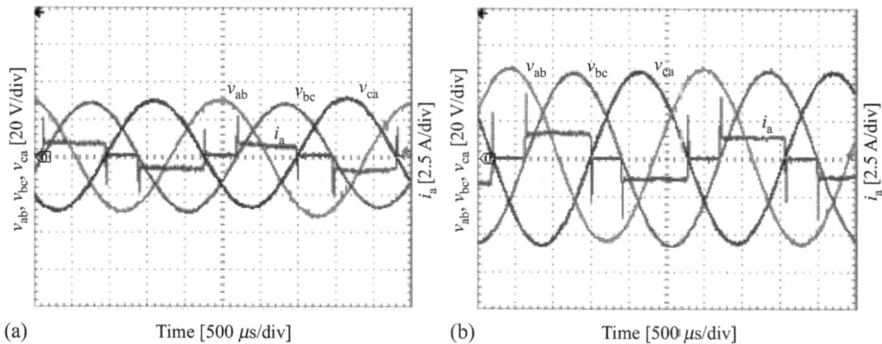

Figure 6.6 Steady-state responses of the three-phase inverter system under a nonlinear load at the reference frequency of 390 Hz using (a) the CRC ($N=26$) and (b) the FACRC ($N=25.6$)

Table 6.2 Performance of the three-phase inverter system with the CRC or the FACRC under different fundamental frequencies and loading conditions

Fundamental frequency (Hz)	**Control scheme**	**Tracking error (RMS) (V)**		**Total harmonic distortion (%)**	
		Linear load	**Nonlinear load**	**Linear load**	**Nonlinear load**
50	CRC ($N=200$)	0.813	1.733	1.258	1.482
46	FACRC ($N=217.4$)	0.934	1.916	1.393	1.560
46	CRC ($N=217$)	2.871	6.591	2.243	8.133
400	CRC ($N=25$)	1.510	1.631	0.702	0.707
390	FACRC ($N=25.6$)	1.214	1.751	0.614	0.762
390	CRC ($N=26$)	10.191	11.980	1.061	1.308

under the rectifier load with the harmonic-distorted reference voltage v^*, and Figure 6.7(b) shows the steady-state responses of the inverter system under the same condition with the FACRC scheme. Table 6.3 lists the experimental results shown in Figure 6.7. The results in Figure 6.7 and Table 6.3 indicate that the FACRC-controlled inverter can accurately synthesize the user-defined harmonic-distorted waveforms under the rectifier load, but the CRC fails to accurately track the reference with harmonic components, especially the seventh-order harmonic.

At last, dynamic tests of the FACRC-controlled inverter system with a reference sinusoidal voltage of 46 Hz 104 V (peak) have been conducted under sudden step-change loads. It can be seen in Figure 6.8 that the output voltage recovers from the sudden step-change loads within about five fundamental cycles. Obviously, the FACRC-controlled inverter system offers good performance in terms of fast dynamic responses.

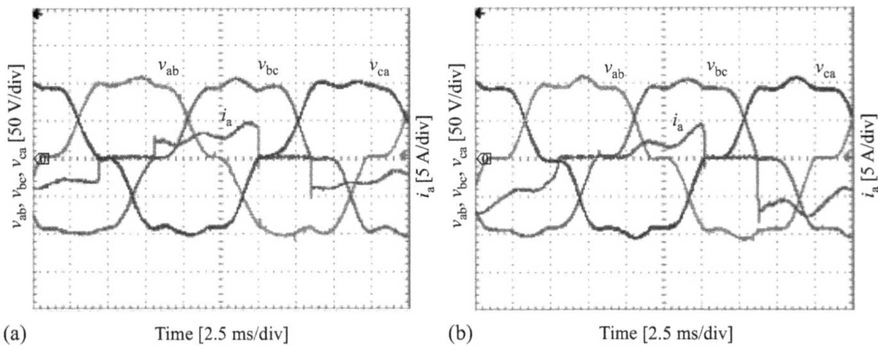

Figure 6.7 Synthesis of the user-composed harmonic-distorted waveform of a 46-Hz sinusoidal voltage with 10% fifth- and seventh-order harmonics using: (a) the CRC and (b) the FACRC

Table 6.3 Synthesis of the user-composed harmonic-distorted waveforms with the three-phase inverter systems under different control schemes

Fundamental frequency (Hz)	**Control scheme**	**Harmonic (% of fund.)**		
		Fund.	**5th**	**7th**
46	FACRC (N = 217.4)	100.46	9.829	9.820
46	CRC (N = 217)	104.16	10.617	5.443

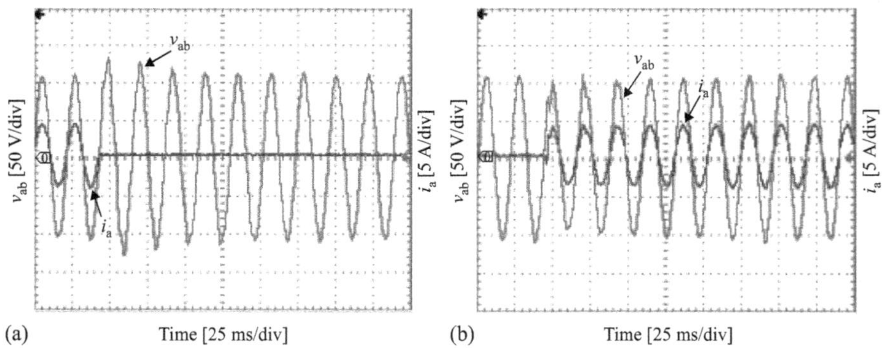

Figure 6.8 Dynamic performance of the FACRC-controlled inverter system under step-change loads: (a) from full load to no load and (b) from no load to full load

6.1.4 Conclusion

The experimental case study of the FACRC-controlled three-phase PWM inverter for programmable AC power sources has been provided in this section. Experimental results indicate that the FACRC enables programmable AC power sources to precisely produce various variable-frequency variable-magnitude periodical output voltages under various loading conditions.

Compared with the CRC, the FACRC scheme with a fixed sampling rate can exactly track or eliminate any periodic signal with a variable frequency. Using the *Lagrange* interpolation-based fractional delay filter to approximate the factional period of varying frequency references, the FACRC offers fast on-line tuning of the fractional delay and the fast update of the coefficients, and then provides a simple but high-performance control solution to high-switching frequency converters. Furthermore, the FACRC can be used in extensive applications, such as the current control of grid-connected converters [23–29].

6.2 FAPC of grid-connected PV inverters

6.2.1 Background

Severe power quality (PQ) problems have been brought by an increase of power electronics-interfaced renewable energy systems, e.g., PV systems and wind turbine systems [30–35]. For grid-connected PV inverter systems, the current distortion level is one important PQ index [34,36]. For instance, it is stated in both the IEEE Std. 1547–2003 and the IEC Standard 61727 that the total harmonic distortion (THD) of the grid current should be lower than 5% to avoid adverse effects on other equipment that is connected to the grid [36]. Moreover, for each individual odd harmonic from third to ninth, the limitation is 4% and at the same time the even harmonics are limited to 25% of the odd harmonic limits.

In practice, power harmonics caused by power converters interfaced loads and distributed generators usually concentrate on some particular frequencies [37,38]. For instance, single-phase H-bridge converters mainly produce $(4k \pm 1)$-order $(k = 1, 2, \ldots)$ power harmonics; n-pulse $(n = 6, 12, \ldots)$ converter-based high-voltage direct current (HVDC) transmission systems mainly produce $(nk \pm 1)$-order $(k = 1, 2, \ldots)$ power harmonics. To deal with power harmonics issues, power converters demand optimal control strategies, which can compensate power harmonics with high control accuracy while maintaining fast dynamic response, guaranteeing robustness, and being feasible for implementation [17,39–44]. Taking the harmonic distribution into consideration, a tailor-made IMP-based optimal harmonic control (OHC) scheme has been introduced to mitigate the power harmonics in [17], and it is further illustrated in Figure 6.9. The OHC scheme includes multiple parallel $(nk \pm m)$-order harmonic RC [14] with independent control gains, a good trade-off among control accuracy, dynamic response, and implementation complexity can be achieved. However, for grid-connected applications, the OHC scheme is sensitive to grid frequency variations due to the mismatch between their embedded integer-period-based internal models and fractional-period grid signals [26,45]. In the case of a time-varying grid frequency, the tracking accuracy of the OHC scheme will be degraded. It means that the OHC scheme cannot accurately compensate current harmonic distortions for grid-connected converters in the case of grid frequency variations.

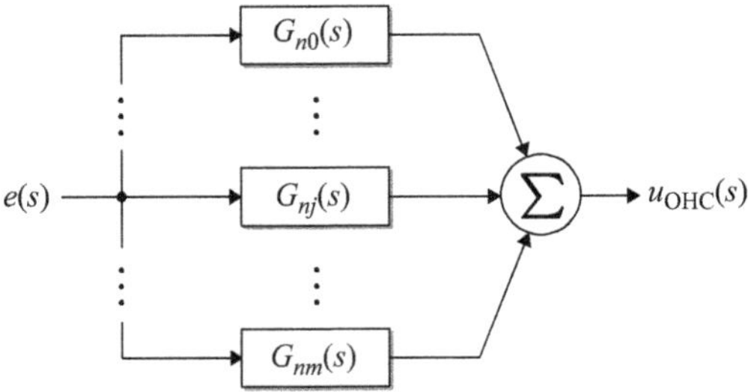

Figure 6.9 Optimal harmonic control scheme, where e(s) is the error input, $u_{OHC}(s)$ is the controller output, and $G_{nj}(s)$ is the corresponding $(nk \pm j)$-order harmonic RC with $j = 0, 1, 2, \ldots, m$

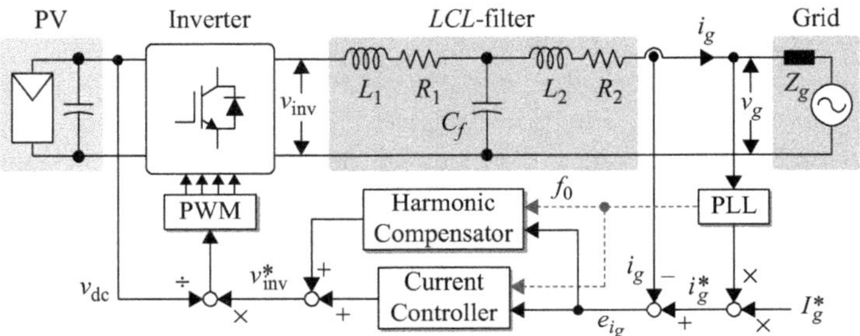

Figure 6.10 A typical single-phase grid-connected PV inverter with an LCL filter

Similar to the FACRC, FD filters can also be employed to enable the OHC to be frequency-adaptive. As a case study in this section, a frequency-adaptive optimal harmonic control (FA-OHC) strategy based on the FD filters will be developed for grid-connected PV inverters to feed sinusoidal currents into grid under grid frequency variations.

6.2.2 Modeling and control of grid-connected PV inverters

Figure 6.10 shows a single-phase grid-connected PV inverter with an *LCL* filter, where the filter-capacitor C_f is used to eliminate high-order harmonics at the switching frequencies. Together with the grid-side inductor L_2, the capacitor C_f can be treated as a "model mismatch" [26]. Therefore, the dynamic model of the single-phase grid-connected system shown in Figure 6.10 is simplified as:

$$L_1 \frac{di_g}{dt} + R_1 i_g = v_{in} - v_g \tag{6.11}$$

where v_g is the grid voltage, i_g is the grid current, L_1 and R_1 are the inductance and resistance of the inductor L_1, respectively.

In this case, a very simple DB current controller is employed, and it can be derived from (6.11) as:

$$v_{\mathrm{inv}}^*(k) = \frac{1}{v_{\mathrm{dc}}(k)}\left[v_g(k) + b_1 i_g^*(k) - (b_1 - b_2) i_g(k)\right] \tag{6.12}$$

in which $b_1 = L_1/T_s$, $b_2 = R_1$, k is the sampling instant number, and v_{dc} is the DC-link voltage. The DB controller in (6.12) will force the feed-in current to track the reference in one sampling period in theory, i.e., $i_g(k+1) = i_g^*(k)$. However, due to system uncertainties (including model mismatch), the control parameters, b_1 and b_2, should be adjusted to enhance the robustness of the controller [31].

One of the main functions of the grid-connected PV inverter is to feed clean sinusoidal current into the grid. To compensate the inherent odd-order harmonics in the feed-in current from single-phase inverters [37,38] under system uncertainties and frequency variations, a FA-OHC scheme is added into the DB current controller as the harmonic compensator shown in Figure 6.10. Such a hybrid structure control scheme can achieve both zero steady-state tracking error and good dynamics. According to Figure 6.9 and the discussion in Section 5.3 in Chapter 5, the digital FA-OHC scheme can be written as:

$$\begin{aligned} G_{\mathrm{aOHC}}(z) &= \sum_{m\in N_m} G_{\mathrm{asm}}(z) \\ &= \sum_{m\in N_m} \frac{k_m G_f(z)\left(\cos\left(\frac{2\pi m}{n}\right)\left(z^{N_i}\sum_{k=0}^{n} A_k z^k\right)Q(z) - Q^2(z)\right)}{\left(z^{N_i}\sum_{k=0}^{n} A_k z^k\right)^2 - 2\cos\left(\frac{2\pi m}{n}\right)\left(z^{N_i}\sum_{k=0}^{n} A_k z^k\right)Q(z) + Q^2(z)} \end{aligned} \tag{6.13}$$

where $N = T_0/T_s \in \mathbb{N}$ with T_s being the sampling period, $N_i = [N/n] \in \mathbb{N}$ is the integer part of N/n, F is the fractional part of N/n, k_m is control gain, and N_m is the considered number of harmonic sets. The control module $G_{\mathrm{asm}}(z)$ for $(nk \pm m)$-order harmonics in (6.13) is given as:

$$G_{\mathrm{asm}}(z) = \frac{k_m G_f(z)\left(\cos\left(\frac{2\pi m}{n}\right)\left(z^{N_i}\sum_{k=0}^{n} A_k z^k\right)Q(z) - Q^2(z)\right)}{\left(z^{N_i}\sum_{k=0}^{n} A_k z^k\right)^2 - 2\cos\left(\frac{2\pi m}{n}\right)\left(z^{N_i}\sum_{k=0}^{n} A_k z^k\right)Q(z) + Q^2(z)} \tag{6.14}$$

which is also called the frequency-adaptive selective harmonic control (FA-SHC) module. The fractional delay filter is given as:

$$\sum_{k=0}^{n} A_k z^{-k} = \sum_{k=0}^{n}\left(\prod_{\substack{i=0\\ i\neq k}}^{n} \frac{F-i}{k-i}\right) z^{-k} \tag{6.15}$$

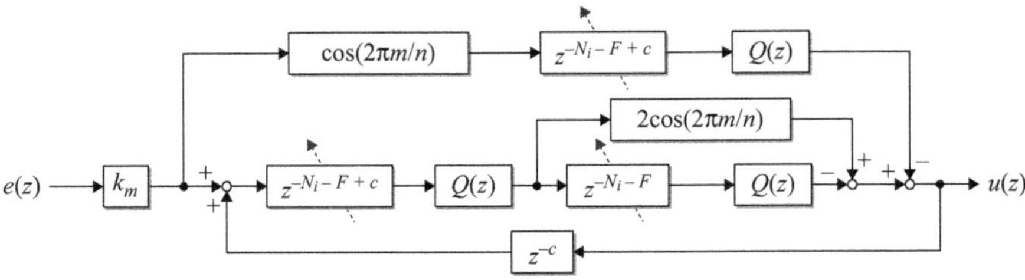

Figure 6.11 Frequency-adaptive digital selective harmonic control scheme with a linear phase-lead compensator z^c, which is a control module of the FA-OHC scheme at ($nk \pm m$)-order harmonics

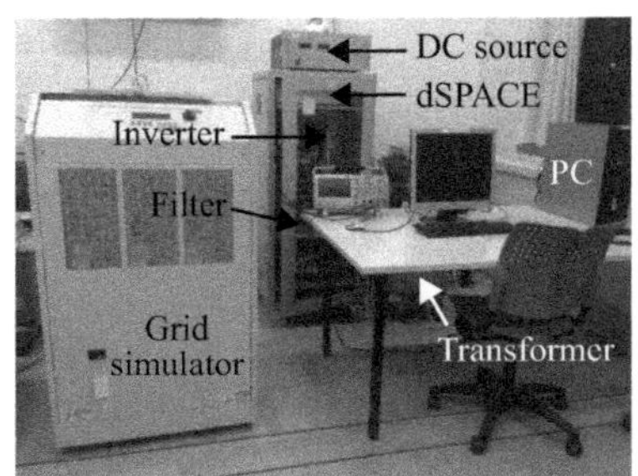

Figure 6.12 Experimental setup of a single-phase grid-connected inverter system

It is used to approximate the fractional part z^{-F} of the time-varying signal period. Obviously, if $F = 0$, the FA-OHC in (6.13) is equivalent to the OHC in (3.38).

Also, for simplicity, a linear phase-lead compensator $G_f(z) = z^c$ is used to compensate the system delay and reinforce the system stability [21,26]. The implementation of a FA-SHC module with the linear phase compensator $G_f(z) = z^c$ in the FA-OHC can be found in Figure 6.11.

When the bandwidth of the fractional delay filter in (6.15) is larger than the bandwidth of the LPF $Q(z)$, the FA-OHC in (6.13) will be equivalent to the conventional OHC within the bandwidth of LPF $Q(z)$ due to $|Q(z)|\left|\sum_{k=0}^{n} A_k z^{-k}\right| \to 1$. Under such conditions, the design and synthesis of the FA-OHC system shown in Figure 6.11 are the same as those of the conventional OHC system that are presented in Chapter 3.

6.2.3 Experimental validation

In order to verify the theoretical study, experimental tests have been performed on a single-phase grid-connected inverter system shown in Figure 6.12, where the controllers are implemented in a dSPACE DS1103-based system, a California Instruments Programmable AC Power Source is adopted as the simulated power grid, a DC power source is utilized instead of PV panels, and a commercial full-bridge PV

Table 6.4 System parameters of the single-phase grid-connected inverter system

Parameters	Value
Nominal grid voltage amplitude	$v_{gn} = 220$ V (rms)
Nominal grid frequency	$f_0 = 50$ Hz
Reference current amplitude	$I_g^* = 5$A
Transformer leakage impedance	$L_g = 2$ mH, $R_g = 0.2\ \Omega$
LCL-filter	$L_1 = L_2 = 3.6$ mH $R_1 = R_2 = 0.04\ \Omega$ $C_f = 2.35\ \mu$F
Sampling and switching frequency	$f_s = f_{sw} = 10$ kHz
DC bus voltage	$v_{dc} = 400$ V
Repetitive control gain	$k_{rc} = k_0 + k_1 + k_2 = 1.8$

inverter with an *LCL*-filter connected to the grid through an isolation transformer has been used as the power conversion stage. System parameters are listed in Table 6.4. As mentioned previously, the main control objective of the grid-connected inverter is to feed unity-power-factor clean sinusoidal currents (e.g., THD < 5%) into the grid.

Figure 6.13(a) shows the steady-state performance of the single-phase grid-connected inverter system only with the DB control. The corresponding harmonic spectrum of the feed-in current i_g is shown in Figure 6.13(b), which indicates that the feed-in current i_g contains dominant $(4k \pm 1)$-order harmonics. Specifically, the ratio of all $4k$-order ($k = 0, 1, 2, ...$) harmonics to the total harmonics is nearly 18%, the ratio of all $(4k \pm 1)$-order ($k = 1, 2, 3, ...$) harmonics to the total harmonics is nearly 62%, and the ratio of all $(4k \pm 2)$-order ($k = 1, 2, 3, ...$) harmonics to the total harmonics is nearly 20%. Accordingly, the following FA-OHC is adopted:

$$G_{aOHC}(z) = \sum_{m \in N_m} G_{asm}(z) = G_{as0}(z) + G_{as1}(z) + G_{as2}(z) \tag{6.16}$$

where the FA-SHC modules $G_{as0}(z)$, $G_{as1}(z)$, and $G_{as2}(z)$ are correspondingly used to compensate the $4k$-order harmonics, $(4k \pm 1)$-order harmonics, and $(4k \pm 2)$-order harmonics. Their corresponding control gains k_0, k_1, and k_2 will be proportional to the ratios of harmonics to the total. Moreover, a CRC scheme with identical system parameters and the conventional OHC are also developed for comparison.

As discussed in Chapters 2, 3, and 5, the stability range of the control gain(s) for the CRC or the FACRC is $0 < k_{rc} < 2$, and for the OHC or the FA-OHC, it is $0 < k_0 + k_1 + k_2 < 2$. Since the error convergence rate is proportional to the control gain, the CRC control gain is chosen as $k_{rc} = k_0 + k_1 + k_2$ to facilitate the comparison between the CRC and the FA-OHC. In the following experiments the control gain $k_{rc} = 1.8$ is chosen for the CRC, and thus the control gains $k_0 = 0.2$, $k_1 = 1.4$, and $k_2 = 0.2$ are chosen for the FA-OHC. Notably, the control gains for the FA-OHC scheme are weighted according to the harmonic distribution of the feed-in current as shown in Figure 6.13(b).

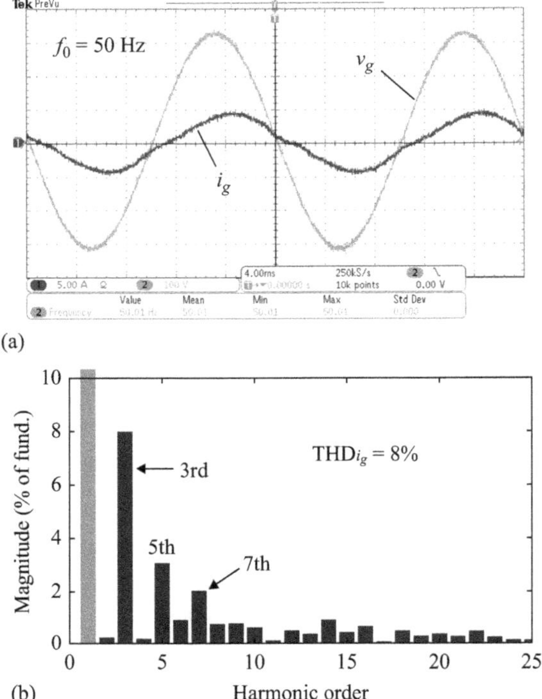

Figure 6.13 Steady-state performance of the DB-controlled single-phase grid-connected inverter system: (a) grid voltage v_g: [100 V/div], grid current i_g: [5 A/div], and time: [4 ms/div] and (b) harmonic spectrum of the grid current i_g

To further evaluate the harmonic of the grid current i_g, a harmonic ratio $h_r(j)$ is defined as:

$$h_r(j) = \left(\sum_{i=1}^{j} M_i\right) \Big/ \left(\sum_{i=1}^{199} M_i\right) \times 100\% \tag{6.17}$$

where M_i is the magnitude of the ith-order harmonic. Correspondingly, the harmonic ratio $h_r(j)$ for the current spectrum shown in Figure 6.13(b) can be represented in Figure 6.14(a), which indicates that over 83% of the harmonics are within a certain range (0~3.5 kHz). Consequently, if the cutoff frequency f_{cutoff} of the LPF $Q(z)$ for the CRC and $Q^2(z)$ for the OHC or the FA-OHC are chosen as $f_{\text{cutoff}} >$ 3.5 kHz, most of the harmonics can be removed by the CRC or the FA-OHC. Figure 6.15(b) shows that the cutoff frequencies of the designed $Q(z) = 0.1z + 0.8 + 0.1z^{-1}$ and $Q^2(z) = (0.05z + 0.9 + 0.05z^{-1})^2$ are 3.3 kHz and 3.5 kHz, respectively, which approximately meet the bandwidth requirement.

In the case of a nominal grid frequency $f_0 = 50$ Hz, both $N = f_s/f_0 = 200$ and $p = N/4 = 50$ are integers. In those cases, the CRC and the conventional

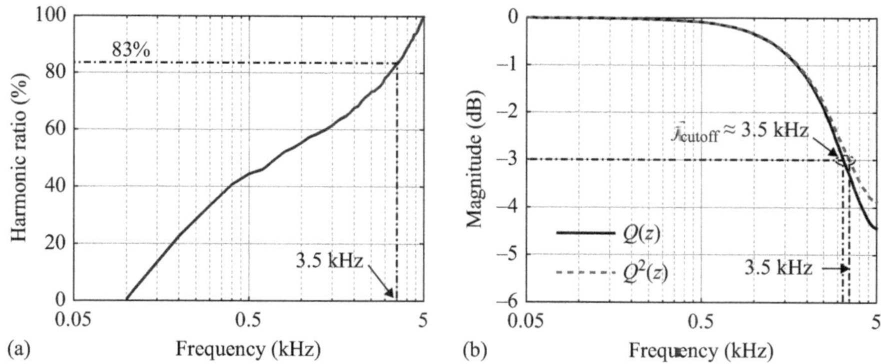

Figure 6.14 Harmonic spectrum analysis under the nominal grid frequency $f_0 = 50$ Hz of the DB-controlled inverter system: (a) harmonic ratio $h_r(j)$ of the grid current i_g and (b) magnitude responses of the LPFs $Q(z)$ and $Q^2(z)$

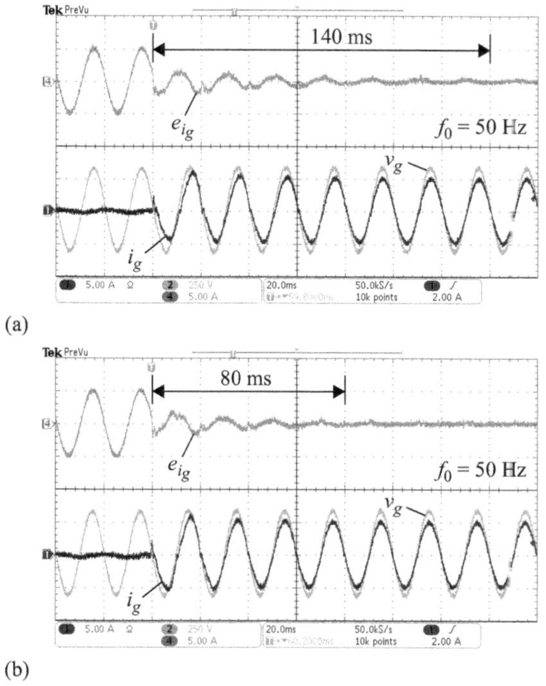

Figure 6.15 Start-up transient tracking error of the grid current using different control schemes (grid voltage v_g: [250 V/div], grid current i_g: [5A/div], tracking error e_{ig}: [5 A/div], time: [20 ms/div]): (a) DB with the CRC and (b) DB with the conventional OHC

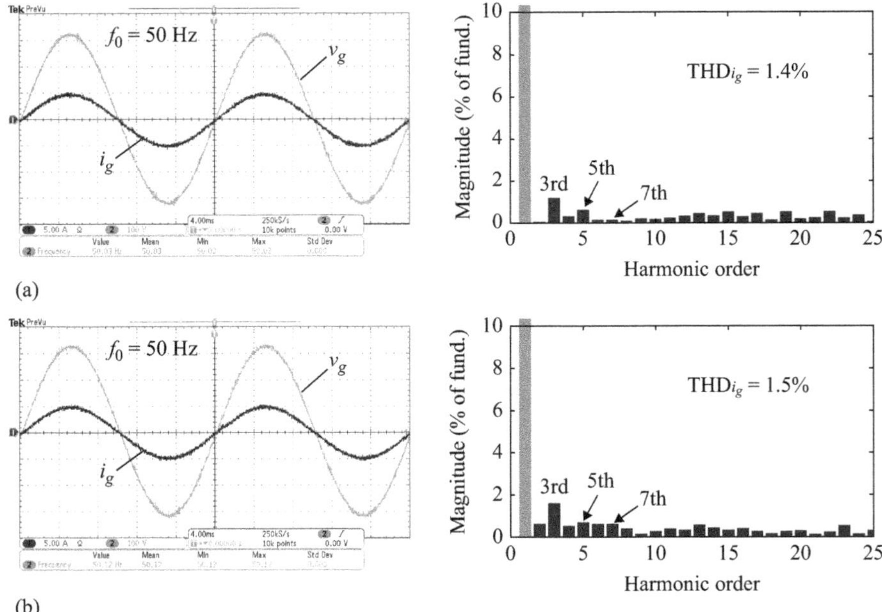

Figure 6.16 Steady-state responses of the single-phase grid-connected inverter system with different control schemes in case of $f_0 = 50$ Hz (left: grid voltage v_g: [100 V/div], grid current i_g: [5 A/div], time: [4 ms/div]; right: harmonic spectrum of the grid current i_g): (a) DB with the CRC and (b) DB with the conventional OHC

OHC schemes are plugged into the DB-controlled system to enhance the harmonic elimination. Figure 6.15 demonstrates that, within the stability range for the control gain(s), the dynamic response of the OHC can be much faster (up to $n/2$ times) than that of the CRC. Figure 6.16 shows that the plug-in CRC, and the OHC-controlled single-phase inverter can feed good quality current i_g into the grid and their corresponding harmonic distortion levels of the feed-in grid current i_g are $\text{THD}_{ig} = 1.4\%$ and $\text{THD}_{ig} = 1.5\%$, respectively. It is clearly indicated in Figures 6.15 and 6.16 that, compared with the CRC, the OHC scheme with optimized control gains can significantly improve the dynamic response with almost the same tracking accuracy. It means that the OHC scheme can offer an optimal and efficient PC solution to the grid-connected inverters.

However, in the case of frequency variations, (e.g., if $f_0 = 50.2$ Hz, $N/4 = 49.8$ is fractional, and if $f_0 = 49.8$ Hz, $N/4 = 50.2$ is also fractional), the tracking accuracy of the CRC or the OHC schemes will be affected. Table 6.5 lists the current THDs under various grid frequencies using different control schemes, which are also plotted in Figure 6.17. It can be seen from Table 6.5 that the DB controller with almost constant feed-in current THDs of 8% ~ 8.65% (higher than 5%) can be treated as irrelevant to frequency variations, while the THDs of the CRC-controlled feed-in current will be increased with the frequency deviation $\Delta f_0 = f_0 - 50$ Hz.

Table 6.5 Current THDs under various grid frequencies using different control schemes

Grid frequency (Hz)		**49**	**49.5**	**49.6**	**49.7**	**49.8**	**49.9**	
	DB	8.39	8.12	8.09	8.07	8.05	8.01	
THD of i_g (%)	DB + CRC	6.25	4.5	4	3.38	2.7	2	
	DB + FACRC	3.02	1.9	1.78	1.63	1.48	1.42	
	DB + FA-OHC	3.08	2.02	1.85	1.73	1.63	1.52	
Grid frequency (Hz)		**50**	**50.1**	**50.2**	**50.3**	**50.4**	**50.5**	**51**
	DB	8	8.02	8.05	8.11	8.15	8.2	8.65
THD of i_g (%)	DB + CRC	1.55	2.12	2.8	3.4	3.95	4.44	6.5
	DB + FACRC	1.4	1.43	1.52	1.63	1.82	2	3.16
	DB + FA-OHC	1.49	1.52	1.63	1.77	1.95	2.13	3.16

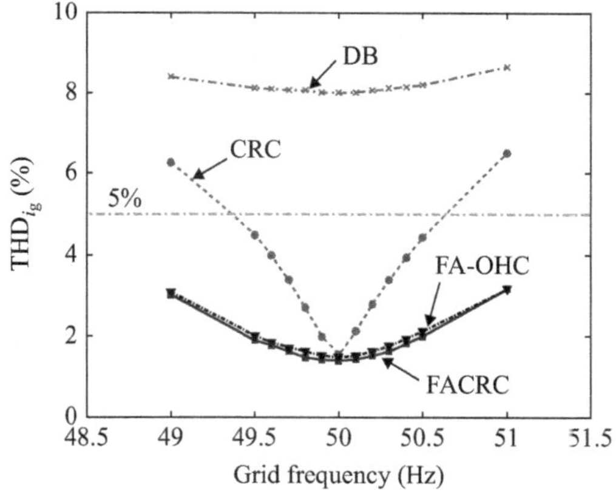

Figure 6.17 Feed-in current THD levels in a single-phase grid-connected inverter system with various current controllers under a varying grid frequency (49 Hz $\leq f_0 \leq$ 51 Hz)

Based on the FD filters, the FACRC and FA-OHC can effectively suppress the current THDs to below 3.5% (lower than 5%), as observed in Table 6.5 and also in Figure 6.17. It means that the FAPC schemes are much less sensitive to grid frequency variations. In addition, as shown in Figure 6.18, the emerging phase shifts between the grid voltage and the grid current will lead to reactive power injections or absorptions into the grid, if the frequency-adaptive capabilities of the periodic control schemes are deactivated.

In order to further evaluate the robustness and convergence rate of the FA-OHC scheme and also the FACRC scheme in response to frequency step

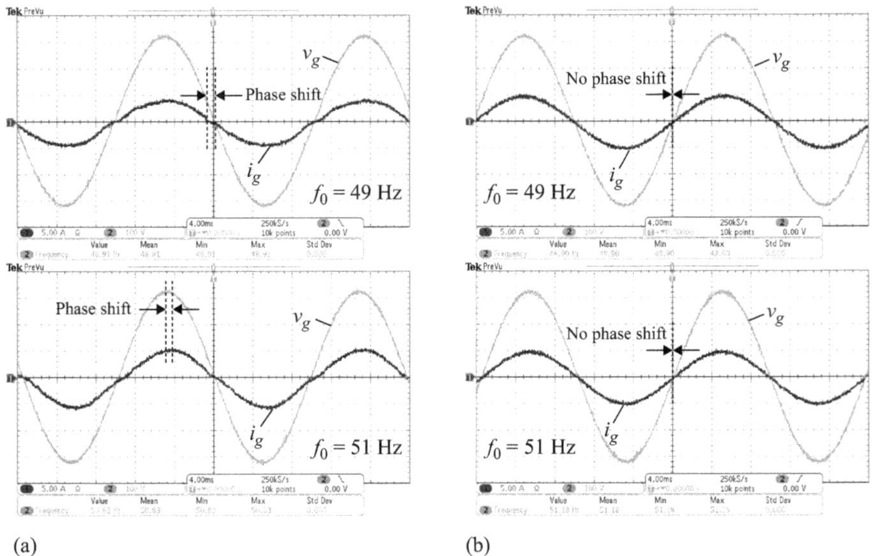

Figure 6.18 Steady-state performance of the CRC and the frequency adaptive RC current controllers under various grid frequencies (grid voltage v_g: [100 V/div] and grid current i_g: [5 A/div], time: [4 ms/div]): (a) the CRC scheme and (b) the frequency-adaptive RC scheme

changes, more experiments have been carried out. Figure 6.19 shows the start-up dynamic responses of the FACRC and the FA-OHC in case of a nonnominal grid frequency (i.e., $f_0 = 51$ Hz). In contrast to the start-up response of the two controllers without frequency-adaptive schemes shown in Figure 6.15, the dynamic performances of both the FACRC and the FA-OHC are almost the same. It means that the frequency-adaptive schemes by means of FD filters will not change the dynamic performance of the PC schemes. Finally, the dynamic performances of the FACRC and the FA-OHC schemes under grid frequency step changes are presented in Figure 6.20. It is indicated in Figure 6.20 that the two FAPC schemes are robust to grid frequency variations as well.

6.2.4 Conclusion

In general, seen from the above representative cases, the following can be concluded: in grid-connected applications, the FAPC schemes presented in Chapter 5 are quite competitive to other current controllers in terms of high tracking accuracy, fast dynamic performance, good robustness, and being easy for implementation.

Specifically, for grid-connected converters, the FAPC schemes, especially the FA-OHC, are developed to remove uneven distributed power harmonics from the feed-in current under grid frequency variations. Series of experiments demonstrate that the FA-OHC, which can achieve accurate, fast, robust, and cost-effective feed-in

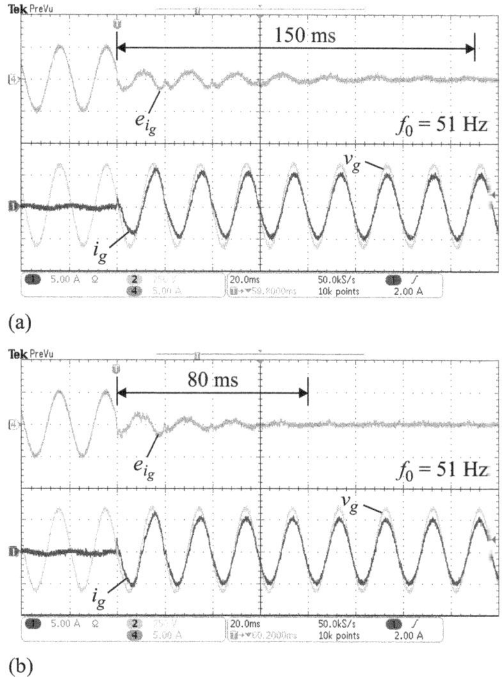

Figure 6.19 Start-up transient tracking error of the grid current with different PC schemes (grid voltage v_g: [250 V/div], grid current i_g: [5 A/div], tracking error e_{ig}: [5 A/div], time: [20 ms/div]): (a) DB with the frequency-adaptive RC scheme and (b) DB with the FA-OHC scheme

current regulation, provides a tailor-made optimal current control solution to grid-connected converters. Furthermore, since various periodic controllers can be treated as special cases of the FA-OHC, the FA-OHC actually provides a general form of the FAPC for power converters [17,26].

6.3 FAPC of shunt active power filters

6.3.1 Background

The rapidly growing use of nonlinear loads causes considerable power quality degradation in power distribution networks, such as power harmonic distortions and resonances. Shunt active power filters (APFs) [46–53], which operate as controllable power sources and offer fast responses to load dynamic changes, are widely used to cancel the current harmonics produced by the nonlinear loads. The digital CRC provides a simple but accurate current control solution to shunt APFs

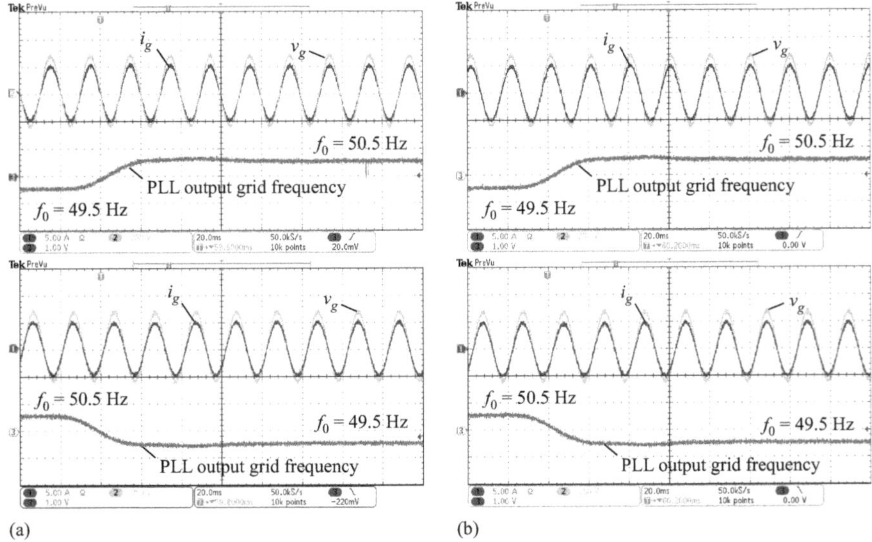

Figure 6.20 Transient responses of the single-phase grid-connected system under grid frequency changes with different PC schemes (grid voltage v_g: [250 V/div], grid current i_g: [5 A/div], grid frequency [1 Hz/div], time: [20 ms/div]): (a) DB with the FACRC and (b) DB with the FA-OHC scheme

for grid current distortion compensation. It can achieve zero steady-state error tracking of any periodic signal with a known integer period of $N = f_s/f_0$, where f_0 is the fundamental frequency of reference signal and f_s is the sampling rate.

Obviously, a time-varying grid frequency f_0 will result in a fractional period of N. However, in practical applications, the grid-connected shunt APFs usually operate under a time-varying grid frequency f_0. As discussed in Chapter 5, the digital CRC with a fixed sampling frequency f_s is sensitive to such grid frequency variations. In cases of fractional-period grid voltages, the CRC cannot enable shunt APFs to accurately compensate the current harmonics. Addressing this issue, an FACRC strategy is developed for a single-phase shunt APF to accurately remove harmonic distortions from the grid current [25].

6.3.2 Modeling and control of shunt active power filters

Figure 6.21(a) shows the typical configuration of a single-phase PWM inverter-based APF system, which is used to compensate the current harmonic distortions induced by the nonlinear rectifier loads with L_r and C_r being the load filter inductor and capacitor, respectively, and R being the load resistor. While keeping the DC bus voltage v_{dc} constant at a preset level, the shunt APF inverter is thus regulated to accurately generate the compensation current i_c to compensate the harmonic components of the load current i_L, and to achieve sinusoidal grid current i_g with a unity

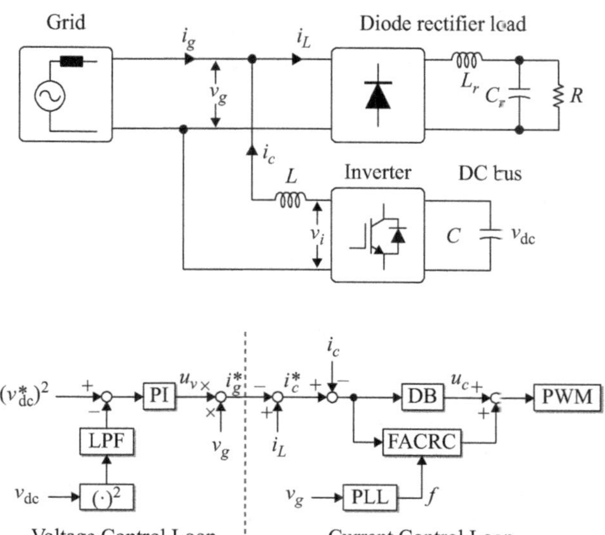

Figure 6.21 Single-phase shunt active power filter system connected to the grid with a nonlinear load: (a) the hardware schematic and (b) the dual-loop control scheme

power factor. The dynamics of the shunt APF in Figure 6.21(a) can be written as [25]:

$$\dot{i}_c = \frac{1}{L}(v_g - v_i) \tag{6.18}$$

$$\dot{v}_{dc} = \frac{1}{C} i_c S(t) \tag{6.19}$$

where v_{dc}, v_g, and v_i are the DC bus voltage, the grid voltage and the output voltage of the inverter, i_c is the compensation current fed by the APF, L, and C are the filter inductor and DC bus capacitor, and $S(t)$ denotes the switching function with value being 1 or -1.

Figure 6.21(b) shows the diagram of the control scheme for the shunt APF, where it can be seen that the control scheme includes two control loops. Namely, a proportional–integral (PI) controller-based outer voltage loop is to force the DC bus voltage v_{dc} to track the reference voltage v_{dc}^*, and a DB plus FACRC scheme-based inner current loop is to force the compensation current i_c to track the reference current i_c^*. The compensation current i_c is expected to be the inverse of the harmonic components of and the reactive power components in the load current i_L. Since the bandwidth of the inner current loop is usually much larger than that of the outer voltage loop, the reference current i_c^* can be treated as a constant input for the inner control loop. Consequently, outer voltage loop controller and inner current loop controller can be separately designed.

A PI controller is employed to force the DC bus voltage v_{dc} to track the constant reference value v_{dc}^* [25,53], and its output can be given as:

$$u_v = k_p\left[\left(v_{dc}^*\right)^2 - v_{dc}^2\right] + k_i\int\left[\left(v_{dc}^*\right)^2 - v_{dc}^2\right]dt \tag{6.20}$$

with u_v being the output of the voltage loop PI controller and k_p, k_i being the proportional and integral gains of the PI controller, respectively. Note that, due to the power exchange between the grid and the APF, the DC bus voltage for the APF will contain voltage ripples denoted as $\tilde{v}_{dc}$. The voltage ripples $\tilde{v}_{dc}$ in the DC bus voltage will lead to unexpected harmonics in the reference compensation current i_c^* and thus in the compensation current i_c [37,38]. As shown in Figure 6.21(b), an LPF is thus used to remove unexpected harmonics due to the DC bus voltage ripples, which can be passed to the reference compensation current i_c^*.

Since the main control objective of the APF is to achieve a sinusoidal grid current i_g with a unity power factor, the reference compensation current i_c^* for the inner current loop can be generated by:

$$i_c^*(t) = i_L(t) - u_v(t)v_g(t) \tag{6.21}$$

The sampled-data form of (6.18) can be expressed as:

$$i_c(k+1) = i_c(k) + \frac{T_s}{L}(v_g(k) - v_i(k)) \tag{6.22}$$

where k and $k+1$ represent the k-th and $(k+1)$-th sampling instants, and $T_s = 1/f_s$ is the sampling period. Let $i_c(k+1) = i_c^*(k)$, and the transfer function for the inner control loop will be $H(z) = z^{-1}$. That is to say, a DB current control loop is achieved. Accordingly, we have:

$$v_i(k) = v_g(k) - \frac{L}{T_s}(i_c^*(k) - i_c(k)) \tag{6.23}$$

where $v_i(k)$ is the output voltage of the inverter and

$$v_i(k) = u_c(k)v_{dc}(k) \tag{6.24}$$

with $u_c(k) \in [-1, 1]$ being the output of inner loop DB current controller.

Therefore, the DB current control law can be obtained as:

$$u_c(k) = \frac{1}{v_{dc}(k)}\left[v_g(k) - \frac{L}{T_s}(i_c^*(k) - i_c(k))\right] \tag{6.25}$$

However, the DB controller is sensitive to various system uncertainties (e.g., the time delay and parameter variations). To ensure that the APF can achieve an exact compensation of current harmonics in the presence of various system uncertainties and frequency variations, an FACRC $G_{arc}(z)$ is plugged into the DB current controller as shown in Figure 6.21(b). The FACRC can be expressed as:

$$G_{arc}(z) = k_{rc}\frac{(z^{-N_i}z^{-F})Q(z)}{1 - (z^{-N_i}z^{-F})Q(z)}G_f(z) \tag{6.26}$$

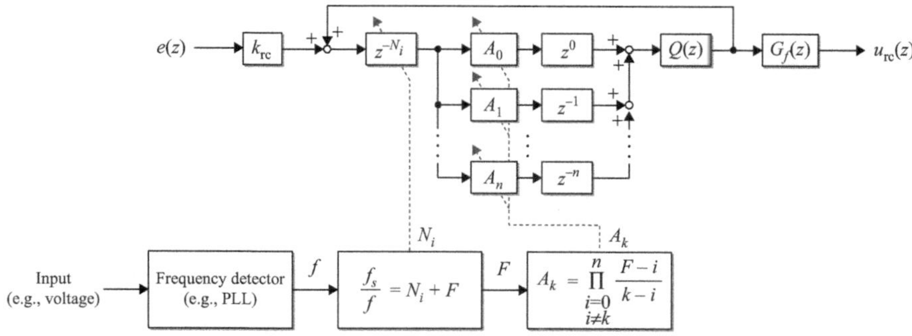

Figure 6.22 Implementation of the FACRC given in (6.26)

where $G_f(z) = z^c$ is a linear phase-lead filter with phase-lead step $c \in \mathbb{N}$ being determined by experiments [19,21], $Q(z)$ is an LPF, and k_{rc} is the control gain. In (6.26), the *Lagrange* interpolation-based FD filter z^{-F} can be written as:

$$z^{-F} = \sum_{k=0}^{n} A_k z^{-k} = \sum_{k=0}^{n} \prod_{\substack{i=0 \\ i\neq k}}^{n} \frac{F-i}{k-i} z^{-k} \quad k, i = 0, 1, \ldots, n \tag{6.27}$$

with F being the fractional part of the fractional number $N = f_s/f = N_i + F$, f_s being the sampling frequency and f being the phase-locked loop (PLL)-detected grid frequency. In the following experiments, $n = 3$ is chosen to be the *Lagrange* polynomial degree.

As shown in Figure 6.22, for online update of the fractional delay of the FACRC, a frequency detector which is performed by a PLL system shown in Figure 6.21, is used to provide the real-time measurement of the grid frequency f. The frequency detector should be insensitive to grid disturbances and harmonics, and can operate within a wide frequency range (i.e., 47~52 Hz) [25]. The dynamic response of the frequency detector with the settling time being 50 ms is much slower than that of the current loop, and is much faster than the maximum change rate of the grid frequency. For example, if the grid frequency f changes from 50 to 50 ± 1 Hz, the fundamental signal period $N = f_s/f = N_i + F$ with the sampling frequency $f_s = 5$ kHz will change from 100 to about 100 ± 2. If the change rate of the grid frequency is 1 Hz/s (a typical maximum rate value), it will take 500 ms to increase or decrease the integer part N_i by 1, and the fractional period N for FACRC would only change $\pm 4 \times 10^{-4}$ over one sampling cycle. Correspondingly, online updating coefficients A_k will not change significantly, and can be treated as constants in the system stability analysis. It implies that the FACRC is quite robust to the grid frequency change. Furthermore, as long as the bandwidth of the FD filter in (6.27) is larger than that of $Q(z)$, the FACRC in (6.26) is identical to the CRC in (2.5). Hence the synthesis of the FACRC can be fully compatible to that of the CRC.

Table 6.6 System parameters of a single-phase shunt active power filter

Symbol	Description	Nominal value
v_{dc}^*	DC bus voltage reference	250 V
v_g	Grid voltage amplitude	120 V
L	Filter inductor	5 mH
L_r	Rectifier inductor	5 mH
C	DC bus capacitor	1100 μF
C_r	Rectifier capacitor	4400 μF
R	Rectifier resistor	12 Ω
f_c	Switching frequency	5 kHz
f_s	Sampling frequency	5 kHz
f	Grid frequency	50 Hz
k_p	Proportional gain	1.8
k_i	Integral gain	25
k_{rc}	Repetitive gain	0.8

6.3.3 Experimental validation

A dSPACE DS1104-based single-phase shunt APF system has been built up to verify the effectiveness of the plug-in FACRC. A programmable AC power source is used to simulate the grid voltage. The system parameters are listed in Table 6.6. The phase-lead compensation filter $G_f(z) = z^2$ is employed in the FACRC, and the LPF $Q(z)$ is chosen as $Q(z) = 0.1z + 0.8 + 0.1z^{-1}$. According to Figure 6.3, the bandwidths of $Q(z)$ and the FD filter in (6.28) can be correspondingly calculated as about $0.66f_s/2 = 1650$ Hz and $0.75f_s/2 = 1875$ Hz, respectively. Thus, it enables the FACRC to efficiently compensate up to 30th-order current harmonics.

As two examples, when the grid frequency changes from 50 Hz to 50 ± 0.2 Hz, the period N will change from integer $N = 100$ to fractional $N = 100 \pm 0.4$. Figure 6.23 shows the Bode plots of the CRC and the FACRC under such conditions. It can be seen in Figure 6.23 that the CRC with an integer delay z^{-100} and the FACRC with approximated fractional delays $z^{-99.6}$ and $z^{-100.4}$ have large gains at the harmonic frequencies of $50n$ Hz and $(50 \pm 0.2)n$ Hz ($n = 1, 2, \ldots$), respectively. In addition, the phase shifts at these frequencies are zero. It is indicated that if the fundamental grid frequency changes from 50 Hz to 50 ± 0.2 Hz, the CRC cannot exactly compensate the harmonic of $(50 \pm 0.2)n$ Hz with $n = 1, 2, \ldots$, due to its reduced gains at these frequencies. This means that the CRC is sensitive to grid frequency variations.

Figure 6.24 shows the performance of the grid-connected system in Figure 6.21 when the APF is deactivated. It can be observed in Figure 6.24 that, without the APF, the THD of the grid current i_g under the nominal grid frequency of 50 Hz is up to 44.4%. Then, the APF control is enabled. Figure 6.25 shows the steady-state performance of the system with the CRC-controlled APF ($N = 100$, i.e., $F = 0$ for the FACRC) under the nominal grid frequency of 50 Hz. It can be seen that the THD of the grid current i_g is significantly reduced to 3.7%.

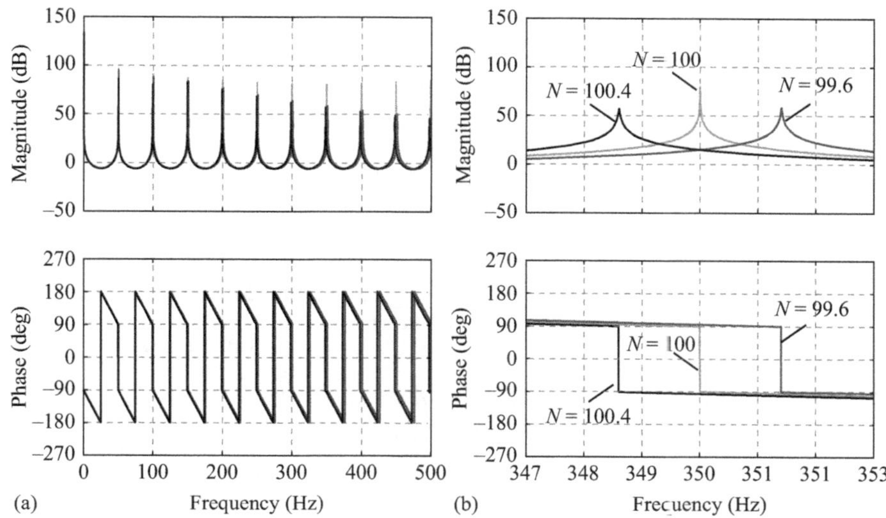

Figure 6.23 Bode diagrams of the CRC with N = 100 and the FACRC with N = 100.4 and 99.6 within: (a) a low frequency band and (b) a neighboring frequency band around the seventh-order harmonic

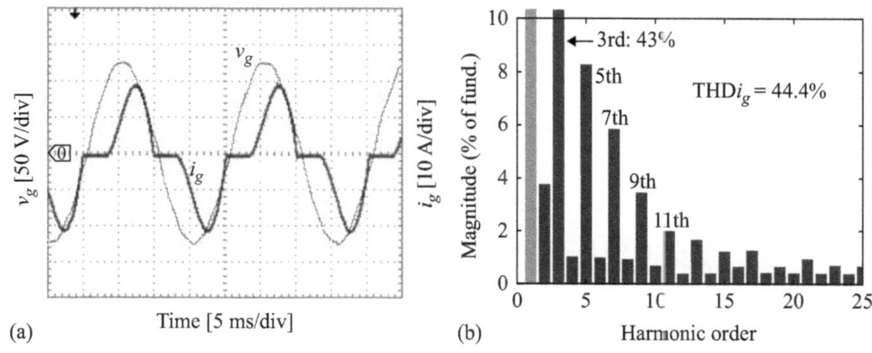

Figure 6.24 Steady-state performance of the system at 50 Hz without APF: (a) grid voltage v_g and grid current i_g and (b) harmonic spectrum of the grid current i_g

Figures 6.24 and 6.25 indicate that the CRC-controlled APF can efficiently compensate the majority of the harmonic distortions in the grid current i_g under a grid voltage with an integer period.

However, when the change of grid frequency leads to the grid voltage with a fractional period, the compensation performance of CRC will be degraded. Figure 6.26 shows the steady-state performance of the system with the CRC-controlled APF under a grid frequency of 49.8 Hz (N is round to 100). It can be

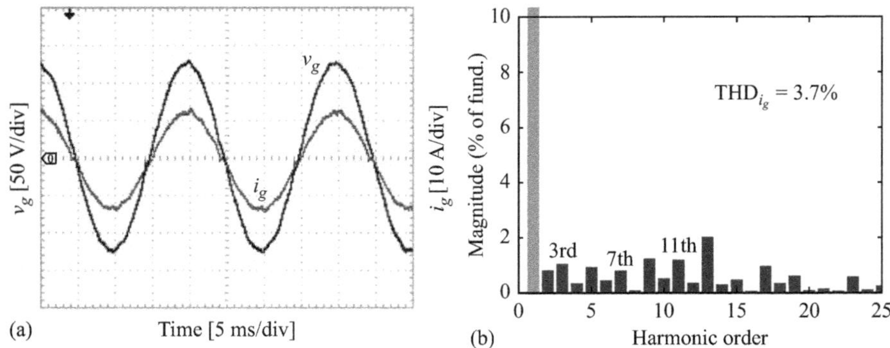

Figure 6.25 Steady-state performance of the system at 50 Hz with the CRC-controlled APF: (a) grid voltage v_g and grid current i_g and (b) harmonic spectrum of the compensated grid current i_g

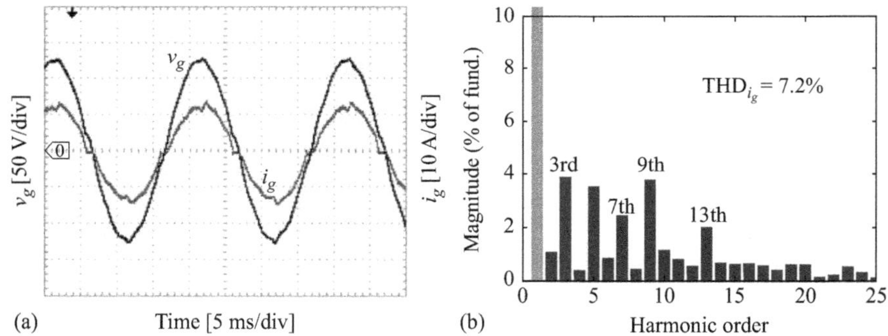

Figure 6.26 Steady-state performance of the system at 49.8 Hz with the CRC-controlled APF: (a) grid voltage v_g and grid current i_g and (b) harmonic spectrum of the compensated grid current i_g

observed in Figure 6.26 that the THD of the grid current i_g is around 7.2%. By contrast, Figure 6.27 shows the steady-state performance of the system with the FACRC-controlled APF under the same grid frequency (49.8 Hz with $N = 100.4$). Clearly, the THD of the grid current i_g is reduced to 3.0%. Figures 6.26 and 6.27 confirm that, in the case of a fractional grid period, the FACRC offers much higher current harmonic compensation accuracy than what the CRC does.

In addition, Figure 6.28 shows the steady-state performance of the system with the CRC-controlled APF under a grid frequency of 50.2 Hz (N is rounded to 100). It is shown that the THD of the grid current i_g is around 8.7%. For comparison, the FACRC control scheme with $N = 100.4$ is also applied to the APF system. As presented in Figure 6.29, under the grid frequency of 50.2 Hz, the FACRC-controlled system can reduce the THD of the grid current i_g to around 2.8%.

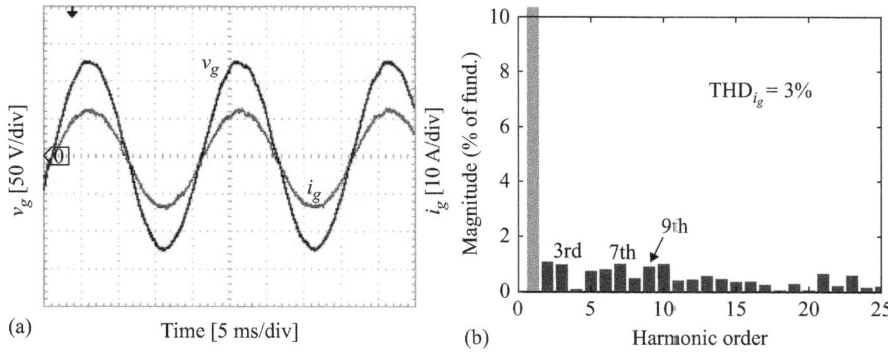

Figure 6.27 Steady-state performance of the system at 49.8 Hz with the FACRC-controlled APF: (a) grid voltage v_g and grid current i_g and (b) harmonic spectrum of the compensated grid current i_g

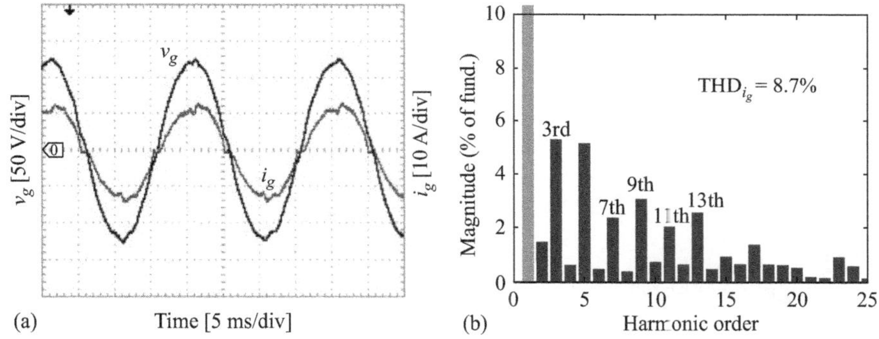

Figure 6.28 Steady-state performance of the system at 50.2 Hz with the CRC-controlled APF: (a) grid voltage v_g and grid current i_g and (b) harmonic spectrum of the compensated grid current i_g

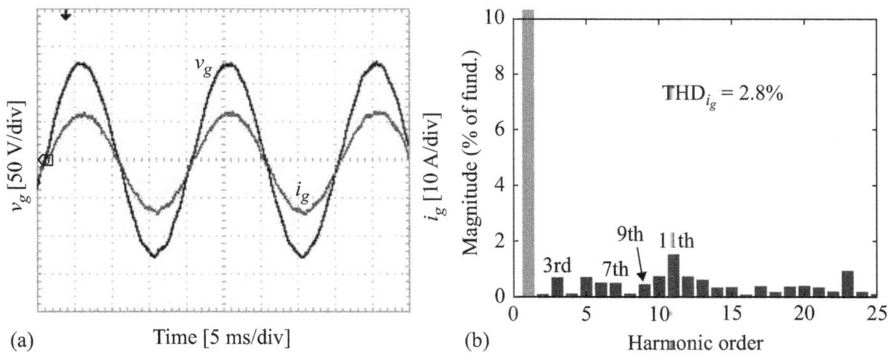

Figure 6.29 Steady-state performance of the system at 50.2 Hz with the FACRC-controlled APF: (a) grid voltage v_g and grid current i_g and (b) harmonic spectrum of the compensated grid current i_g

Table 6.7 Performance of the APF system under different grid frequencies

Grid frequency (Hz)	THD (%)	
	CRC	FACRC
49.5	3.587	2.739
49.6	4.357	2.958
49.7	8.102	2.920
49.8	7.179	2.987
49.9	4.795	2.838
50.0	3.719	3.719
50.1	4.866	2.706
50.2	8.749	2.795
50.3	8.039	3.135
50.4	4.908	2.908
50.5	3.649	3.166

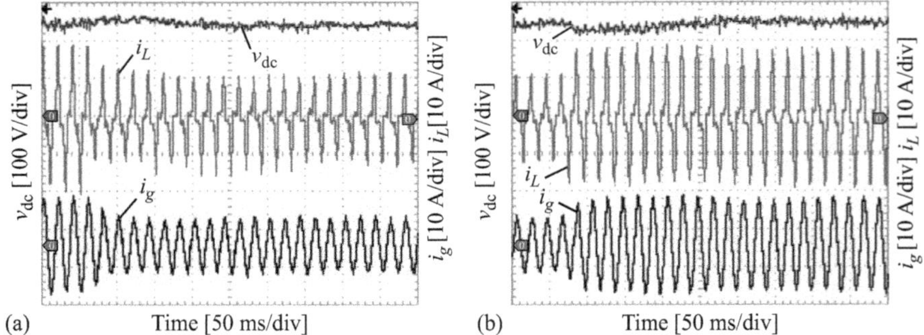

Figure 6.30 System performance of the FACRC-controlled APF system in the case of load changes at a grid frequency of 49.8 Hz: (a) R changed from 15 Ω to 30 Ω and (b) R changed from 30 Ω to 15 Ω

Figures 6.28 and 6.29 also confirm that, in the case of a fractional grid period, the FACRC-controlled APF system performs much better than what the CRC does.

Table 6.7 further summarizes the THDs of the compensated grid current i_g under various grid frequencies from 49.5 Hz to 50.5 Hz. The results in Table 6.7 indicate that, in the case of various grid frequencies the FACRC-controlled APF always can offer very high current harmonic compensation accuracy, while the CRC-controlled APF fails to provide a satisfactory current harmonic compensation. Obviously, the CRC is sensitive to frequency variations, while the FACRC is much more immune to frequency changes.

In order to evaluate the robustness of the FACRC scheme for the shunt APF in response to load disturbances, more experiments have been carried out. As shown in Figure 6.30, the grid current i_g compensated by the FACRC-controlled APF can keep sinusoidal under the step-change loads. Moreover, in the load disturbance tests, the settling time of the DC bus voltage is about 250 ms. The above results thus verifies that the FACRC-controlled APF system is robust to load changes.

6.3.4 Conclusion

The experimental tests of the FACRC-controlled single-phase shunt APF system indicate that the FACRC enables the shunt APF to precisely and rapidly compensate the harmonic distortions in the grid current under grid frequency variations. Moreover, the FACRC scheme is robust to step frequency changes and load disturbances.

6.4 Summary

The FD filter-based FAPC with a fixed sampling rate provides power converters (especially grid-connected converters) with a simple, accurate, fast, and robust control solution. The synthesis and design of such FAPC systems are compatible to those of the corresponding non-FAPC schemes, which have been explored in Chapters 2 and 3. All the above three application examples verified the effectiveness of the FD filter-based FAPC schemes.

References

[1] Tzou, Y.Y., Ou, R.S., Jung, S.L., and Chang, M.Y., "High-performance programmable AC power source with low harmonic distortion using DSP-based repetitive control technique," *IEEE Trans. Power Electron.*, vol. 12, no. 4, pp. 715–725, 1997.

[2] Tamyurek, B., "A high-performance SPWM controller for three-phase UPS systems operating under highly nonlinear loads," *IEEE Trans. Power Electron.*, vol. 28, no. 8, pp. 3689–3701, 2013.

[3] Cardenas, R., Juri, C., Pena, R., Wheeler, P., and Clare, J., "The application of resonant controllers to four-leg matrix converters feeding unbalanced or nonlinear loads," *IEEE Trans. Power Electron.*, vol. 27, no. 3, pp. 1120–1129, 2012.

[4] Mattavelli, P., "An improved deadbeat control for UPS using disturbance observers," *IEEE Trans. Ind. Electron.*, vol. 52, no. 1, pp. 206–212, 2005.

[5] Komurcugil, H., "Rotating-sliding-line-based sliding-mode control for single-phase UPS inverters," *IEEE Trans. Ind. Electron.*, vol. 59, no. 10, pp. 3719–3726, 2012.

[6] Zhou, K. and Wang, D., "Zero-tracking error controller for three-phase CVCF PWM inverter," *Electronics Letters*, vol. 36, no. 10, pp. 864–865, 2000.

[7] Zhou, K. and Wang, D., "Digital repetitive learning controller for three-phase CVCF PWM inverter," *IEEE Trans Ind. Electron.*, vol. 48, no. 4, pp. 820–830, 2001.

[8] Chen, S., Lai Y.M., Tan S.C., and Tse C.K., "Analysis and design of repetitive controller for harmonic elimination in PWM voltage source inverter systems," *IET Power Electron.*, vol. 1, no. 4, pp. 497–506, 2008.

[9] Loh, P.C., Tang, Y., Blaabjerg, F., and Wang, P., "Mixed-frame and stationary-frame repetitive control schemes for compensating typical load and grid harmonics," *IET Power Electron.*, vol. 4, no. 2, pp. 218–226, 2011.

[10] Ye, Y., Zhang, B., Zhou, K., Wang, D., and Wang, Y., "High-performance cascade-type repetitive controller for CVCF PWM inverter: Analysis and design," *IET Electric Power Appl.*, vol. 1, no. 1, pp. 112–118, 2007.

[11] Yang, Y., Zhou, K., and Lu, W., "Robust repetitive control scheme for three-phase CVCF PWM inverters," *IET Power Electron.*, vol. 5, no. 6, pp. 669–677, 2011.

[12] Lu, W., Zhou, K., Wang, D., and Cheng, M., "A general parallel structure repetitive control scheme for multiphase DC–AC PWM converters," *IEEE Trans. Power Electron.*, vol. 28, no. 8, pp. 3980–3987, 2013.

[13] Lu, W., Zhou, K., and Wang, D., "General parallel structure digital repetitive control," *Int. J. Control*, vol. 86, no. 1, pp. 70–83, Jan. 2013.

[14] Lu, W., Zhou, K., Wang, D., and Cheng, M., "A generic digital nk ± m-order harmonic repetitive control scheme for PWM converters," *IEEE Trans. Ind. Electron.*, vol. 61, no. 3, pp. 1516–1527, 2014.

[15] Zhou, K., Lu, W., Yang, Y., and Blaabjerg, F., "Harmonic control: A natural way to bridge resonant control and repetitive control," *in Proc. Amer. Control Conf.*, pp. 3189–3193, Jun. 2013.

[16] Zhou, K., Low, K.S., Wang, Y., Luo, F.L., Zhang, B., and Wang, Y., "Zero-phase odd-harmonic repetitive controller for a single-phase PWM inverter," *IEEE Trans. on Power Electron.*, vol. 21, no. 1, pp. 193–201, 2006.

[17] Zhou, K., Yang, Y., Blaabjerg, F., and Wang, D., "Optimal selective harmonic control for power harmonics mitigation," *IEEE Trans. Ind. Electron.*, vol. 62, no. 2, pp. 1220–1230, 2015.

[18] Zhu, W., Zhou, K., and Cheng, M., "A bidirectional high-frequency-link single-phase inverter: Modulation, modeling, and control," *IEEE Trans. Power Electron.*, vol. 29, no. 8, pp. 4049–4057, 2014.

[19] Zou, Z., Zhou, K., Wang, Z., and Cheng, M., "Fractional-order repetitive control of programmable AC power sources," *IET Power Electron.*, vol. 7, no. 2, pp. 431–438, 2014.

[20] Zhang, B., Zhou, K., Wang, Y., and Wang, D., "Performance improvement of repetitive controlled PWM inverters: A phase-lead compensation solution," *Int. J. Circuit Theory Appl.*, vol. 38, no. 5, pp. 453–469, 2010.

[21] Zhang, B., Wang, D., Zhou, K., and Wang, Y., "Linear phase-lead compensation repetitive control of a CVCF PWM inverter," *IEEE Trans. Ind. Electron.*, vol. 55, no. 4, pp. 1595–1602, Apr. 2008.

[22] Laakso, T.I., Valimaki, V., Karjalainen, M., and Laine, U.K., "Splitting the unit delay," *IEEE Signal Processing Mag.*, vol. 13, no. 1 pp. 30–60, 1996.

[23] Chen, D., Zhang, J., and Qian, Z., "An improved repetitive control scheme for grid-connected inverter with frequency-adaptive capability," *IEEE Trans. Ind. Electron.*, vol. 60, no. 2, pp. 814–823, 2013.

[24] Rashed, M., Klumpner, C., and Asher, G., "Repetitive and resonant control for a single-phase grid-connected hybrid cascaded multilevel converter," *IEEE Trans. Power Electron.*, vol. 28, no. 5, pp. 2224–2234, 2013.

[25] Zou, Z.-X., Zhou, K., Wang, Z., and Cheng, M., "Frequency adaptive fractional order repetitive control of shunt active power filters," *IEEE Trans. Ind. Electron.*, vol. 62, no. 3, pp. 1659–1668, 2015.

[26] Yang, Y., Zhou, K., Wang, H., Blaabjerg, F., Wang, D., and Zhang, B., "Frequency-adaptive selective harmonic control for grid-connected inverters," *IEEE Trans. Power Electron.*, vol. 30, no. 7, pp. 3912–3924, 2015.

[27] Wang, Y., Wang, D., Zhang, B., and Zhou, K., "Fractional delay-based repetitive control with application to PWM DC/AC Converters," *16th IEEE Int. Conf. Control Appl.*, pp. 928–933, 2007.

[28] Liu T. and Wang D., "Parallel structure fractional repetitive control for PWM inverters," *IEEE Trans. Ind. Electron.*, vol. 62, no. 8, pp. 5045–5054, Feb. 2015.

[29] Liu, T., Wang, D., and Zhou, K., "High-performance grid simulator using parallel structure fractional repetitive control," *IEEE Trans. Power Electron.*, vol. 31, no. 3, pp. 2669–2679, 2016.

[30] Blaabjerg, F., Teodorescu, R., Liserre, M., and Timbus, A. V., "Overview of control and grid synchronization for distributed power generation systems," *IEEE Trans. Ind. Electron.*, vol. 53, no. 5, pp. 1398–1409, Oct. 2006.

[31] Timbus, A.V., Liserre, M., Teodorescu, R., Rodriguez, P., and Blaabjerg, F., "Evaluation of current controllers for distributed power generation systems," *IEEE Trans. Power Electron.*, vol. 24, no. 3, pp. 654–664, Mar. 2009.

[32] Teodorescu, R., Liserre, M., and Rodriguez, P., *Grid Converters for Photovoltaic and Wind Power Systems*. Hoboken, NJ, USA: Wiley, 2011.

[33] Kjaer, S.B., Pedersen, J.K., and Blaabjerg, F., "A review of single-phase grid-connected inverters for photovoltaic modules," *IEEE Trans. Ind. Appl.*, vol. 41, no. 5, pp. 1292–1306, Sep./Oct. 2005.

[34] Yang, Y., Enjeti, P., Blaabjerg, F., and Wang, H., "Wide-scale adoption of photovoltaic energy: Grid code modifications are explored in the distribution grid," *IEEE Ind. Appl. Mag.*, vol. 21, no. 5, pp. 21–31, 2015.

[35] Yang, Y. and Blaabjerg, F., "Overview of single-phase grid-connected photovoltaic systems," *Electric Power Components and Systems*, vol. 43, no. 12, pp. 1352–1363, 2015.

[36] IEEE Application Guide for IEEE Std. 1547, "IEEE Standard for Interconnecting Distributed Resources with Electric Power Systems," *IEEE Standard 1547. 2–2008*, 2009.

[37] Zhou, K., Qiu, Z., Watson, N.R., and Liu, Y., "Mechanism and elimination of harmonic current injection from single-phase grid-connected PWM converters," *IET Power Electron.*, vol. 6, no. 1, pp. 88–95, Jan. 2013.

[38] Yang, Y., Zhou, K., and Blaabjerg, F., "Harmonics suppression for single-phase grid-connected PV systems in different operation modes," in *Proc. of IEEE APEC'13*, pp. 889–896, Mar. 2013.
[39] Zhou, K., Wang, D., Zhang, B., Wang, Y., Ferreira, J.A., and de Haan, S.W.H., "Dual-mode structure digital repetitive control," *Automatica*, vol. 43, no. 3, pp. 546–554, 2007.
[40] Zhou, K., Lu, W., Yang, Y., and Blaabjerg, F., "Harmonic control: A natural way to bridge resonant control and repetitive control," *American Control Conf.*, 3189-3193, 2013
[41] Teodorescu, R., Blaabjerg, F., Liserre, M., and Loh, P. C., "Proportional-resonant controllers and filters for grid-connected voltage-source converters," *Proc. IEE*, vol. 153, no. 5, pp. 750–762, Sep. 2006.
[42] Liserre, M., Teodorescu, R., and Blaabjerg, F., "Multiple harmonics control for three-phase grid converter systems with the use of PI-RES current controller in a rotating frame," *IEEE Trans. Power Electron.*, vol. 21, no. 3, pp. 836–841, May 2006.
[43] Yang, Y., Zhou, K., Cheng, M., and Zhang, B., "Phase compensation multi-resonant control of CVCF PWM converters," *IEEE Trans. Power Electron.*, vol. 28, no. 8, pp. 3923–3930, Aug. 2013.
[44] Yang, Y., Zhou, K. and Cheng, M., "Phase compensation resonant controller for single-phase PWM converters," *IEEE Trans. Ind. Informatics*, vol. 9, no. 2, pp. 957–964, 2013.
[45] Yang, Y., Zhou, K., and Blaabjerg, F., "Enhancing the frequency adaptability of periodic current controllers with a fixed sampling rate for grid-connected power converters," *IEEE Trans. Power Electron.*, vol. 31, no. 10, pp. 7273–7285, 2016.
[46] Akagi, H., "Active harmonic filters," *Proc. IEEE*, vol. 93, no. 12, pp. 2128–2141, Dec. 2005.
[47] Le Roux, A.D., Mouton, H., and Akagi, H., "DFT-based repetitive control of a series active filter integrated with a 12-pulse diode rectifier," *IEEE Trans. Power Electron.*, vol. 24, no. 6, pp. 1515–1521, Jun. 2009.
[48] Mattavelli P. and Marafao, F.P., "Repetitive-based control for selective harmonic compensation in active power filters," *IEEE Trans. Ind. Electron.*, vol. 51, no. 5, pp. 1018–1024, Oct. 2004.
[49] Miret, J., Castilla, M., Matas, J., Guerrero, J.M., and Vasquez, J.C., "Selective harmonic-compensation control for single-phase active power filter with high harmonic rejection," *IEEE Trans. Ind. Electron.*, vol. 56, no. 8, pp. 3117–3127, Aug. 2009.
[50] Costa-Castell´o, R., Grin´o, R., and Fossas, E., "Odd-harmonic digital repetitive control of a single-phase current active filter," *IEEE Trans. Power Electron.*, vol. 19, no. 4, pp. 1060–1068, Jul. 2004.
[51] Angulo, M., Ruiz-Caballero, D.A., Lago, J., Heldwein, M.L., and Mussa, S.A., "Active power filter control strategy with implicit closed-loop current control and resonant controller," *IEEE Trans. Ind. Electron.*, vol. 60, no. 7, pp. 2721–2730, Jul. 2013.

[52] Lascu, C., Asiminoaei, L., Boldea, I., and Blaabjerg, F., "High-performance current controller for selective harmonic compensation in active power filters," *IEEE Trans. Power Electron.*, vol. 22, no. 5, pp. 1826–1835, 2007.

[53] Miret, J., Castilla, J., Matas, J., Guerrero, J.M., and Vasquez, J.C., "Selective harmonic-compensation control for single-phase active power filter with high harmonic rejection," *IEEE Trans. Ind. Electron.*, vol. 56, no. 8, pp. 3117–3127, Aug. 2009.

Chapter 7
Continuing developments of periodic control

Abstract

This chapter will discuss the most recent developments of the periodic control (PC) technology, such as the PC for multi-period signals and the periodic signal filtering problem. All these developments would extend the PC to more extensive engineering practice and can enable PC to further improve system performance.

7.1 Periodic control for multi-period signals

All PC laws developed in the previous chapters apply to periodic signals with a single fundamental period [1–16], which can be decomposed into a linear combination of a DC component, a fundamental frequency component of the given period and all harmonics. Periodic controllers [17–19] can also be used to deal with signals with multiple periods, for example, harmonics and inter-harmonics in electrical power systems.

7.1.1 Digital multi-period repetitive control

Let us consider the digital RC system shown in Figure 7.1, which can be described as:

$$y(z) = \frac{(1 + G_R(z))H(z)}{1 + G_R(z)H(z)} r(z) + \frac{1}{1 + G_R(z)H(z)} d(z) \tag{7.1}$$

where $G_p(z)$ is the plant, $G_c(z)$ is the feedback controller, $G_R(z)$ is the multi-period RC scheme, $y(z)$ is the output, $r(z)$ is the reference input, and $d(z)$ is the multi-period disturbance, and:

$$H(z) = \frac{G_c(z)G_p(z)}{1 + G_c(z)G_p(z)}$$

Signals $r(z)$ and/or $d(z)$ are a linear combination of frequencies of multiple given periods. In [17,18], a novel multi-period RC to simplify the design and

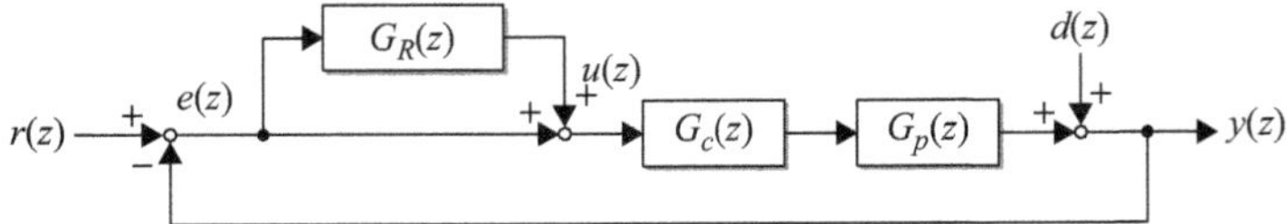

Figure 7.1 Repetitive control system with multi-period signals

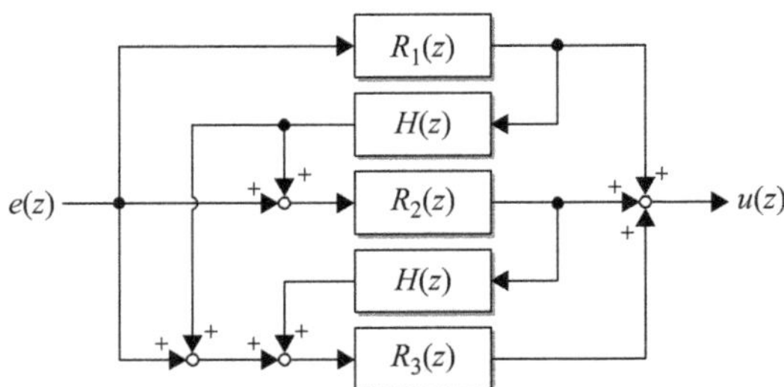

Figure 7.2 Multi-period repetitive controller $G_R(z)$ in (7.2) with three given periods (i.e., $p = 3$)

analysis of the corresponding RC systems has been developed. The multi-period RC $G_R(z)$ [17–19] can be written as:

$$G_R(z) = \sum_{j=1}^{p} R_j(z) + \sum_{j,k=1,j\neq k}^{p} R_j(z)R_k(z)H(z) + \prod_{j=1}^{p} R_j(z)H^2(z) \tag{7.2}$$

where

$$R_j(z) = k_j \frac{Q(z)z^{-N_j}}{1 - Q(z)z^{-N_j}} G_f(z) \tag{7.2a}$$

is a digital CRC scheme for the periodic component with a given period $N_j (j = 1, 2, \ldots, p)$ with p being the number of periods considered, $Q(z)$ is the frequency-selection low pass filter (LPF), and $G_f(z)$ is the phase-lead compensator. An example of a multi-period RC with three given periods is given in Figure 7.2.

Furthermore, we can have:

$$\frac{1}{1 + G_R(z)H(z)} = \frac{1}{\prod_{j=1}^{p} (1 + R_j H)} = \prod_{j=1}^{p} \frac{1 - Q(z)z^{-N_j}}{1 - (1 - k_j G_f(z)H(z))Q(z)z^{-N_j}} \tag{7.3}$$

From (7.1) and (7.3), it can be seen that the characteristic polynomial for such a multi-period RC system is equivalent to the product of every characteristic polynomial for each separate period CRC system. Hence, the design process of each period CRC component in the multi-period RC is decoupled. That is to say, the control gain k_j for each CRC of $R_j(z)$ can be separately tuned.

Therefore, it is clear that the plug-in multi-period RC system is stable if the two following sufficient stability conditions are met [17–19]:

1. Roots of $1 + G_c(z)G_p(z) = 0$ are inside the unit circle, i.e., without the plug-in multi-period RC scheme, the closed-loop feedback control system $H(z)$ is stable.
2. Roots of the characteristic equation $1 - \left(1 - k_j G_f(z)H(z)\right)Q(z)z^{-N_j} = 0$ with $j = 1, 2, \ldots p$ for each period RC, are inside the unit circle, i.e.:

$$\left|\left(1 - k_j G_f(z)H(z)\right)Q(z)\right| < 1, \quad \text{for } z = e^{j\omega} \text{ with } \omega < \frac{\pi}{T_s} \tag{7.4}$$

The above stability criteria for the multi-period RC system indicates that the design and analysis of the multi-period RC is almost the same as that of the CRC system discussed in Chapter 2. Furthermore, if $G_f(z)$ is designed to enable

$$0 < G_f(z)H(z) \leq 1, \tag{7.5}$$

the stability criterion of (7.4) can be simplified into:

$$0 < k_j < 2, \quad j = 1, 2, \ldots, p \tag{7.6}$$

As shown in Figure 7.3, in order to deal with multi-period inputs/disturbances, another straightforward way is to plug the following paralleled digital RC compensators into the feedback control system:

$$G_R(z) = \sum_{i=1}^{p} R_j(z) = \sum_{i=1}^{p}\left(k_j \frac{Q(z)z^{-N_j}}{1 - Q(z)z^{-N_j}} G_f(z)\right) \tag{7.7}$$

If the plug-in multi-period RC of (7.2) is replaced by the one of (7.7), the stability condition of (7.6) for the plug-in multi-period RC system in Figure 7.1 will be changed into:

$$0 < \sum_{j=1}^{p} k_j < 2 \tag{7.8}$$

The stability condition of (7.8) can be proved in the same way as that of (3.17) for the parallel-structure RC system discussed in Chapter 3. Unlike the inequality in

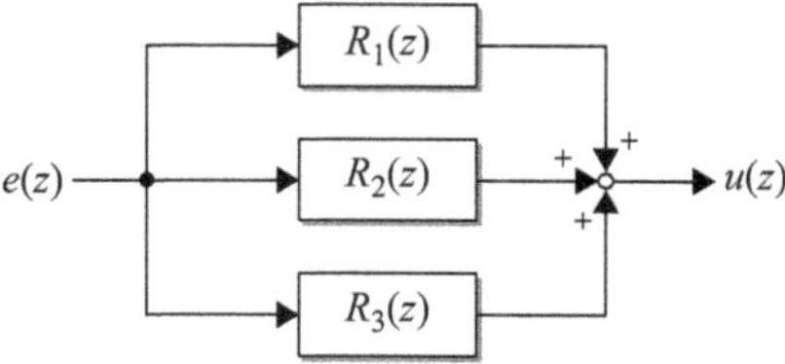

Figure 7.3 Alternative multi-period repetitive controller $G_R(z)$ with three given periods (i.e., $p = 3$)

(7.6) for the multi-period RC of (7.2), the inequality in (7.8) indicates that the design selection of the control gains for the parallel-structure multi-period RC of (7.8) is not completely independent.

Furthermore, an advanced multi-period PC scheme can be obtained by replacing the CRC components of the multi-period RC in either (7.2) or (7.7) with the advanced periodic controllers (such as the parallel-structure RC, selective harmonic control, and optimal harmonic control) explored in Chapter 3. The multi-period periodic control schemes can be adopted to improve the control system performance at the cost of higher complexity in their implementation and synthesis.

7.1.2 Multi-period resonant control

For each period component in multi-period signals, one multiple resonant control (MRSC) scheme can be employed to track or eliminate major harmonics. For the j-th period component with fundamental frequency ω_j ($j = 1, 2, \ldots p$), the MRSC scheme can simply be written as:

$$R_{Mj}(s) = \sum_{h \in N_{jh}} R_{jh}(s) = \sum_{h \in N_{jh}} \mathcal{L}\{k_{jh} \cos(\omega_{jh} t + \phi_{jh})\}$$

$$= \sum_{h \in N_{jh}} \left(k_{hj} \frac{s \cos \phi_{hj} - \omega_{hj} \sin \phi_{hj}}{s^2 + \omega_{jh}^2} \right) \tag{7.9}$$

where N_{jh} represents the set of selected harmonic orders, ω_{jh} is the h-th order harmonic frequency, k_{jh} is the control gain for tuning the error convergence rate, ϕ_{jh} is the phase-lead angle for system delay compensation at the resonant frequency ω_{jh}.

To compensate multi-period signals, multiple MSRC schemes can be directly connected in parallel to construct the multi-period resonant control (RSC) scheme as follows:

$$G_{\mathrm{MR}}(s) = \sum_{j=1}^{p} R_{Mj}(s) = \sum_{j=1}^{p} \left(\sum_{h \in N_{jh}} R_{jh}(s) \right) \tag{7.10}$$

which is shown in Figure 7.4.

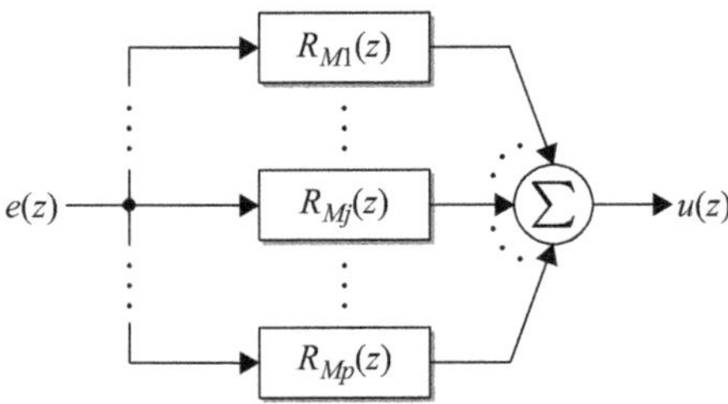

Figure 7.4 Multi-period resonant control shown in (7.10)

Like the MRSC system, it can be derived that the overall closed-loop multi-period RSC system is stable if the following conditions hold:

1. Roots of $1 + G_c(s)G_p(s) = 0$ are inside the unit circle (i.e., $H(s)$ is stable).
2. Roots of $1 + G_{\mathrm{MR}}(s)H(s) = 0$ are inside the unit circle.

Unlike the multi-period RC system, the stability range of the control gains k_{jh} for the multi-period RSC system cannot directly be derived from the above conditions. Empirical evidences show that the less selected harmonics are, the larger the stability range of the control gains k_{jh} are.

7.1.3 Frequency-adaptive periodic control for multi-period signals

For multi-period signals with components of coprime periods N_j ($j = 1, 2, \ldots, p$), it is not easy to choose a suitable system sampling rate to enable all the periods N_j to be integers. For example, considering a multi-period reference signal that contains three fixed-frequency components as

$$x(t) = \sin(50 \times 2\pi t) + \sin(51 \times 2\pi t) + \sin(49 \times 2\pi t),$$

the minimum value of the system sampling rate will be chosen as $f_s = 49 \times 50 \times 51 = 124950$ Hz to enable every period N_j for the CRC component in (7.2a) of the multi-period RC in (7.2) or (7.7) to be an integer. $N_j \in [50 \times 51, 49 \times 51, 49 \times 50]$ is a very large number for implementing the CRC. Moreover, if the frequency of any one signal component changes to be a fractional value, the minimum value of the system sampling rate may become incredibly large to ensure that every period N_j is an integer. Therefore, in practice, in the presence of frequency variations, it is almost impossible to choose an appropriate system sampling rate to guarantee all the CRC components:

$$R_j(z) = k_j \frac{Q(z)z^{-N_j}}{1 - Q(z)z^{-N_j}} G_f(z) \tag{7.11}$$

with integer periods N_j ($j = 1, 2, \ldots, p$) whether the sampling rate is fixed or variable. That is to say, whether the sampling rate is fixed or variable, it is hard to guarantee that the multi-period RC can achieve zero error tracking/elimination of the multi-period signals in the presence of frequency variations.

Fortunately, the fractional delay (FD) filters with a fixed sampling rate can be used to enable the multi-period RC in (7.2) or (7.7) to be frequency-adaptive. Like (5.14), if all the CRC components $R_j(z)$ in (7.2) and (7.7) are replaced with the *Lagrange* interpolation-based fractional-order adaptive ones [20–28], the following equation will be obtained:

$$R_{aj}(z) = k_j \frac{\left(z^{-N_{ji}} \sum_{k=0}^{n} A_k z^{-k}\right) Q(z)}{1 - \left(z^{-N_{ji}} \sum_{k=0}^{n} A_k z^{-k}\right) Q(z)} G_f(z) \tag{7.12}$$

where $N_{ji} = [N_j]$ ($j = 1, 2, \ldots, p$). As a result, the multi-period RC of (7.2) or (7.7) with a fixed sampling rate will become frequency-adaptive. It should be noted that, a frequency detector is needed to separately measure the fundamental frequency of each signal component.

Furthermore, as mentioned in Section 5.1, the frequency-adaptive multi-period RSC scheme [29–31] can also be obtained as:

$$G_{\mathrm{aMR}}(s) = \sum_{j=1}^{p} R_{\mathrm{a}Mj}(s) = \sum_{j=1}^{p}\left(\sum_{h\in N_{jh}}\left(k_{jh}\frac{s\cos\phi_{jh} - \hat{\omega}_{jh}\sin\phi_{jh}}{s^2+\hat{\omega}_{jh}^2}\right)\right) \tag{7.13}$$

where N_{jh} represents the set of the selected harmonic orders for the j-th period signal, $\hat{\omega}_{jh} = h\hat{\omega}_{j0}$ denotes the detected h-th order interested harmonic frequency for the j-th period signal with $\hat{\omega}_{j0}$ being the detected fundamental harmonic frequency and h being the order of harmonics, k_{jh} is the control gain for tuning the error convergence rate, ϕ_{jh} is the phase-lead angle for system delay compensation at the resonant frequency $\hat{\omega}_{jh}$. A frequency detector (such as a phased-locked loop-PLL system) is commonly employed to detect $\hat{\omega}_{j0}$ and enables the RSC in the multi-period RSC to directly adapt to the varying frequency.

7.2 Periodic signal filtering

Figure 7.5 shows a typical feedback control system, where $G_c(z)$ is the feedback controller, $G_p(z)$ is the control plant, $d(z)$ is the disturbance, $r(z)$ is the reference signal, and $y(z)$ is the output signal. The error transfer function for such a control system can be written as:

$$e(s) = r(s) - y(z) = \frac{1}{1+G_c(z)G_p(z)}r(z) - \frac{1}{1+G_c(z)G_p(z)}d(z) \tag{7.14}$$

From (7.14), for a given control plant $G_p(s)$, if the feedback controller $G_c(z) \to \infty$, we will have $y(z) \to r(z)$ even in the presence of disturbances $d(z)$. That is also why periodic controllers can be used to force the output $y(z)$ to exactly track the periodic reference signal $r(z)$ at the interested harmonic frequencies.

However, in practical power conversion applications [32,33], the reference signal $r(z)$ may suffer from unexpected harmonic distortions. For example, for the grid-connected single-phase PWM inverter shown in Figure 7.6, the DC-AC power conversion of the PWM inverter [34,35] injects the double-grid frequency voltage

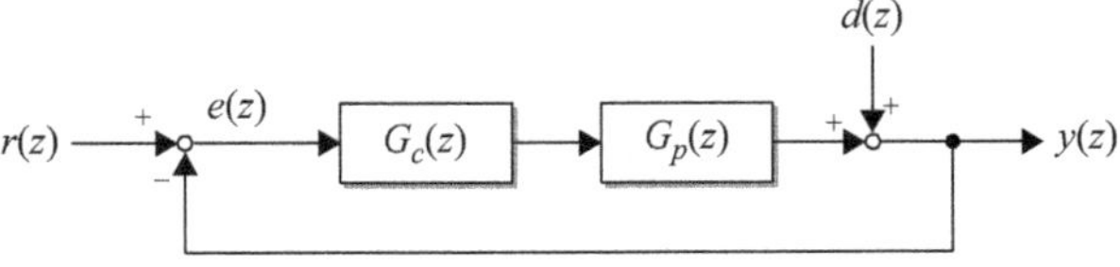

Figure 7.5 A typical digital feedback control system

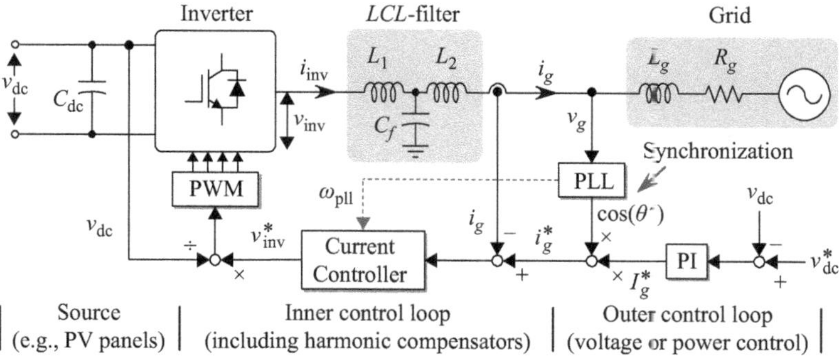

Figure 7.6 Grid-connected single-phase PWM inverter system

ripples into the DC bus voltage v_{dc}, and then it leads to significant twice the grid fundamental frequency components in the output I_g^* of the proportional–integral (PI) controller. Moreover, under distorted grid voltage v_g, the PLL-detected synchronization signal $\cos\theta'$ may contain unfiltered harmonic distortions. Since the reference AC feed-in current signal i_g^* for the inner current control loop is the product of the output I_g^* of the outer loop voltage PI controller and the PLL-detected synchronization signal $\cos\theta'$, the reference cosine current signal i_g^* will contain unexpected odd-order harmonics distortions due to the interaction between the polluted grid frequency $\cos\theta'$ and the double-grid-frequency voltage ripples in I_g^*. Therefore, in order to feed clean sine/cosine currents into the grid, filtering of both the DC bus voltage v_{dc} and the grid AC voltage v_g are needed to remove harmonic distortions from the reference current signal i_g^*. Hereby, such a problem is called *periodic signal filtering*.

7.2.1 Notch and comb filters

Low pass filters (LPFs) can be used to remove periodic harmonics. However, the bandwidth of the LPFs must be below the frequency of the lowest order harmonics, which then often leads to slow dynamic system responses. Moreover, the LPFs may not be able to significantly attenuate low-frequency harmonics due to their nonideal transition band.

To deal with the harmonics filtering problem, engineers would often resort to notch and comb filters, which pass most frequencies unaltered, but attenuate those at specific frequencies to very low levels. A notch filter is a band-stop filter with a narrow notch at a specified frequency. The frequency response of a comb filter consists of a series of regularly spaced notches, giving the appearance of a comb. By observing their frequency responses in the Bode diagram, it can be found that a notch/comb filter actually can be treated as the "reciprocal" of a proportional item "1" plus a RSC/a RC, respectively. In other words, a notch or comb filter can be obtained by taking the denominator of a RSC or a RC, since the poles at the resonant frequencies become zeros at the notch frequencies.

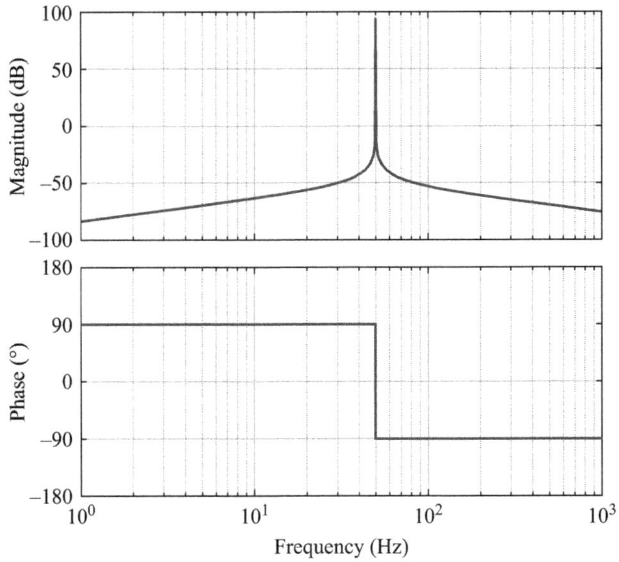

Figure 7.7 Bode plot of the normalized resonant controller in (7.15) with $\omega_h = 2\pi \times 50$ rad/s and $k_h = 1$

As mentioned in Section 2.2.1 in Chapter 2, a typical RSC can be expressed as:

$$\hat{G}_h(s) = k_h \frac{s}{s^2 + \omega_h^2} \tag{7.15}$$

with k_h being the control gain and ω_h being the resonant frequency. As shown in Figure 7.7, for the RSC $\hat{G}_h(s)$ in (7.15), $|\hat{G}_h(j\omega)| \rightarrow \infty$ at the resonant frequency of $\omega = \omega_h$, and $|G_h(j\omega)| \rightarrow 0$ at other frequencies of $\omega \neq \omega_h$.

On the contrary, for a typical notch filter $G_{\text{notch}}(s)$, $|G_{\text{notch}}(j\omega)| \rightarrow 0$ at the notch frequency of $\omega = \omega_h$, and $|G_{\text{notch}}(j\omega)| \rightarrow 1$ at other frequencies of $\omega \neq \omega_h$. Thus, based on the RSC in (7.15), a notch filter can be derived as [36,37]:

$$G_{\text{notch}}(s) = \frac{1}{1 + \hat{G}_h(s)} = \frac{s^2 + \omega_h^2}{s^2 + k_h s + \omega_h^2} = \frac{s^2 + \omega_h^2}{s^2 + \frac{\omega_h}{Q_n} s + \omega_h^2} \tag{7.16}$$

where $k_h = k\omega_h$, $Q_n = 1/k$ denotes the quality factor, and the resonant frequency ω_h in (7.15) represents the notch frequency in (7.16). The quality factor Q_n determines the sharpness of the filter response, which affects the stop-band width as well as the phase-shift characteristics. A larger value of Q_n will lead to a narrower stop-band width as well as a less phase delay for frequencies that lie below the notch frequency ω_h. However, due to the finite numerical resolution, too narrow notch width caused by an overlarge Q_n would result in difficulties in accurate

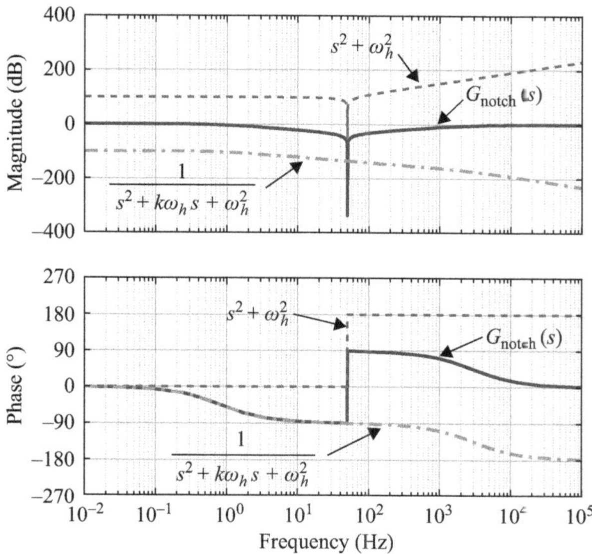

Figure 7.8 Frequency response of the notch filter in (7.16) with $\omega_h = 2\pi \times 50$ rad/s and $k = 60$

implementation of the digital notch filter. Figure 7.8 gives an example of the frequency responses of the notch filter in (7.16) with $k = 60$ and $\omega_h = 2\pi \times 50$ rad/s.

From (7.15), (7.16), and the frequency response in Figure 7.8, it is clear that the item $(s^2 + \omega_h^2)$ results in the notch at the harmonic frequency ω_h. Moreover, the denominator polynomial of the notch filter in (7.16) is adopted to shape the frequency response of the notch filter.

Nevertheless, a more general notch filter can be given as:

$$G_{\text{notch}}(s) = \frac{s^2 + k_1\omega_h s + \omega_h^2}{s^2 + k_2\omega_h s + \omega_h^2} \quad \text{with} \quad k_1 \ll k_2 \tag{7.17}$$

where the addition of $k_1\omega_h s$ can change both the stop-band width and the phase shift characteristics near the notch frequency. For example, Figure 7.9 shows the frequency responses of the notch filter in (7.17) with the notch frequency $\omega_h = 10$ rad/s, $k_2 = 5$ for various values of k_1. Obviously, as shown in Figure 7.9, if $k_1 > k_2$, the filter becomes a band-pass filter; the notch filter with $k_1 = 0$, i.e., the notch filter in the form of (7.16), achieves the sharpest notch and the steepest phase change at the notch frequency, and it thus can completely filter out the interested harmonic component.

As mentioned in Section 2.1.1 in Chapter 2, for periodic signals with the fundamental period of T_0, the simplest RC scheme can be written as:

$$\hat{G}_{\text{rc}}(s) = \frac{e^{-sT_0}}{1 - e^{-sT_0}} \tag{7.18}$$

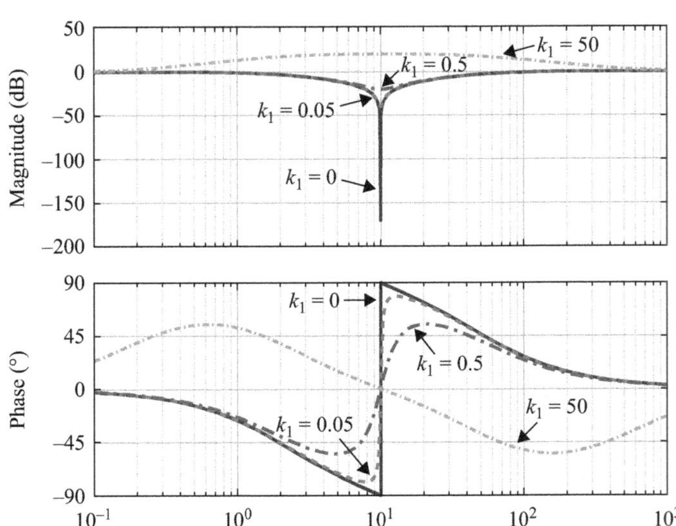

Figure 7.9 Frequency response of the notch filter in (7.17) with $\omega_h = 2\pi \times 10$ rad/s and $k_2 = 5$

which has poles at $s = \pm jn\omega_0$ with $n \in \mathbb{N}$ and $\omega_0 = 2\pi/T_0$. As shown in Figure 2.1, for the RC controller $\hat{G}_{\rm rc}(s)$ in (7.18), $\left|\hat{G}_{\rm rc}(j\omega)\right| \to \infty$ at all harmonic frequencies of $\omega = n\omega_0$, and $\left|\hat{G}_{\rm rc}(j\omega)\right| \to 0$ at other frequencies of $\omega \neq n\omega_0$.

Similarly, the comb filter can be obtained as the "reciprocal" of a proportional item "1" plus the RC in (7.18). Hence the resultant full harmonic comb filter is given as:

$$G_{\rm fhc}(s) = \frac{1}{1 + \hat{G}_{\rm rc}(s)} = 1 - e^{-sT_0} \tag{7.19}$$

which gives notches at full harmonic frequencies $n\omega_0$. As shown in Figure 7.10, for the comb filter in (7.19), $\left|\hat{G}_{\rm fhc}(j\omega)\right| \to 0$ at all notch frequencies of $\omega = n\omega_0$, and $\left|\hat{G}_{\rm fhc}(j\omega)\right| \to 1$ at other frequencies of $\omega \neq n\omega_0$.

For periodic signals with the fundamental period of T_0, a simplest odd-order harmonic RC scheme can be written as:

$$\hat{G}_{\rm or}(s) = \frac{-e^{-sT_0/2}}{1 + e^{-sT_0/2}} \tag{7.20}$$

which has poles at $s = \pm j(2k + 1)\omega_0$, $k \in \mathbb{N}$, and $|\hat{G}_{\rm or}(j\omega)| \to \infty$ at the odd-order harmonic frequencies $(2k + 1)\omega_0$, and $|\hat{G}_{\rm or}(j\omega)| \to 0$ at other frequencies.

Following the same way, the odd-order harmonic comb filter can be directly derived as:

$$G_{\rm ohc}(s) = \frac{1}{1 + \hat{G}_{\rm or}(s)} = 1 + e^{-sT_0/2} \tag{7.21}$$

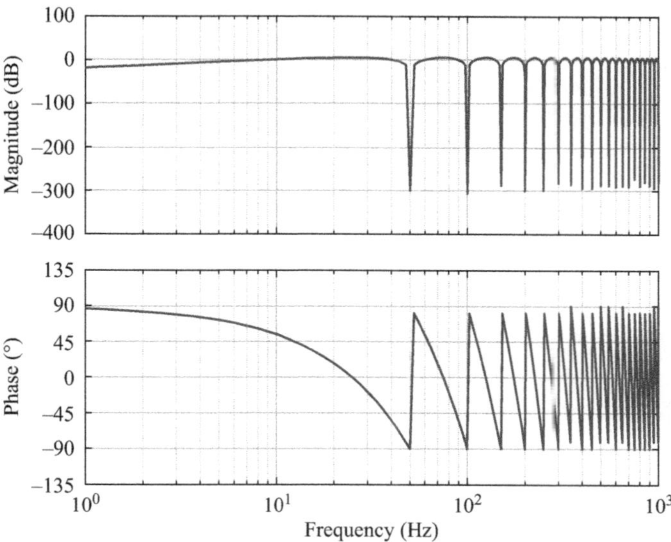

Figure 7.10 Frequency response of the full harmonic comb filter in (7.19) with $T_0 = 0.02$ s

which results in notches at the odd-order harmonic frequencies $(2k + 1)\omega_0$, as demonstrated in Figure 7.11.

Likewise, a selective harmonic comb filter can be developed by simply taking the denominator of the selective harmonic control scheme in (3.24) as follows:

$$G_{\text{shc}}(s) = \frac{1}{1 + \hat{G}_{\text{sm}}(s)} = \frac{e^{-\frac{2sT_0}{n}} - 2\cos(2\pi m/n)e^{-\frac{sT_0}{n}} + 1}{-\cos(2\pi m/n)e^{-\frac{sT_0}{n}} + 1} \tag{7.22}$$

where $k = 0, 1, 2, \ldots$ and $m < n$. Obviously (7.19) and (7.21) are special cases of (7.22) with appropriate m and n.

Likewise, more general comb filters can be given as follows:

$$G_{\text{fhc}}(s) = 1 - \alpha e^{-sT_0} \tag{7.23}$$

$$G_{\text{ohc}}(s) = 1 + \alpha e^{-sT_0/2} \tag{7.24}$$

$$G_{\text{shc}}(s) = \frac{2}{\dfrac{1}{1 - \alpha e^{\frac{-sT_0}{n} - j\frac{2\pi m}{n}}} + \dfrac{1}{1 - \alpha e^{\frac{-sT_0}{n} + j\frac{2\pi m}{n}}}} = \frac{\alpha^2 e^{-\frac{2sT_0}{n}} - 2\alpha\cos(2\pi m/n)e^{-\frac{sT_0}{n}} + 1}{-\alpha\cos(2\pi m/n)e^{-\frac{sT_0}{n}} + 1} \tag{7.25}$$

where the tuning of coefficient $\alpha \in [-1, 1]$ could change both the stop-band width and the phase-shift characteristics near the notch frequencies. As shown in

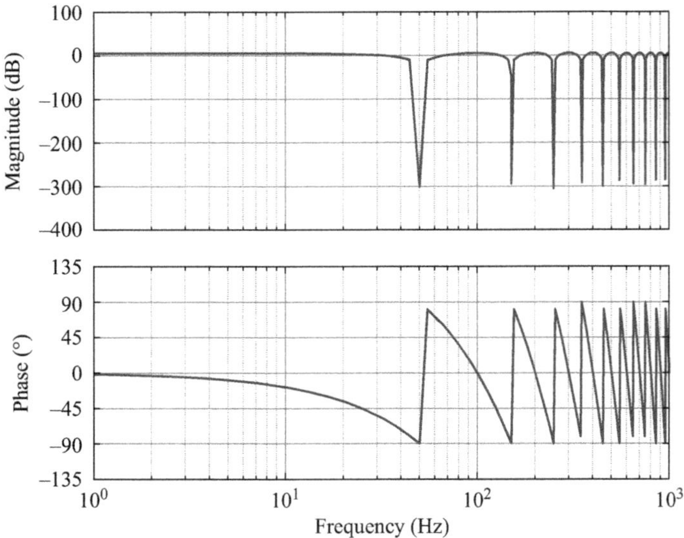

Figure 7.11 Frequency response of the odd harmonic comb filter in (7.21) with $T_0 = 0.02$ s

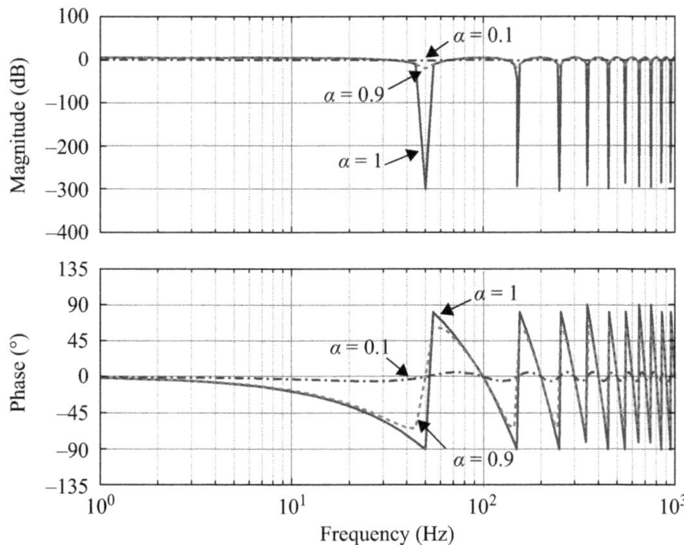

Figure 7.12 Frequency response of the comb filter in (7.24) with $T_0 = 0.02$ s and $\alpha = 1, 0.9, 0.1$

Figure 7.12, for the odd-order harmonic comb filter in (7.24), both harmonics attenuation and phase shift at the notch frequencies will be significantly reduced with the decrease of $|\alpha|$. Obviously, the comb filter in (7.24) with $|\alpha| = 1$ achieves the sharpest attenuation at the notch frequencies, and thus it is widely used in practical applications.

7.2.2 Digital notch and comb filters

Since the control system components are often designed in the continuous time, they must generally be discretized for implementation on digital computers and embedded processors. Common methods for discretizing filters include zero-order hold, first-order hold, impulse invariant, Tustin (bilinear approximation), Tustin with frequency pre-warping, matched poles, and zeros. Among these, the zero- and first-order hold methods and the impulse-invariant method are well-suited for discrete approximations in the time domain. By contrast, the Tustin and matched methods tend to perform better in the frequency domain because they introduce less gain and phase distortion near the Nyquist frequency. Moreover, the higher the sampling rate is, the closer the match between the continuous and discretized responses is.

In practice, the continuous-time notch filter can be converted into a discrete-time one with the matched frequency response at the notch frequency ω_h by simply substituting the following Tustin without/with frequency pre-warping transform into the continuous notch filter transfer function in (7.16) or (7.17):

$$\text{Tustin: } s = \frac{2}{T_s}\frac{z-1}{z+1} \tag{7.26a}$$

$$\text{Tustin with frequency pre-warping: } s = \frac{\omega_h}{\tan(\omega_h T_s/2)}\frac{z-1}{z+1} \tag{7.26b}$$

where T_s is the system sampling period.

Referring to the digital forms of corresponding periodic controllers in Chapters 2 and 3, the corresponding discrete-time comb filters for the continuous-time comb filters in (7.23), (7.24), and (7.25) can be discretized into:

$$G_{\text{fhc}}(z) = 1 - \alpha z^{-N} \tag{7.27}$$

$$G_{\text{ohc}}(z) = 1 + \alpha e^{-N/2} \tag{7.28}$$

$$G_{\text{shc}}(z) = \frac{\alpha^2 z^{-\frac{2N}{n}} - 2\alpha\cos(2\pi m/n) z^{-\frac{N}{n}} + 1}{-\alpha\cos(2\pi m/n) z^{-\frac{N}{n}} + 1} \tag{7.29}$$

where $N = T_0/T_s \in \mathbb{N}$ with T_s being the sampling period, and $-1 \le \alpha \le 1$.

In practice, the comb filter with full frequency coverage as shown in (7.27) is more widely used. When $N = 1$, the comb filter in (7.27) becomes a DC notch filter with the notch frequency at $\omega_h = 0$ as below:

$$G_{Ic}(z) = 1 - \alpha z^{-1} \quad \text{with} \quad 0 \le \alpha \le 1 \tag{7.30}$$

which is the opposite to the discrete-time integrator:

$$G_I(z) = \frac{\alpha z^{-1}}{1 - \alpha z^{-1}} \tag{7.31}$$

The frequency response of a discrete-time notch filter or comb filter can be obtained by making the substitution $z = e^{j\omega}$. For example, for the comb filter in (7.27), its frequency response can be written as:

$$G_{\text{fhc}}(j\omega) = 1 - \alpha e^{-jN\omega} \tag{7.32}$$

Using the Euler's identity, the frequency response of (7.27) can be rewritten as:

$$G_{\text{fhc}}(j\omega) = [1 - \alpha\cos(N\omega)] + j\alpha\sin(N\omega) \tag{7.33}$$

Therefore the corresponding magnitude response can be written as:

$$|G_{\text{fhc}}(j\omega)| = \sqrt{(1+\alpha^2) - 2\alpha\cos(N\omega)} \tag{7.34}$$

7.2.3 Discrete Fourier transform-based comb filter

As discussed in Section 2.3 in Chapter 2, the discrete Fourier transform (DFT) can be adopted to create the finite-impulse-response (FIR) bandpass filter in (2.42) at the selected harmonic frequencies as [38,39]:

$$F_{\text{DFT}}(z) = \sum_{h\in N_h} F_{\text{dh}}(z) = \frac{2}{N}\sum_{i=1}^{N}\left[\sum_{h\in N_h}\cos\left(\frac{2\pi}{N}hi\right)\right]z^{-i} \tag{7.35}$$

in which $F_{\text{dh}}(z)$ denotes the bandpass filter for the h-th order harmonic, N is the number of samples per fundamental period, h is the harmonic order, and N_h represents the set of selected bandpass frequencies. Such a DFT-based FIR filter actually produces unity gains at all selected harmonic frequencies (i.e., $h \in N_h$) and zero gains at other harmonic frequencies, as shown in Figure 2.10.

On the contrary, a DFT-based comb filter $G_{\text{DFTC}}(z)$ with the selected notch frequencies, can be given as:

$$G_{\text{DFTC}}(z) = 1 - F_{\text{DFT}}(z) = 1 - \sum_{h\in N_h} F_{\text{dh}}(z) = 1 - \frac{2}{N}\sum_{i=1}^{N}\left[\sum_{h\in N_h}\cos\left(\frac{2\pi}{N}hi\right)\right]z^{-i} \tag{7.36}$$

which produces zero gains at all selected notch frequencies (i.e., $h \in N_h$) and unity gains at other harmonic frequencies.

In the same way of (7.19), the DFT-based comb filter in (7.36) can also be developed by using the DFT-based RC in (2.44) as follows:

$$G_{\text{DFTC}}(z) = \frac{1}{1 + \hat{G}_{\text{DFT}}(z)} = \frac{1}{1 + \dfrac{F_{\text{DFT}}(z)}{1 - F_{\text{DFT}}(z)}} = 1 - F_{\text{DFT}}(z) \tag{7.36a}$$

The DFT-based comb filter $G_{\text{DFTC}}(z)$ in (7.36) provides a flexible selective harmonic comb filter. Its computational complexity is independent of the number of selected harmonics to be filtered. However, the implementation of $G_{\text{DFTC}}(z)$ will involve a large amount of parallel computations, which is proportional to the fundamental period N. Its implementation is suitable for high-performance fixed-point digital signal processors (DSP).

7.2.4 Frequency-adaptive notch and comb filters

In practical applications, in the presence of frequency variations, the real fundamental frequency can be written as:

$$\hat{\omega}_0 = \omega_0 + \Delta\omega \tag{7.37}$$

where ω_0 is the nominal fundamental frequency, and $\Delta\omega$ is the frequency deviation.

The magnitude response of the notch filter in (7.16) around the harmonic frequency (i.e. $s = j\hat{\omega}_h = jh\hat{\omega}_0$) can be obtained as:

$$\begin{aligned}|G_{\text{notch}}(jh\hat{\omega}_0)| &= \left|\frac{(jh\hat{\omega}_0)^2 + (h\omega_0)^2}{(jh\hat{\omega}_0)^2 + k(h\omega_0)(jh\hat{\omega}_0) + (h\omega_0)^2}\right| \\ &= \left|\frac{-h^2(\omega_0+\Delta\omega)^2 + h^2(\omega_0)^2}{-h^2(\omega_0+\Delta\omega)^2 + k(h\omega_0)jh(\omega_0+\Delta\omega) + (h\omega_0)^2}\right| \\ &= \left|\frac{\delta^2 + 2\delta}{\delta^2 + 2\delta + jk(1+\delta)}\right| == 1/\sqrt{1 + k^2\left(\frac{1+\delta}{\delta^2+2\delta}\right)^2}\end{aligned} \tag{7.38}$$

where $\delta = \Delta\omega/\omega_0$ [20]. Equation (7.38) indicates that the gain of the notch filter at the nominal harmonic frequency ω_h will not reach its minimum value unless $\delta = 0$ (i.e., $\Delta\omega = 0$). That is to say, the gains of the notch filter in (7.16) at the notch frequencies are sensitive to frequency variations. As shown in Figure 7.13, it can be observed that even a small frequency variation (i.e., $\pm 0.02\%$) will result in a significant magnitude increase near the notch frequency.

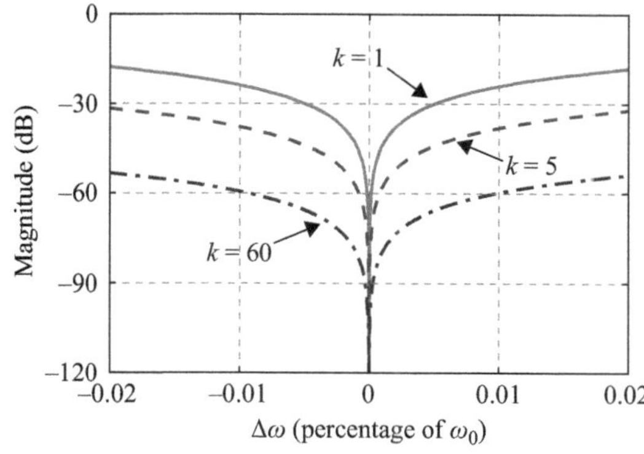

Figure 7.13 Magnitude response of the notch filter in (7.38) as a function of the frequency variation

If the interested harmonic frequency ω_h can be detected and updated in real time, the notch filter in (7.16) can always produce zero or very small gain at notch frequencies $\pm\omega_h$. The frequency adaptive notch filter can accurately filter out varying-frequency harmonics. To be differentiated from the notch filter in (7.16), the frequency-adaptive notch filter can be written as:

$$G_{\mathrm{anotch}}(s) = \frac{s^2 + \hat{\omega}_h^2}{s^2 + k\hat{\omega}_h s + \hat{\omega}_h^2} \tag{7.39}$$

where $\hat{\omega}_h = h\hat{\omega}_0$ denotes the detected interested harmonic frequency, with $\hat{\omega}_0$ being the detected fundamental harmonic frequency and h being the order of harmonics.

In the same way, the actual fundamental frequency is again expressed as $\hat{\omega}_0 = \omega_0 + \Delta\omega = \omega_0(1+\delta)$, where $\Delta\omega$ is the frequency deviation and $\delta = \Delta\omega/\omega_0$. Using the Euler's identity, the magnitude response of the comb filter in (7.23) around the actual harmonic frequency (i.e., $s = jh\hat{\omega}_0$) can be obtained as [32]:

$$\begin{aligned} |G_{\mathrm{fhc}}(jh\hat{\omega}_0)| &= \left|1 - \alpha e^{-j2\pi h(\hat{\omega}_0/\omega_0)}\right| = \left|1 - \alpha e^{-j2\pi h(1+\delta)}\right| \\ &= \sqrt{1 + \alpha^2 - 2\alpha\cos(2\pi h\delta)} \end{aligned} \tag{7.40}$$

which implies that the comb filter no longer approaches its minimum gain at the corresponding harmonic frequency when there is a frequency variation, i.e., $\delta \neq 0$. It means that the comb filter in (7.40) is sensitive to the frequency variation. Figure 7.14 shows that, the higher the harmonic order h is, the more sensitive the magnitude response of the comb filter in (7.40) to the frequency variation is.

For all delay-based digital comb filters with a fixed sampling rate, such as the digital comb filters in (7.27) to (7.29), the mismatch between the integer order delay z^{-N} and the time-varying fundamental period T_0 will lead to the deviation on harmonics rejection ability of the comb filter in the presence of harmonic frequency

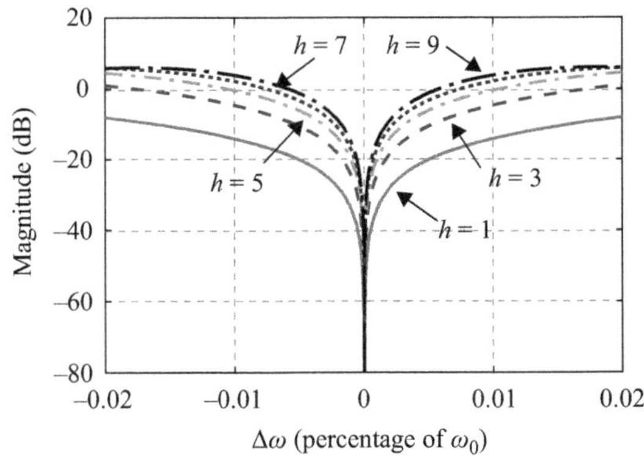

Figure 7.14 Magnitude response of the comb filter with $\alpha = 1$ in (7.40) as a function of the frequency variation

variations. Actually, they are sensitive to the harmonic frequency variations. In addition, the digital DFT-based comb filter is also sensitive to the harmonic frequency variations since it assumes that the integer order N of its polynomial is equal to the number of samples per period of the fundament harmonic frequency. The DFT-based comb filter in (7.36) can be equivalent to one kind of time-delay-based periodic controllers.

Similar to the delay-based periodic controllers in Chapter 5, the frequency-adaptive digital comb filters with a fixed sampling rate can be obtained by adopting FD filters-based fractional time-delay elements. For example, in the presence of frequency variations, the number of delay elements $N_F = f_s/f$ becomes fractional. Such fractional digital delay elements can be written as [23–25]:

$$z^{-N_F} = z^{-N_i - F} \tag{7.41}$$

with $N_i = [N_F]$ being the integer part of N_F and $F = N_F - N_i$ $(0 \leq F < 1)$ being the fractional part of N_F and the fractional digital delay z^{-F} can be approximated by:

$$z^{-F} \approx \sum_{k=0}^{p} A_k z^{-k} \tag{7.42}$$

where p is the order of the *Lagrange* polynomial, $k = 0, 1, \ldots, p$, and the coefficient A_k can be obtained as:

$$A_k = \prod_{\substack{i=0 \\ i \neq k}}^{p} \frac{F - i}{k - i} \quad k, i = 0, 1, \ldots, n \tag{7.43}$$

Therefore, the frequency-adaptive comb filters in (7.26–7.28) can be written as:

$$G_{\text{afhc}}(z) = 1 - \alpha z^{-N_i} \left(\sum_{k=0}^{p} A_k z^{-k} \right) \quad \text{with} \quad N_i = [N_F] \tag{7.44}$$

$$G_{\text{aohc}}(z) = 1 + \alpha z^{-N_i} \left(\sum_{k=0}^{p} A_k z^{-k} \right) \quad \text{with} \quad N_i = \left[\frac{N_F}{2} \right] \tag{7.45}$$

$$G_{\text{ashc}}(z) = \frac{\alpha^2 z^{-2N_i} \left(\sum_{k=0}^{p} A_k z^{-k} \right)^2 - 2\alpha \cos\left(\frac{2\pi m}{n} \right) z^{-N_i} \left(\sum_{k=0}^{p} A_k z^{-k} \right) + 1}{-\alpha \cos\left(\frac{2\pi m}{n} \right) z^{-N_i} \left(\sum_{k=0}^{p} A_k z^{-k} \right) + 1}$$

$$\text{with} \quad N_i = \left[\frac{N_F}{n} \right] \tag{7.46}$$

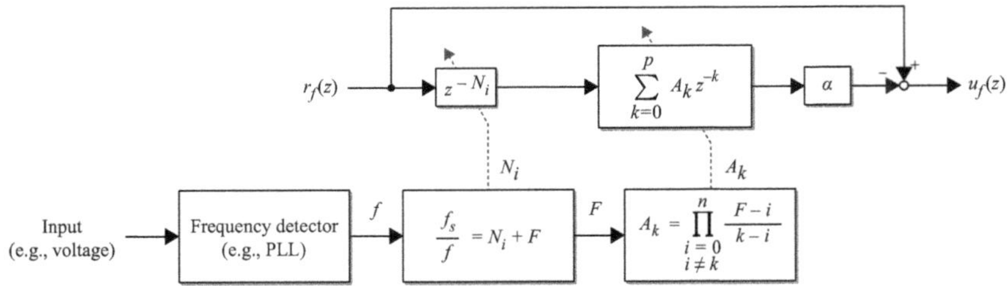

Figure 7.15 Implementation of the frequency-adaptive digital comb filter in (7.44) based on the Lagrange interpolation polynomial

Figure 7.15 exemplifies the implementation of the frequency-adaptive digital comb filter as given in (7.44), where $r_f(z)$ represents the filter input and $u_f(z)$ is the filter output. As discussed in Chapter 5, the frequency adaptability is achieved by feeding the detected frequency into the digital filters. Thus, the entire system performance is dependent on the frequency detector, which is usually performed by a PLL system in practice.

In addition, similar to the frequency-adaptive DFT-based RC, in the presence of frequency variations, a frequency-adaptive DFT-based comb filter with a fixed sampling rate can be obtained by replacing the unit delay z^{-1} with the virtual unit delay z_v^{-1} introduced in Chapter 5. The resultant frequency adaptive DFT-based comb filter is then expressed as:

$$G_{\text{aDFTC}}(z) = 1 - \frac{2}{N}\sum_{i=1}^{N}\left[\sum_{h\in N_h}\cos\left(\frac{2\pi}{N}hi\right)\right]z_v^{-i} \tag{7.47}$$

For convenience, the virtual unit delay z_v^{-1} as a frequency-adaptive approximated FD [53] is given again by:

$$z_v^{-1} = z^{-(1+F_N)} \approx \begin{cases} |F_N|z^0 + (1-|F_N|)z^{-1} & -1 < F_N < 0 \\ (1-|F_N|)z^{-1} + |F_N|z^{-2} & 0 \le F_N < 1 \end{cases}$$

and

$$\frac{N_F}{N} = \frac{N+F}{N} = 1 + F_N$$

where $N = f_s/f_0$ is the nominal integer number with f_0 being the nominal frequency and $N_F = f_s/f$ is the actual number of samples per fundamental period with F being the fractional part of N_F and $|F_N| = |F/N| \ll 1$.

The frequency-adaptive DFT-based comb filter with a fixed sampling rate is approximately equivalent to a frequency-adaptive DFT-based comb filter with a variable sampling rate.

7.2.5 Periodic signal filtering for grid-connected converters

For grid-connected power converters, the PLLs are widely used to detect fundamental frequency and phase angle of power grid voltages for generating proper reference signals for synchronizing their output voltages or currents with the grid voltages. Under ideal grid conditions without any harmonics, unbalance, or DC offset, the three-phase grid voltages only contain positive-sequence components. With the clockwise rotating operation $e^{-j\omega t}$, the ideal balanced three-phase voltages (v_a, v_b, v_c) can be transformed into two constant DC voltages (v_d, v_q) in the dq synchronous reference frame (SRF). In that case, a typical SRF-PLL shown in Figure 7.16 can achieve a fast and precise detection of the phase angle and frequency of the grid voltages, which has been widely used due to its ease of operation and robust behavior.

Under unbalanced grid voltages, the three-phase grid voltages (v_a, v_b, v_c) can be decomposed into symmetrical positive-sequence components ($v_{a(+)}$, $v_{b(+)}$, $v_{c(+)}$), negative-sequence components ($v_{a(-)}$, $v_{b(-)}$, $v_{c(-)}$), and zero-sequence components ($v_{a(0)}$, $v_{b(0)}$, $v_{c(0)}$) as follows [40,41]:

$$\begin{bmatrix} \mathbf{v}_{(+)} \\ \mathbf{v}_{(-)} \\ \mathbf{v}_{(0)} \end{bmatrix} = \frac{1}{3} \cdot \begin{bmatrix} 1 & \alpha & \alpha^2 \\ 1 & \alpha^2 & \alpha \\ 1 & 1 & 1 \end{bmatrix} \begin{bmatrix} \mathbf{v}_a \\ \mathbf{v}_b \\ \mathbf{v}_c \end{bmatrix} \tag{7.48}$$

where ($\mathbf{v}_a$, $\mathbf{v}_b$, $\mathbf{v}_c$) are the phasors of three-phase grid voltages (v_a, v_b, v_c), ($\mathbf{v}_{(+)}$, $\mathbf{v}_{(-)}$, $\mathbf{v}_{(0)}$) are the phasors of symmetrical positive-sequence components, negative-sequence components, and zero-sequence components, and $\alpha = e^{j120^\circ}$ is a unit complex operator that shifts a phasor by an angle of 120° counter-clockwise. Furthermore, for three-phase three-wire systems, the zero-sequence components will be omitted.

Unbalanced grid voltages usually occur in the presence of unsymmetrical grid faults or heavy unbalanced loads, e.g., one or two phases shorted to ground or to each other. For the grid-connected three-phase converters under unbalanced grid voltages, the conventional control scheme in the positive-sequence SRF can only regulate positive-sequence feed-in currents into the grid, and would worsen the unexpected power oscillations due to the interaction between the grid voltages and the negative-sequence feed-in current components. If the three-phase PLL system

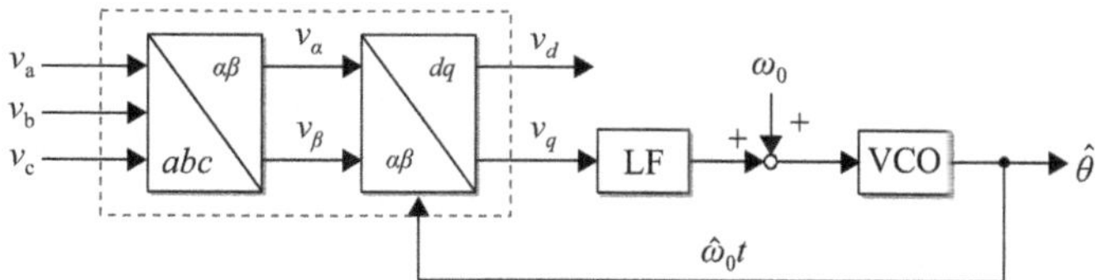

Figure 7.16 Schematic diagram of the synchronous reference frame-based phase-locked loop system, where the loop filter (LF) is usually performed by a PI controller and VCO is the voltage controlled oscillator

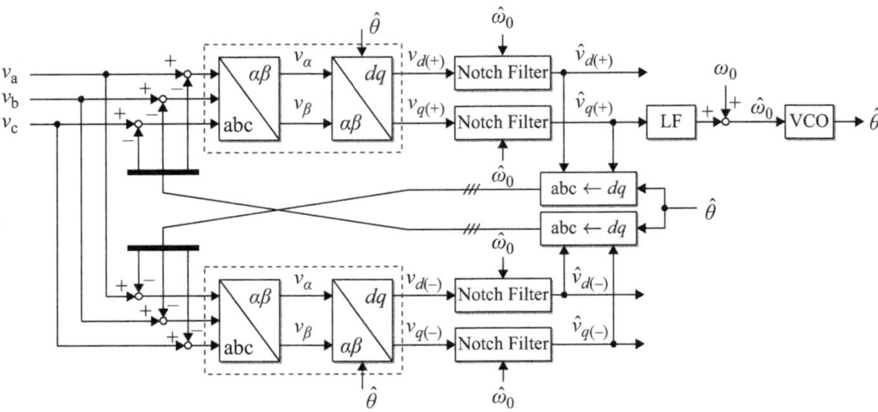

Figure 7.17 Dual synchronous reference frame based phase-locked loop for unbalanced grid voltages

is not designed to be robust to the unbalance, second-order harmonic oscillations will appear in the phase angle signal, thus in the current references. In addition, the second-order harmonic ripple component in the DC-link voltage would also bring pollutions into the current references for the current control loop. In such cases, positive- and negative-sequence feed-in currents must be controlled independently to avoid injecting power oscillations into the grid. For example, grid codes put more and more stress on the ability of power converter interface distributed generation units to ride through such short grid disturbances.

Under unbalanced conditions, dual synchronous reference frame (DSRF) current controllers [42–45], one for the positive-sequence and the other for the negative-sequence reference frames are needed to independently regulate the positive- and negative-sequence currents. Correspondingly, DSRF-PLL systems are demanded to produce clean synchronization signals for the positive-sequence and the negative-sequence reference frames, respectively. Under unbalanced grid voltages, the PLL system must be designed to filter out the negative-sequence, and produce a clean synchronization signal. An example of the DSRF-PLL systems is given in Figure 7.17, where the interactions between grid voltages and the opposite-sequence rotating reference frame give rise to ripples at twice the fundamental grid frequency in both dq voltages $(v_{d(+)}, v_{q(+)})$ and $(v_{d(-)}, v_{q(-)})$ obtained from the Park transformation. Fortunately, proper frequency-adaptive notch filters in (7.39) can be employed to filter out the second-order harmonic ripples in $(v_{d(+)}, v_{q(+)})$ and $(v_{d(-)}, v_{q(-)})$, and then remove the unexpected second-order harmonic ripples from the produced synchronous signals – grid fundamental frequency $\hat{\omega}_0$ and phase angle $\hat{\theta}$.

In common cases, a large amount of grid-connected nonlinear/unsymmetrical loads will inject high level current harmonics distortions into the grid, and then will distort the grid voltages at the PCC (point of common coupling) due to the line impedance. Grid voltage distortions will also inject unexpected harmonics into

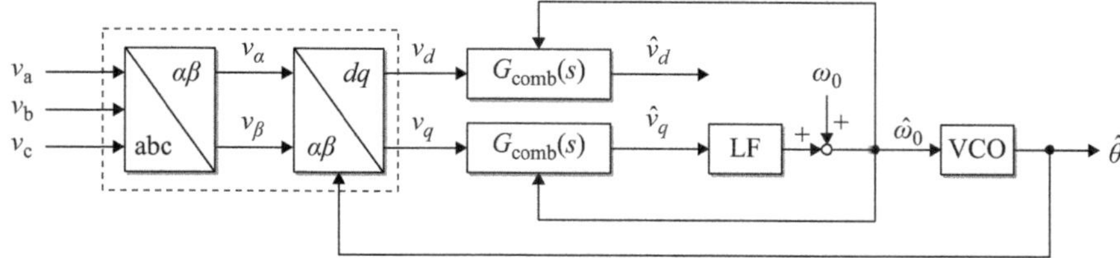

Figure 7.18 Synchronous reference frame-based phase-locked loop system for distorted grid voltages, where the comb filter shown in (7.49) has been adopted

the PLL. Similarly, proper comb filters can be employed to cancel out such harmonics distortions to produce clean synchronization signals [46–48]. For example, as shown in Figure 7.18, $(6k \pm 1)$-order harmonics with $k = 1, 2, 3, \ldots$, in the distorted three-phase grid voltages (v_a, v_b, v_c) will lead to 6th, 8th, 12th, 14th, ... harmonics in the dq-voltages (v_d, v_q) after the Park transformation. An even-order harmonic comb filter given below can thus be used to eliminate the harmonics in the dq-voltages (v_d, v_q):

$$G_{\text{comb}}(s) = \frac{1}{T_0 s}\left(1 - e^{-s\frac{T_0}{2}}\right) \tag{7.49}$$

where the integrator $1/s$ is added to compensate the attenuation at the DC frequency and will allow DC signals to pass [49,50], as compared in Figure 7.19. Notably, the original comb filter in (7.49) (i.e., $1-e^{-sT_0/2}$) can be derived according to (7.22) by setting $n = 2$ and $m = 0$.

It can be observed from Figure 7.19 that the comb filter in (7.49) will allow the DC component to pass and completely block the AC components with frequencies of integer multiples of $2/T_0$. Obviously, even under unbalanced and distorted grid voltages, the comb filter can also remove the second-order harmonics from (v_d, v_q), and as a consequence, the PLL would produce clean synchronization signals.

The discrete-time form of the comb filter in (7.49) can be written as:

$$G_{\text{comb}}(z) = \frac{1}{N} \cdot \frac{1}{1 - z^{-1}} \cdot \left(1 - z^{-N/2}\right) \tag{7.50}$$

Compared with the basis function in (2.56), it can be found that the comb filter in (7.50) is actually a special case of the basis function in (2.56) with the harmonic order $m = 0$.

Furthermore, it is known that delay-based digital comb filters are sensitive to grid frequency variations. In the presence of frequency variations, the following frequency-adaptive comb filter could be employed:

$$G_{\text{acomb}}(z) = \frac{1}{N} \cdot \frac{1}{1 - z_v^{-1}} \cdot \left(1 - z_v^{-N/2}\right) \tag{7.51}$$

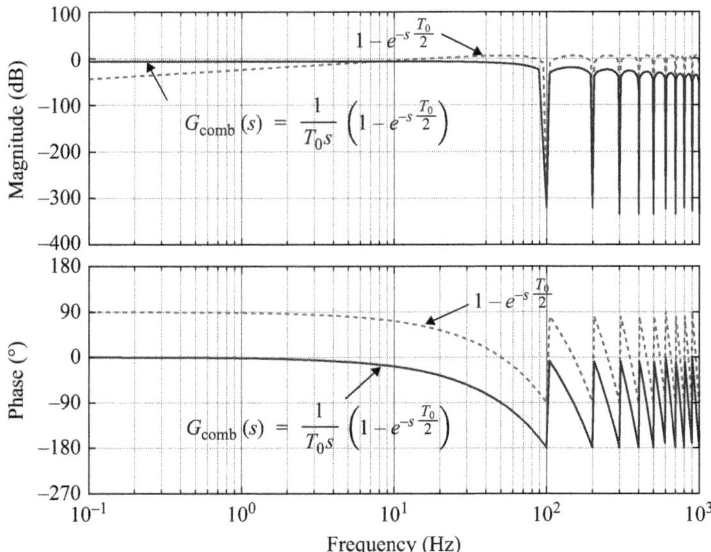

Figure 7.19 Frequency responses of the comb filter in (7.49) with $T_0 = 0.02$ s, which are also compared with the original comb filter derived according to (7.22) by setting $n = 2$ and $m = 0$

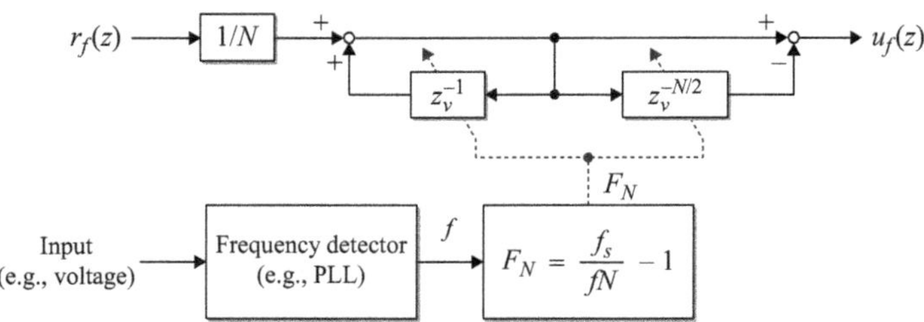

Figure 7.20 Digital implementation of the frequency-adaptive comb filter shown in (7.51)

or

$$G_{\text{acomb}}(z) = \frac{1}{N} \cdot \frac{1}{1 - z_v^{-1}} \cdot \left[1 - z^{-N_i}\left(\sum_{k=0}^{p} A_k z^{-k}\right)\right] \tag{7.52}$$

where the definitions of z_v^{-1} , N_i, N, and A_k can be found in (7.41) and (7.47). Accordingly, the implementation of the frequency-adaptive comb filter given in (7.51) is shown in Figure 7.20, where $r_f(z)$ represents the filter input and $u_f(z)$ is the filter output. Such a comb filter shown in Figure 7.20 can cancel out unexpected varying-frequency harmonics distortions in the output synchronization signals in the PLL.

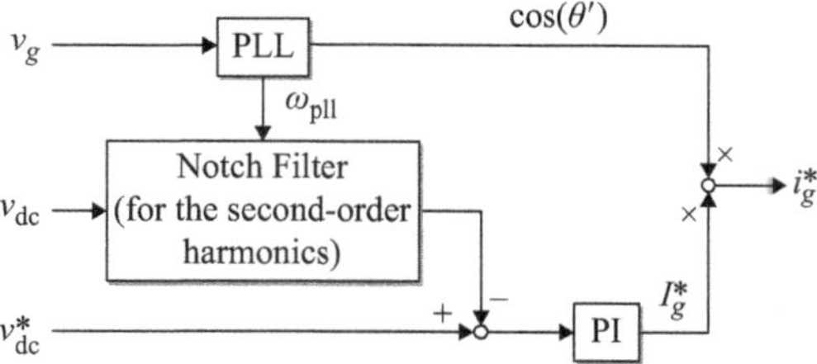

Figure 7.21 Improved outer voltage control loop for the single-phase converter shown in Figure 7.6, where ω_{pll} is the PLL output frequency

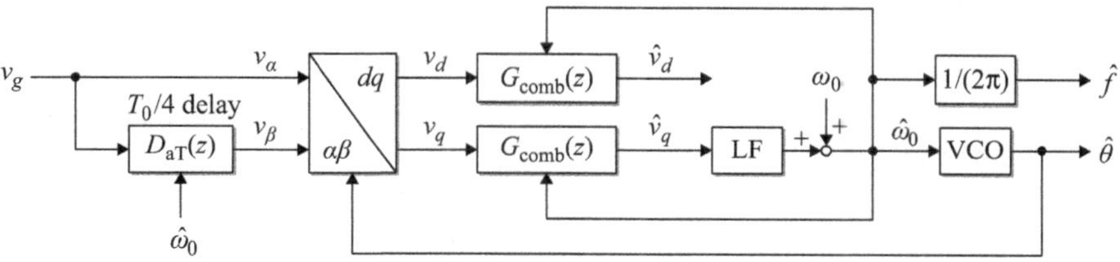

Figure 7.22 Frequency-adaptive $T_0/4$ delay-based PLL for single-phase systems

As shown in Figure 7.6, since current references for the inner current control loop is the production of synchronized cosine signal $\cos\theta'$ from the PLL and the output DC signal I^*_g from the outer voltage controller, both the synchronized sinusoidal signal and the DC signal must be clean. However, due to the commutation of the converter, harmonics ripples will appear in the DC-link voltage of the power converter, e.g., the second-order ripple for single-phase H-bridge converters, and the sixth-order ripple for three-phase bridge converters. It is clear that the outer loop voltage PI controller cannot remove these ripple harmonics. Therefore, proper notch filters can be added to eliminate these ripple harmonics in the detected feedback DC-link voltage v_{dc}, or resonant controllers are employed at the outer voltage loop to suppress the unexpected ripples in the output I^*_g, and then the produced current reference signal i^*_g is clean. Moreover, as shown in Figure 7.21, in case of the power grid with a varying frequency, frequency-adaptive notch filters can be used to deal with the resultant harmonics.

What's more, some other delay-based components might be needed in the control system for the power converters. For example, as shown in Figure 7.22, the $T_0/4$ delay-based PLL [51,52,53] provides a very simple grid synchronization method to single-phase grid-connected converters. The $T_0/4$ delay-based PLL adds a phase shift of $-90°$ to the single-phase grid voltage v_g, and then creates orthogonal signals (v_α, v_β) for the PLL to detect synchronization signals. The digital implementation of the $T_0/4$ delay can be written as:

$$D_T(z) = z^{-N_T} \tag{7.53}$$

Table 7.1 Parameters of the single-phase grid-connected system shown in Figure 7.6

Parameter	Value	Unit
Grid nominal voltage (peak) v_g	325	V
Grid nominal frequency f_0	50	Hz
Inverter-side inductor L_1	3.6	mH
Grid-side inductor L_2	4	mH
Capacitor of the *LCL* filter C_f	2.35	μF
Sampling frequency f_s	10	kHz
Proportional gain of the LF (PI controller)	0.24	–
Integral gain of the LF (PI controller)	4.81	–

where $N_T = N/4 \in \mathbb{N}$, $N = f_s/f_0 \in \mathbb{N}$ with f_s being the sampling frequency and f_0 being the nominal grid frequency.

Obviously the $T_0/4$ delay in (7.53) is sensitive to grid frequency variations. In the same manner of the frequency-adaptive delay-based comb filters, a frequency-adaptive $T_0/4$ delay can be obtained as:

$$D_{aT}(z) = z^{-N_i}\left(\sum_{k=0}^{p} A_k z^{-k}\right) \tag{7.54}$$

or

$$D_{aT}(z) = z_v^{-N_T} \tag{7.55}$$

where $N_i = [f_s/(4f)]$ with f being the actual grid frequency. In addition, as shown in Figure 7.22, in order to remove the even-order harmonics appearing in the dq-voltages (v_d, v_q), a frequency-adaptive even-order comb filter in (7.51) or (7.52) can be added into the $T_0/4$ delay-based PLL to enhance the detection accuracy for synchronization signals.

In order to demonstrate the necessity and effectiveness of the virtual unit delay in advancing the digital comb filters, simulations have been carried out on a single-phase grid-connected inverter system referring to Figure 7.6. The system parameters are shown in Table 7.1. In this case study, the virtual unit delays have been adopted to enable frequency-adaptive $T_0/4$ delay-based PLL, according to Figure 7.22.

First, steady-state performance tests are conducted under abnormal grid conditions. Simulation results are shown in Figure 7.23. It can be observed that the conventional $T_0/4$ delay-based PLL will present frequency variations of twice the grid fundamental frequency in the case of an abnormal grid frequency or a highly harmonics-distorted grid voltage. It means that the $T_0/4$ delay-based PLL is sensitive to grid frequency variations and also has a poor harmonic rejection capability. Hence, the virtual unit delay system $D_{aT}(z)$ in (7.55) is applied. As can be seen in Figure 7.23(a), under an abnormal grid frequency ($f = 50$ Hz), the PLL system can

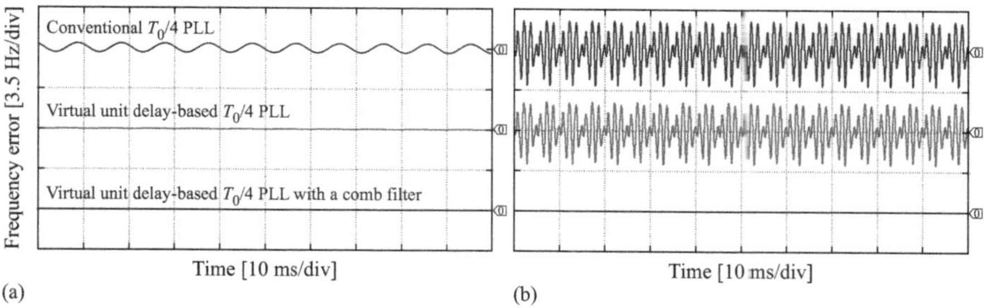

Figure 7.23 Simulation results (estimated frequency error $\Delta f = \hat{f} - f$) of the $T_0/4$ delay-based PLL system under abnormal grid conditions: (a) grid frequency $f = 52$ Hz and (b) harmonics-distorted grid voltage with the voltage THD being 14.5%

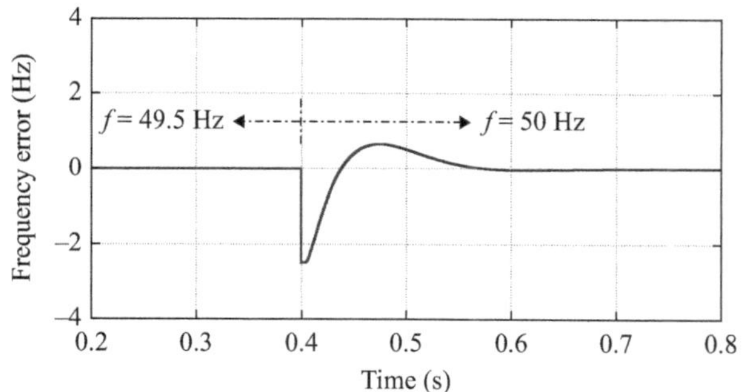

Figure 7.24 Dynamic performance (estimated frequency error $\Delta f = \hat{f} - f$) of the frequency adaptive $T_0/4$ delay-based PLL system with the comb filter given in (7.50) and the virtual unit delay system given in (7.55) under a grid frequency step-up change (49.5 Hz → 50 Hz)

estimate it accurately, being a frequency-adaptive $T_0/4$ delay-based PLL. However, its harmonic rejection capability is still poor, as shown in Figure 7.23(b). According to the above discussion, the comb filter shown in (7.50) can be adopted to enhance the harmonic immunity of the $T_0/4$ delay-based PLL. Observations in Figure 7.23 have confirmed that the $T_0/4$ delay-based PLL with a comb filter can estimate the grid frequency accurately in the presence of abnormal conditions. Additionally, the dynamic performance of the frequency-adaptive $T_0/4$ delay-based PLL is shown in Figure 7.24, where the grid frequency experienced a step change from 49.5 Hz to 50 Hz. It can be seen in Figure 7.24 that, the frequency-adaptive PLL system can estimate the change within 150 ms. This settling time can be further tuned considering the PI-based LF settling time and the comb filter delay. Nevertheless, the case study on the $T_0/4$ delay-based PLL has demonstrated the effectiveness of the virtual unit delay and the comb filters in advancing digital control technology for grid-connected power converters.

7.3 Summary

Control and filtering of periodic signals play critical roles in the control system for power converters, especially the grid-connected converters. Good understanding and full use of specific properties of periodic signals – the waveform shape and the period – enable us to develop simple but very effective controllers and/or filters to produce expected periodic signals or eliminate periodic disturbances. The internal model principle (IMP) provides an excellent universal frame for housing various periodic control technologies, and it enables us to gain insights into complex periodic control problems. Regarding most recent progress of periodic control technology, two emerging topics on the PC of power converters are investigated in this chapter as follows:

1. Periodic control for multi-period signals –
   - Multi-period repetitive control
   - Multi-period resonant control
   - Frequency-adaptive periodic control of multi-period signals
2. Periodic signal filtering –
   - Notch and comb filters and their discrete-time implementations
   - DFT-based comb filters
   - Frequency-adaptive notch and comb filters
   - Periodic signal filtering for grid-connected converters.

As mentioned in the previous chapters, the proportional–integral–derivative (PID) control is a special case of the IMP-based periodic control. The PID control offers a simple but effective approach to DC signal regulation, and it is the most popular control method in industrial applications. However, being a generalized PID control for periodic signals, the PC is still far away from the center of industrial control stage, and it is an emerging topic. This book is written to motivate peers to recognize the great potentials of the periodic control in extensive engineering practice, and promote relevant research and applications.

References

[1] Wang, Y., Gao, F., and Doyle III, F.J., "Survey on iterative learning control, repetitive control, and run-to-run control," *J. Process Control*, vol. 19, no. 10, pp. 1589–1600, 2009.

[2] Tomizuka, M., Tsao, T.-C., and Chew, K.-K., "Analysis and synthesis of discrete-time repetitive controllers," *J. Dyn. Syst., Meas. Control*, vol. 111, no. 3, pp. 353–358, 1989.

[3] Lu, W., Zhou, K., Wang, D., and Cheng, M., "A general parallel structure repetitive control scheme for multiphase DC–AC PWM converters," *IEEE Trans. Power Electron.*, vol. 28, no. 8, pp. 3980–3987, 2013.

[4] Lu, W., Zhou, K., and Wang, D., "General parallel structure digital repetitive control," *Int. J. Control*, vol. 86, no. 1, pp. 70–83, Jan. 2013.

[5] Lu, W., Zhou, K., Wang, D., and Cheng, M., "A generic digital nk ± m order harmonic repetitive control scheme for PWM converters," *IEEE Trans. Ind. Electron.*, vol. 61, no. 3, pp. 1516–1527, 2014.

[6] Zhou, K., Yang, Y., Blaabjerg, F., and Wang, D., "Optimal selective harmonic control for power harmonics mitigation," *IEEE Trans. Ind. Electron.*, vol. 62, no. 2, pp. 1220–1230, 2015.

[7] Zhou, K., Low, K.S., Wang, Y., Luo, F.L., Zhang, B., and Wang, Y., "Zero-phase odd-harmonic repetitive controller for a single-phase PWM inverter," *IEEE Trans. on Power Electron.*, vol. 21, no. 1, pp. 193–201, 2006.

[8] Zhang, B., Wang, D., Zhou, K., and Wang, Y., "Linear phase-lead compensation repetitive control of a CVCF PWM inverter," *IEEE Trans. Ind. Electron.*, vol. 55, no. 4, pp. 1595–1602, Apr. 2008.

[9] Zhang, B., Zhou, B., Wang, Y., and Wang, D., "Performance improvement of repetitive-controlled PWM inverters: A phase-lead compensation solution," *Int. J. Circuit Theory Appl.*, vol. 38, no. 5, pp. 453–469, Jun. 2010.

[10] Zhou, K., Lu, W., Yang, Y., and Blaabjerg, F., "Harmonic control: A natural way to bridge resonant control and repetitive control," *American Control Conference*, pp. 3189–3193, 2013.

[11] Lu, W., Zhou, K., and Wang, D., "General parallel structure digital repetitive control," *Int J. Control*, vol. 86, no. 1, pp. 70–83, 2013.

[12] Liserre, M., Teodorescu, R., and Blaabjerg, F., "Multiple harmonics control for three-phase grid converter systems with the use of PI-RES current controller in a rotating frame," *IEEE Trans. Power Electron.*, vol. 21, no. 3, pp. 836–841, May 2006.

[13] Lascu, C., Asiminoaei, L., Boldea, I., and Blaabjerg, F., "High-performance current controller for selective harmonic compensation in active power filters," *IEEE Trans. Power Electron.*, vol. 22, no. 5, pp. 1826–1835, 2007.

[14] Lascu, C., Asiminoaei, L., Boldea, I., and Blaabjerg, F., "Frequency response analysis of current controllers for selective harmonic compensation in active power filters," *IEEE Trans. Ind. Electron.*, vol. 56, no. 2, pp. 337–347, 2009.

[15] Yang, Y., Zhou, K., Cheng, M., and Zhang, B., "Phase compensation multi-resonant control of CVCF PWM converters," *IEEE Trans. Power Electron.*, vol. 28, no. 8, pp. 3923–3930, Aug. 2013.

[16] Yang, Y., Zhou, K., and Cheng, M., "Phase compensation resonant controller for single-phase PWM converters," *IEEE Trans Ind. Inform.*, vol. 9, no. 2, pp. 957–964, 2013.

[17] Yamada, M., Riadh, Z., and Funahashi, Y., "Discrete-time repetitive control system with multiple periods," *6th International Workshop on Advanced Motion Control*, pp. 228–233, 2000.

[18] Yamada, M., Riadh, Z., and Funahashi, Y., "Design of robust repetitive control system for multiple periods," *Proc. 39th IEEE Conf. Decision and Control*, vol. 4, pp. 3739–3744, 2000.

[19] Longman, R. W., "On the theory and design of linear repetitive control systems," *European J. Control*, vol. 5, no. 5, pp. 447–496, 2010.

[20] Yang, Y., Zhou, K., and Blaabjerg, F., "Frequency adaptability of harmonics controllers for grid-interfaced converters," *Int. J. Control*, (doi:10.1080/00207179.2015.1022957) (Early Online Publication), 2015.

[21] Liu, T. and Wang, D., "Parallel structure fractional repetitive control for PWM inverters," *IEEE Trans. Ind. Electron.*, vol. 62, no. 8, pp. 5045–5054, Feb. 2015.

[22] Liu, T., Wang, D., and Zhou, K., "High-performance grid simulator using parallel structure fractional repetitive control," *IEEE Trans. Power Electron.*, vol. 31, no. 3, pp. 2669–2679, Mar. 2016.

[23] Laakso, T.I., Valimaki, V., Karjalainen, M., and Laine, U.K., "Splitting the unit delay [FIR/all pass filters design]," *IEEE Signal Proc. Mag.*, vol. 13, no. 1, pp. 30–60, Jan. 1996.

[24] Zou, Z., Zhou, K., Wang, Z., and Cheng, M., "Fractional-order repetitive control of programmable AC power sources," *IET Power Electron.*, vol. 7, no. 2, pp. 431–438, 2014.

[25] Zou, Z.-X., Zhou, K., Wang, Z., and Cheng, M., "Frequency-adaptive fractional order repetitive control of shunt active power filters," *IEEE Trans. Ind. Electron.*, vol. 62, no. 3, pp. 1659–1668, 2015.

[26] Nazir, R., Zhou, K., Watson, N., and Wood, A., "Analysis and synthesis of fractional order repetitive control for power converters," *Electric Power Systems Research*, vol. 124, pp. 110–119, 2015.

[27] Yang, Y., Zhou, K., Wang, H., Blaabjerg, F., Wang, D., and Zhang, B., "Frequency-adaptive selective harmonic control for grid-connected inverters," *IEEE Trans. Power Electron.*, vol. 30, no. 7, pp. 3912–3924, 2015.

[28] Yang, Y., Zhou, K., and Blaabjerg, F., "Enhancing the frequency adaptability of periodic current controllers with a fixed sampling rate for grid-connected power converters," *IEEE Trans. Power Electron.*, vol. 31, no. 10, pp. 7273–7285, 2016.

[29] Gonzalez-Espin, F., Garcera, G., Patrao, I., and Figueres, E., "An adaptive control system for three-phase photovoltaic inverters working in a polluted and variable frequency electric grid," *IEEE Trans. Power Electron.*, vol. 27, no. 10, pp. 4248–4261, Oct. 2012.

[30] Teodorescu, R., Blaabjerg, F., Liserre, M., and Loh, P.C., "Proportional-resonant controllers and filters for grid-connected voltage-source converters," *Proc. IEE*, vol. 153, no. 5, pp. 750–762, Sep. 2006.

[31] Teodorescu, R., Liserre, M., and Rodriguez, P., *Grid Converters for Photovoltaic and Wind Power Systems*. Hoboken, NJ, USA: Wiley, 2011.

[32] Blaabjerg, F., Teodorescu, R., Liserre, M., and Timbus, A.V., "Overview of control and grid synchronization for distributed power generation systems," *IEEE Trans. Ind. Electron.*, vol. 53, no. 5, pp. 1398–1409, 2006.

[33] Rocabert, J., Luna, A., Blaabjerg, F., and Rodríguez, P., "Control of power converters in AC microgrids," *IEEE Trans. Power Electron.*, vol. 27, no. 11, pp. 4734–4749, 2012.

[34] Yang, Y., Zhou, K., and Blaabjerg, F., "Current harmonics from single-phase grid-connected inverters – examination and suppression," *IEEE Jo. Emerging and Selected Topics in Power Electron.*, vol. 4, no. 1, pp. 221–233, 2016.

[35] Zhou, K., Qiu, Z., Watson, N. R., and Liu, Y., "Mechanism and elimination of harmonic current injection from single-phase grid-connected PWM converters," *IET Power Electron.*, vol. 6, no. 1, pp. 88–95, 2013.

[36] Rodriguez, P., Luna, A., Ciobotaru, M., Teodorescu, R., and Blaabjerg, F., "Advanced grid synchronization system for power converters under unbalanced and distorted operating conditions," *32nd Annual Conf. IEEE Ind.l Electron. (IECON)*, pp. 5173–5178, 2006.

[37] Rodriguez, P., Pou, J., Bergas, J., Candela, J.I., Burgos, R.P., and Boroyevich, D., "Decoupled double synchronous reference frame PLL for power converters control," *IEEE Trans. Power Electron.*, vol. 22, no. 2, pp. 584–592, 2007.

[38] Wang, L., Chai, S., and Rogers, E., "Predictive repetitive control based on frequency decomposition, *2010 American Control Conference*, Baltimore, MD, USA, Jun. 30–Jul. 02, 2010.

[39] Mattavelli P. and Marafao, F.P., "Repetitive-based control for selective harmonic compensation in active power filters," *IEEE Trans. Ind. Electron.*, vol. 51, no. 5, pp. 1018–1024, Oct. 2004.

[40] Pillay, P. and Manyage M., "Definitions of voltage unbalance," *IEEE Power Eng. Rev.*, vol. 22, no. 11, pp. 49–50, 2002.

[41] Moran, L., Ziogas, P.D., and Joos, G., "Design aspects of synchronous PWM rectifier-inverter systems under unbalanced input voltage conditions," *IEEE Trans. Ind. Appl.*, vol. 28, no. 6, pp. 1286–1293, 1992.

[42] Song, H.-S. and Nam, K., "Dual current control scheme for PWM converter under unbalanced input voltage conditions," *IEEE Trans. Ind. Elec.*, vol. 46, no. 5, pp. 953–959,1999.

[43] Stankovic, A.V. and Lipo, T.A., "A novel control method for input output harmonic elimination of the PWM boost-type rectifier under unbalanced operating conditions," *IEEE Trans. Power Electron.*, vol. 16, no. 5, pp. 603–611, 2001.

[44] Reyes, M., Rodriguez, P., Vazquez, S., Luna, A., Teodorescu, R., and Carrasco, J.M., "Enhanced decoupled double synchronous reference frame current controller for unbalanced grid-voltage conditions," *IEEE Trans. Power Electron.*, vol. 27, no. 9, pp. 3934–3943, 2012.

[45] Xiao, P., Corzine, K.A., and Venayagamoorthy, G.K., "Multiple reference frame-based control of three-phase PWM boost rectifiers under unbalanced and distorted input conditions," *IEEE Trans. Power Electron.*, vol. 23, no. 4, pp. 2006–2017, 2008.

[46] Gonzalez-Espin, F., Figueres, E., and Garcera, G., "An adaptive synchronous-reference-frame phase-locked loop for power quality improvement in a polluted utility grid," *IEEE Trans. Ind. Electron.*, vol. 59, no. 6, pp. 2718–2731, 2012.

[47] Carugati, I., Maestri, S., Donato, P.G., Carrica, D., and Benedetti, M., "Variable sampling period filter PLL for distorted three-phase systems," *IEEE Trans. Power Electron.*, vol. 27, no. 1, pp. 321–330, 2012.

[48] Freijedo, F.D., Yepes, A.G., López, O., Vidal, A., and Doval-Gandoy, J., "Three-phase PLLs with fast postfault retracking and steady-state rejection of voltage unbalance and harmonics by means of lead compensation," *IEEE Trans. Power Electron.*, vol. 26, no. 1, pp. 85–97, 2011.

[49] Wang, J., Liang, J., Gao, F., Zhang, L., and Wang, Z., "A method to improve the dynamic performance of moving average filter-based PLL," *IEEE Trans. Power Electron.*, vol. 30, no. 10, pp. 5978–5990, Oct. 2015.

[50] Golestan, S., Guerrero, J.M., Vidal, A., Yepes, A.G., and Doval-Gandoy, J., "PLL with MAF-based prefiltering stage: Small-signal modeling and performance enhancement," *IEEE Trans. Power Electron.*, vol. 31, no. 6, pp. 4013–4019, 2016.

[51] Yang, Y. and Blaabjerg, F., "Synchronization in single-phase grid-connected photovoltaic systems under grid faults," *3rd IEEE Int. Symposium Power Electron. for Distributed Generation Systems (PEDG) 2012*, pp. 476–482, 2012.

[52] Hadjidemetriou, L., Kyriakides, E., Yang, Y., and Blaabjerg, F., "A synchronization method for single-phase grid-tied inverters," *IEEE Trans. Power Electron.*, vol. 31, no. 3, pp. 2139–2149, Mar. 2016.

[53] Yang, Y., Zhou, K. and Blaabjerg, F., "Virtual unit delay for digital frequency-adaptive *T*/4 delay phase-locked loop system," in *Proc. IPEMC 2016-ECCE Asia*, pp. 2910–2916, 2016.

Index